AF355971

Assessing Hydrological Drought in a Climate Change: Methods and Measures

Assessing Hydrological Drought in a Climate Change: Methods and Measures

Editor

Alina Barbulescu

Basel • Beijing • Wuhan • Barcelona • Belgrade • Novi Sad • Cluj • Manchester

Editor
Alina Barbulescu
Civil Engineering
Transilvania University
Brasov
Romania

Editorial Office
MDPI
St. Alban-Anlage 66
4052 Basel, Switzerland

This is a reprint of articles from the Special Issue published online in the open access journal *Water* (ISSN 2073-4441) (available at: www.mdpi.com/journal/water/special_issues/Hydrological_Drought).

For citation purposes, cite each article independently as indicated on the article page online and as indicated below:

Lastname, A.A.; Lastname, B.B. Article Title. *Journal Name* **Year**, *Volume Number*, Page Range.

ISBN 978-3-7258-0908-0 (Hbk)
ISBN 978-3-7258-0907-3 (PDF)
doi.org/10.3390/books978-3-7258-0907-3

© 2024 by the authors. Articles in this book are Open Access and distributed under the Creative Commons Attribution (CC BY) license. The book as a whole is distributed by MDPI under the terms and conditions of the Creative Commons Attribution-NonCommercial-NoDerivs (CC BY-NC-ND) license.

Contents

About the Editor

Alina Barbulescu

Alina Bărbulescu obtaineda Ph.D. in Mathematics from Al. I. Cuza University of Iași, Romania, another in Cybernetics and Econometrics from the Academy of Economics Studies of Bucharest, Romania, and yet another in Civil Engineering from the Technical University of Civil Engineering of Bucharest. She obtained a Habilitation in Civil Engineering and another in Cybernetics and Econometrics.

Currently, she is a Professor at Transilvania University of Brașov, Romania. Alina Bărbulescu's main research fields are environmental pollution, hydrological modeling, applied statistics, and time series analysis. Over the last period, most of her research has been developed within the frameworks of UEFISCDI Romania, Horizon Europe, or projects funded by the Zayed University Research Center, the United Arab Emirates. Her publishing activity totals over 200 articles (over half indexed in Web of Science), 32 books, and several chapters. She was the editor of 22 Special Issues of international conferences and is a member of the editorial boards of different journals (three indexed in Web of Science).

water

MDPI

Editorial

Assessing Hydrological Drought in a Climate Change: Methods and Measures

Alina Barbulescu [1,*] and Stefan Mocanu [2]

1 Department of Civil Engineering, Transilvania University of Brașov, 5 Turnului Str., 500152 Brașov, Romania
2 Fcaulty of Mechanical Engineering and Robotics in Constructions, Technical University of Civil Engineering of Bucharest, 59 Calea Plevnei, 010234 Bucharest, Romania
* Correspondence: alina.barbulescu@unitbv.ro

Citation: Barbulescu, A.; Mocanu, S. Assessing Hydrological Drought in a Climate Change: Methods and Measures. *Water* **2023**, *15*, 1978. https://doi.org/10.3390/w15111978

Received: 5 April 2023
Accepted: 18 May 2023
Published: 23 May 2023

Copyright: © 2023 by the authors. Licensee MDPI, Basel, Switzerland. This article is an open access article distributed under the terms and conditions of the Creative Commons Attribution (CC BY) license (https://creativecommons.org/licenses/by/4.0/).

Water is a resource indispensable for human life and activity, significantly affected by climate change (by decreasing the water quantities available for drinking) and anthropogenic activities (by pollution). In recent decades, the frequency and intensity of drought has increased in extended zones of the planet, with an obvious negative impact on the environment, human life, and the economy. In this context, the Special Issue "Assessing Hydrological Drought in a Climate Change: Methods and Measures" addresses the following main topics:

- Designing new drought indices to better quantify drought effects;
- Analyzing drought events from both qualitative and quantitative viewpoints;
- Detecting drought events and their correlations with climate change;
- Estimating the drought frequency and intensity of drought episodes;
- Modeling and forecasting time series related to drought events.

Remote sensing has become a regular tool for assessing hydrological drought, mainly when onsite recorded data are insufficient or absent for specific periods. It was the primary investigation tool used in the case of the Nuntasi-Tuzla Lake, situated in the Danube Delta Natural Reserve, Romania [1]. An event in 2020, when the water level significantly decreased, was the beginning point for finding the causes of this phenomenon and its relationship with climate change and anthropic influence. The 1965–2021 data series and indicators derived from Landsat TM/ETM+/OLI and MODIS datasets (NDVI, MNDWI, WNDWI, and WRI) were used in the investigation. The results showed that hydrological drought and anthropic activities influenced the significant variation in the lake's water level.

A similar study was performed for the Urmia Lake Basin, Iran [2], for 1981–2018, using SPI, SPEI, and SMRI indexes. The decreasing trends of SPEI and SMRI suggested that evaporation and low snowmelt increased the drought after 1995. As in the case of the Nuntasi-Tuzla Lake, the water level diminishing was also the result of defective water management.

Starting from the idea that hazard indexes, built for estimating drought, do not give information on the location and moment when the adverse effects appear, Thomaz et al. [3] proposed a new index to assess drought—the Water Scarcity Risk Index (W-ScaRI). It comprises two sub-indices. The first one, formed by integrating SPI, RDI, and SDI, describes the hazard, whereas the second describes the hazard's consequences. It demonstrated promising results in the presented case study and can be extended by including other subindexes.

In [4], the authors used the SPI drought prediction at different time scales. They also investigated the influence of the data series stationarity/nonstationarity on the SPI computation and the bias introduced and draw attention to the misuse of these indexes.

Botai et al. [5] evaluated hydrological drought in three Cape provinces utilizing the Standardized Streamflow Index computed for trimestral and semiannual accumulation periods in a study aiming to propose solutions to mitigate the adverse drought effects. The results show that drought frequency increased and that there was a spatial inhomogeneity (of the drought events) in the studied zone.

Another series of articles analyzed river discharge events in different countries. After a critical overview of the percentiles and flow duration curve (FDC) use, Raczyński and Dyer [6] introduced a new methodology to compute the low flow threshold (LFT) based on the change point detection. They indicate that the new algorithm has the advantage of objectively identifying the beginning of the low flow events, unifying the approaches for selecting the threshold levels. The algorithm is accompanied by a module written in Python.

To study the low flow patterns, the autocorrelation and partial autocorrelation functions and the Hurst exponent were employed in the article by Raczynski et al. [7]. They helped to make the distinction between the white noise and the seasonal processes.

Climate change has manifested in some regions of Romania through long drought periods followed by high precipitation in a short period, leading to flooding. Ţigăneşti and Brânceni are villages that have been affected many times by floods, as in July 2005, when the water flow was 676 m^3/s. The article [8] investigates the mentioned event and evaluates the produced damages using field observation and recorded data HEC-RAS simulation. It was emphasized that the existence of a levee along the Vedea River would protect the villages from flooding.

In the same vein, Garza-Díaz et al. [9] computed the streamflow drought index in a study related to the landscape modification in a river basin by the anthropogenic drought in Mexico.

In the conditions of the accentuated drought, irrigation is a must for maintaining high crop production and ensures the consumption necessities. With this in mind, Dumitriu et al. [10] propose a new tool—IrrigTool—that facilitates the quick computation of the irrigation rate, having hydro-meteorological variables, crop type, and soil as inputs. It is implemented in Excel and VBA, has a user-friendly interface, and provides the graphical output and possible comparisons between the irrigation rates in two locations. A case study is also provided, together with step-by-step functioning explanations.

We hope that the reader will find useful information on hydrological drought, its quantification, and its analysis.

Conflicts of Interest: The authors declare no conflict of interest.

References

1. Şerban, C.; Maftei, C.; Dobrică, G. Surface Water Change Detection Via Water Indices and Predictive Modeling Using Remote Sensing Imagery: A Case Study of Nuntasi-Tuzla Lake, Romania. *Water* **2022**, *14*, 556. [CrossRef]
2. Habibi, M.; Babaeian, I.; Schöner, W. Changing Causes of Drought in the Urmia Lake Basin—Increasing Influence of Evaporation and Disappearing Snow Cover. *Water* **2021**, *13*, 3273. [CrossRef]
3. Thomaz, F.R.; Miguez, M.G.; de Souza Ribeiro de Sá, J.G.; de Moura Alberto, G.W.; Fontes, J.P.M. Water Scarcity Risk Index: A Tool for Strategic Drought Risk Management. *Water* **2023**, *15*, 255. [CrossRef]
4. Mammas, K.; Lekkas, D.F. Characterization of Bias during Meteorological Drought Calculation in Time Series Out-of-Sample Validation. *Water* **2021**, *13*, 2531. [CrossRef]
5. Botai, C.M.; Botai, J.O.; de Wit, J.P.; Ncongwane, K.P.; Murambadoro, M.; Barasa, P.M.; Adeola, A.M. Hydrological Drought Assessment Based on the Standardized Streamflow Index: A Case Study of the Three Cape Provinces of South Africa. *Water* **2021**, *13*, 3498. [CrossRef]
6. Raczyński, K.; Dyer, J. Development of an Objective Low Flow Identification Method Using Breakpoint Analysis. *Water* **2022**, *14*, 2212. [CrossRef]
7. Raczynski, K.; Dyer, J. Variability of Annual and Monthly Streamflow Droughts over the Southeastern United States. *Water* **2022**, *14*, 3848. [CrossRef]
8. Popescu, C.; Bărbulescu, A. Floods Simulation on the Vedea River (Romania) Using Hydraulic Modeling and GIS Software: A Case Study. *Water* **2023**, *15*, 483. [CrossRef]

9. Garza-Díaz, L.E.; Sandoval-Solis, S. Changes in the Stability Landscape of a River Basin by Anthropogenic Droughts. *Water* **2022**, *14*, 2835. [CrossRef]
10. Dumitriu, C.Ș.; Bărbulescu, A.; Maftei, C.E. IrrigTool—A New Tool for Determining the Irrigation Rate Based on Evapotranspiration Estimated by the Thornthwaite Equation. *Water* **2022**, *14*, 2399. [CrossRef]

Disclaimer/Publisher's Note: The statements, opinions and data contained in all publications are solely those of the individual author(s) and contributor(s) and not of MDPI and/or the editor(s). MDPI and/or the editor(s) disclaim responsibility for any injury to people or property resulting from any ideas, methods, instructions or products referred to in the content.

Article

Surface Water Change Detection via Water Indices and Predictive Modeling Using Remote Sensing Imagery: A Case Study of Nuntasi-Tuzla Lake, Romania

Cristina Șerban [1,*] , Carmen Maftei [2,*] and Gabriel Dobrică [1]

1   Faculty of Mathematics and Computer Science, Ovidius University of Constanta, 900527 Constanta, Romania; gabriel.dobrica@yahoo.com
2   Faculty of Civil Engineering, Transilvania University of Brasov, 900152 Brasov, Romania
*   Correspondence: cgherghina@gmail.com (C.Ș.); cemaftei@gmail.com (C.M.)

Abstract: Water body feature extraction using a remote sensing technique represents an important tool in the investigation of water resources and hydrological drought assessment. Nuntasi-Tuzla Lake, a component of the Danube Delta Natural Reserve, is located on the Romanian Black Sea littoral. On account of an event in summer 2020, when the lake surface water decreased significantly, this study aims to identify the variation of the Nuntasi-Tuzla Lake surface water over a long-term period in correlation with human intervention and climate change. To this end, it provides an analysis in the period 1965–2021 via hydrological drought indices and data mining classification. The latter approach is based on several water indices derived from Landsat TM/ETM+/OLI and MODIS full-time series datasets: Normalized Difference Vegetation Index (NDVI), Normalized Difference Vegetation Index (NDVI), Modified NDWI (MNDWI), Weighted Normalized Difference Water Index (WNDWI), and Water Ratio Index (WRI). The experimental results indicate that the proposed classification methods can extract relevant features from waterbodies using remote sensing imagery with a high accuracy. Moreover, the study shows a similarity in the evolution of surface water cover identified with the data mining classification and the drought periods detected in the flow data series for the Nuntasi and Sacele Rivers that supply the Nuntasi-Tuzla Lake. Overall, the results of our investigation show that human intervention and hydrological drought had an extensive impact on the long-term changes in surface water of the Nuntasi-Tuzla Lake.

Keywords: hydrological drought; remote sensing; water indices; Nuntasi-Tuzla Lake; Romania

Citation: Șerban, C.; Maftei, C.; Dobrică, G. Surface Water Change Detection via Water Indices and Predictive Modeling Using Remote Sensing Imagery: A Case Study of Nuntasi-Tuzla Lake, Romania. *Water* **2022**, *14*, 556. https://doi.org/10.3390/w14040556

Academic Editor: Thomas Meixner

Received: 25 December 2021
Accepted: 8 February 2022
Published: 12 February 2022

Publisher's Note: MDPI stays neutral with regard to jurisdictional claims in published maps and institutional affiliations.

Copyright: © 2022 by the authors. Licensee MDPI, Basel, Switzerland. This article is an open access article distributed under the terms and conditions of the Creative Commons Attribution (CC BY) license (https://creativecommons.org/licenses/by/4.0/).

1. Introduction

Climatic changes can induce modification of lake surfaces and riverbeds with severe implications for agricultural and economic activities [1–3]. The scientific literature demonstrates the "sensitivity" of physical, chemical, and biological lakes' properties to climate change and human intervention [4–6]. Since 1960 permanent surface water has been shrinking, drying or has completely disappeared: Aral Lake, Urmia Lake (70% of its surface between 2012–2017 [7]), Chad Lake, Lop Nur, etc. [4,5]. Lake surface area was used by Benson and Paillet [8] in order to establish an indicator of lakes' hydrological response to climatic change. In comparison with classic methods, Remote Sensing (RS) technology presents a viable alternative to improve the observation of lake water surface in hydrological studies.

Initially, the NDVI (Normalized Difference Vegetation Index) was an indicator used to efficiently monitor vegetation condition [9] and drought activity [10,11]. This index was used also to extract water features [12]. McFetters [13] developed a new method to delineate water bodies based on the spectral characteristics of water which refers to the capability of a water body to absorb near infrared radiation (NIR) and allows visible green light to penetrate the water body. This new indicator, NDWI (Normalized Difference Water

Index) was modified by Xu [14] in order to resolve the issue of the zero (0) threshold agreed by McFetters as the separation between water body and background. Xu [14] proposed a modified NDWI (MNDWI) index using the middle infrared band (MIR) instead of NIR. According to the author [14], MNDWI can more accurately distinguish water from non-water features. One other well-known multiband water index is the water ratio index (WRI). Due to the domination of the spectral characteristics of the green and red bands compared with the NIR and MIR bands, the WRI shows values greater than 1 for water [15]. Guo et al. [16] proposed a new index: the weighted normalized difference water index (WNDWI), that weighted average NIR and SWIR (shortwave infrared) bands. According to the authors, the new index improves the accuracy of water extraction by correctly classifying turbid water as a water body and vegetation in the shadow area as non-water body [16]. Acharya [17,18] proposed a combination of different indices in order to improve water extraction.

In the summer of 2020, Nuntași-Tuzla Lake (1050 ha), located on the Romanian Black Sea littoral (Figure 1), dried out.

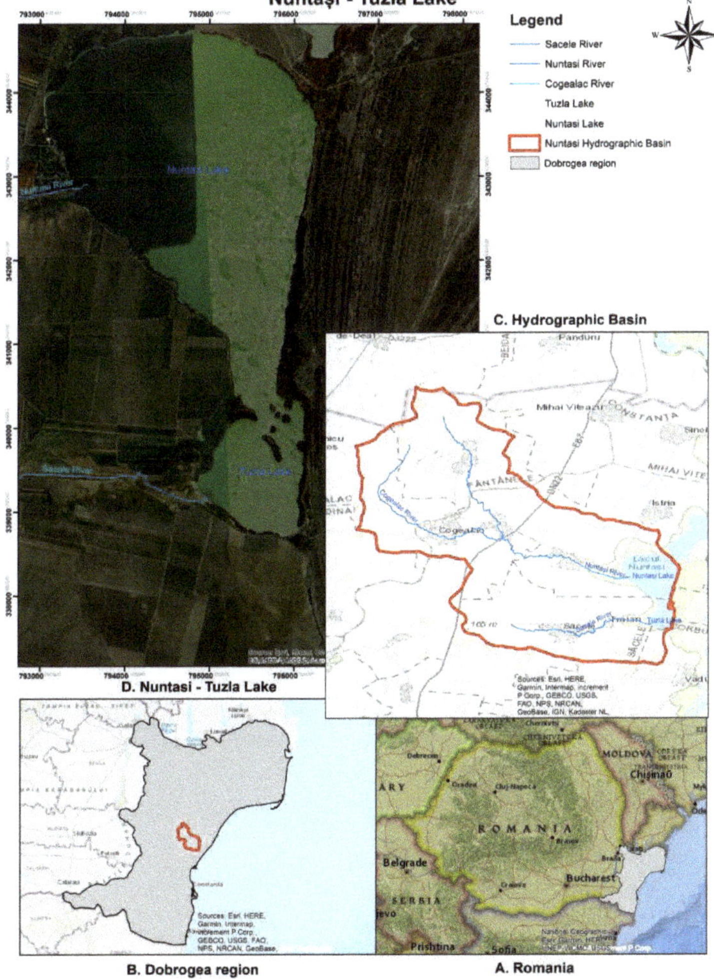

Figure 1. Map of the study area, the Nuntasi-Tuzla Lake.

In this context, the purpose of this paper is to: (i) analyze the hydrological data in order to understand the response of surface water to climate change or/and human intervention and (ii) evaluate the performance of the most commonly used water indices using time series Landsat data in order to capture the variation in surface water during 1984–2021 period. These results will contribute to a better understanding of the lake's evolution in the context of global changes.

2. Materials and Methods

2.1. Study Area

The Nuntasi-Tuzla Lake is situated in the well-known Histria region (Histria or Istros acropolis is the first urban settlement, founded by ancient Milesian in the 7th century BC). From a geomorphological point of view, the Histria region is a coastal lowland developed at the contact between the Central Dobrogea plateau and the Black Sea [19]. The average altitude is 25 m with a slope that decreases towards the sea. Geologically, green schist covered with a loess layer of varying thickness (2–15 m) represents the basement. The climate of the region is temperate continental with marine influences (precipitation of 400 mm and 11 °C average temperature). From a hydrological point of view, the lake is part of the Razim-Sinoe lagunar complex, a component of the Danube Delta Natural Reserve. Today, the Razim-Sinoe complex (Figure 2), initially a marine bay (the former Halmyris gulf), is composed of the following lakes: Razim, Sinoe, Golovita, Zmeica, Babadag, Nuntasi-Tuzla, Istria, and Ceamurlia si Agighiol.

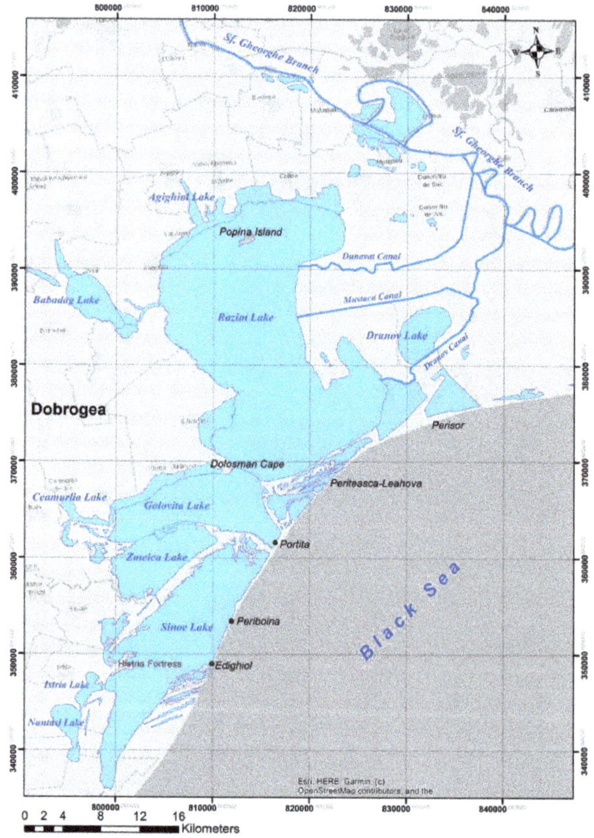

Figure 2. Map of the Razim-Sinoe Lake system.

Concerning the lake's genesis, the latest scientific arguments clearly demonstrate that neo-tectonic activity was the main factor that led to the Nuntasi-Tuzla and Istria lakes' development 1200–1600 years ago [19]. The other lakes were separated from the Black Sea by sand-belts [17], which formerly had outlets allowing the penetration of sea water. The lakes were interconnected by small creeks.

According to Breier [20], Nuntasi-Tuzla Lake has the following morphometric features: the length is 2 km, the breadth varies between 1.7 and a maximum of 3 km, the depth of the lake varies between 2.15 m and 6.15 m, and the surface area is 1050 hectares. The Nuntasi-Tuzla Lake is supplied by two rivers: Nuntasi and Sacele. From the economic point of view, according to town hall documents, agriculture and fisheries are the predominant economic sectors in the region. The mud baths were a great tourist attraction at the beginning of the 20th century due to its therapeutic qualities. Nowadays, the most important touristic attraction in the region is Histria ancient city.

In a recent work [21], we described the irrigation system built in 1971–1975 and operable till 1990. After this year, the demand for irrigation decreased to 20% of its capacity.

In order to ensure water demand, the Razim-Sinoe lagoons were transformed into a large freshwater reservoir. The three greatest hydraulic works were constructed in that period: (i) closure of the natural gateway between the lagoons and Black Sea, (ii) closure—via sluices of the connections between the lakes and (iii) resizing of three canals which supply the reservoir with fresh water from the Danube (St George Arm) so as to ensure a total discharge of 260 mc/s (for a probability of occurrence of 1%) with an average of 80 mc/s [21]. These works led to important changes in the hydrological regime of the region and to a great disturbance of the ecological balance of all the lakes. Nuntasi-Tuzla Lake was connected to Istria Lake by a canal and sluice. This canal was silted up with alluvium.

A detailed description of all human intervention in this area is provided in Section 3.1.

2.2. Data Sets

2.2.1. Hydrological Datasets

Two hydrological parameters are monitored at the two hydrometric stations (Nuntasi and Sacele). This study is based on two time data series: annual average discharge for Nuntasi and Sacele spanning the 1965–2020 period, and daily discharge for the period 2008–2020. The annual average discharge was used in order to investigate changes in the time series. The time series data were collected by the National Administration "Romanian Water" Dobrogea-Littoral Branch at the hydrometric station situated in the Nuntasi-Tuzla lake basin.

2.2.2. Remote Sensing Datasets

The time-series datasets of Landsat surface reflectance are acquired directly from the GEE (Google Earth Engine) platform for the period 1984–2021 in the following way: (i) from 1984 to 1999, and from 2003 to 2011, the data were collected from Landsat 4–5 Thematic Mapper (TM); (ii) for 2000, 2001 and 2002, Landsat 7 Enhanced Thematic Mapper (ETM+) images were collected; and (iii) for 2013–2021, the data were collected from Landsat 8 Operational Land Imager (OLI). The datasets belong to Landsat Collection 1—Level 1 corrected data, which have the highest radiometric and positional quality and are suitable for time-series analysis. The spatial resolution of all Landsat scenes is 30 m.

Due to Landsat 7 ETM+ data gaps from 2012, the time-series dataset of the moderate-resolution imaging spectroradiometer (MODIS) TERRA surface reflectance (MOD09GA) was used to determine the surface water for the seasons of 2012. MODIS images with 500 m spatial resolution were acquired from the GEE platform. The study area is entirely contained within path 181 and row 29 for all Landsat TM/ETM+/OLI and Modis images (Figure 1).

2.3. Methods

The methodology used in this paper consists of three parts: A. a literature investigation to understand human intervention in the Nuntasi-Tuzla Lake basin; B. a hydrological analysis of two tributaries (Nuntasi and Sacele) concerning hydrological drought occurrence, which refers to (i) detecting changes in time series data through KhronoStat software and (ii) calculating hydrological drought; and C. RS analysis, which refers to: (i) processing of the Landsat and MODIS datasets so that only the images with Cloud Cover of less than 30% are investigated (ii) calculation of RS indices (RSI) NDVI, NDWI, MNDWI, WNDWI, WRI and their average for each season of each year of the period considered; (iii) in order to extract the water surface features, a decision tree was built. We chose the Classification and Regression Tree (CART) model developed by Breiman et al. [22], as a very popular statistical learning tool for analyzing large complex data.

2.3.1. Hydrological Analysis

KhronoStat (HydroSciences, Montpellier, France) is statistical software able to identify the sudden changes in a hydro or climatic data series using a set of statistical tests: the Pettitt rank-based test (non-parametric test), Buishand "U" test, Lee and Heghinian Bayesian method (parametric test) and Hubert procedure segmentation. The Buishand "U" and Lee and Heghinian tests are applicable if the time series investigated are normal. The Hubert procedure is the only one able to detect multiple breaks. The series is divided into "m" segments and, in order to limit the segmentation, the averages of two contiguous segments must be significantly different. This constraint is satisfied by Sheffe's test [23,24].

Considering hydrological drought analysis, an overview of the principal scientific literature [25–28] demonstrates that methodologies to characterize hydrological droughts can be divided into two categories (Figure 3): (i) determine the low flow index (a percentile from flow duration curve (FDC); annual minimum flow (MAM); base flow index (BFI); and (ii) deficit characteristics by Threshold Level Method—TLM and SPA—Sequent Peak Algorithm. The biggest disadvantage of the first category of methods is that it cannot give any information about the length of the drought period or about the beginning or end of the drought, while the methods in the second category offer a number of characteristics relating to the drought duration, the volume of severity, the intensity of the event (ratio of the volume and duration of drought), and the minimum of each event.

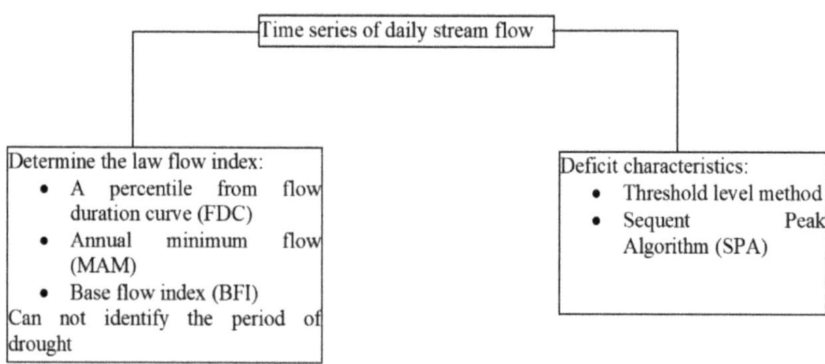

Figure 3. Methods to characterize hydrological drought.

Based on the SPA method, two droughts could be pooled if the reservoir has not totally "recovered" after the first drought event. Our interest is to find minor droughts also and not only major ones. For this study, we decided to use the TLM method. The TLM method, well known as the "method-of-crossing theory", was introduced by Yevjevich in 1967 [29]. First, a threshold level, Q_o, is chosen. The deficit begins where the flow values are below this Q_o level and ends where the flow is above Q_o value. We propose to use as threshold a value in the range of Q70%–Q95% resulting from the FDC analysis. To conclude, in order to evaluate the hydrological drought in the Nuntasi and Sacele Rivers, the methodology proposed is: (i) in order to select the driest period, an analysis of annual discharge time series data

(1965–2020) will be performed with KhronoStat software which detects a sudden change in time series data; (ii) an assessment threshold value with the FDC curve; (iii) hydrological drought assessment with TLM methods using daily discharge data.

2.3.2. RS Analysis

Using the special spectral reflectance (RS) properties of water, it is possible to differentiate between water and other surface materials. A multiple RS indices algorithm based on supervised classification was used to identify the surface water cover of Nuntasi Tuzla Lake. The spectral indices measurements acquired from Landsat TM/ETM+/OLI and MODIS datasets from 1984–2021 were analyzed by the algorithm that trains the randomly chosen samplings to construct a binary decision tree, and finally decodes this in order to classify the surface water (Figure 4). The algorithm was fully implemented and run on the GEE platform.

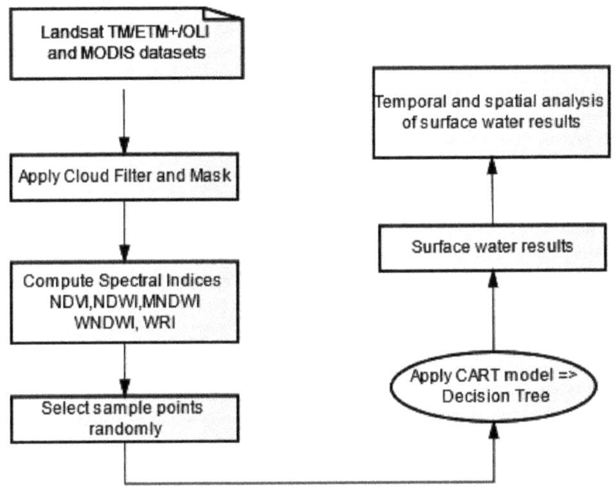

Figure 4. Flowchart of surface water detection based on a Decision Tree Classification.

Cloud filtering and masking of remote sensing data is a necessity to monitor the land surface cover more accurately. We used the cloud flag from the Level 1 Quality Assessment (QA) band of the Landsat datasets to limit the cloud cover at less than 30% in each selected image. Then, cloudy pixels were masked and not taken into consideration during the modeling of the surface water. This method is considered robust against errors [30,31]. After applying the cloud filter, we selected 233 Landsat RS images covering three seasons (spring, summer and autumn) of each year for the period investigated (Table 1). Similarly, the MOD09GA dataset for each season of 2012 was selected. In winter, most of the RS images did not pass the cloud filter, thus we decided to omit this season from our research.

Table 1. Number of Landsat RS images investigated.

Period	Spring	Summer	Autumn
1984–1999	15	50	25
2000–2012			
2000–31 May 2003	9	8	6
1 June 2003–2011	11	29	13
2013–2021	19	27	21

The RSI indices extracted from the full-time Landsat and MODIS surface reflectance dataset (Table 2) were then averaged as seasonal RSI from 1984 to 2021. For example, seasonal RSI in summer 2020 is taken from the RSI indices summed from 1 June to 31 August 2020 divided by the valid data (number of RS images) from the same period.

Table 2. RSI used for water surface extraction [1,12,16].

RSI	Formula	Observation
NDVI	NDVI = (NIR − Red)/(NIR + Red)	Water has negative value
NDWI	NDWI = (Green − NIR)/(Green + NIR)	Water has positive value
MNDWI	MNDWI = (Green − MIR)/(Green + MIR)	Water has positive value
WNDWI	WNDWI = (Green − a · NIR − (1 − a) · SWIR)/(Green + a · NIR + (1 − a) · SWIR)	a [0;1] [1]
WRI	WRI = (Green + Red)/(NIR + MIR)	Water is >1

[1] weighted coefficient; if a = 0 than WNDWI = MNDWI, if a = 1 than WNDWI = NDWI. For our study, the weighted coefficient a was set to 0.50, as the tests show that this value allows a high overall accuracy of the index performance [17].

In the formula, Green denotes the reflectance of the green band (Band 3 of the Landsat OLI data, Band 2 of the Landsat TM/ETM+ imagery, Band 4 of the MODIS data), Red represents the reflectance of the red band (Band 4 of the Landsat OLI data, Band 3 of the Landsat TM/ETM+ imagery, Band 1 of the MODIS data), NIR designates the reflectance of the near-infrared band (Band 5 for Landsat OLI, Band 4 for Landsat TM/ETM+, Band 2 of the MODIS data), MIR indicates the reflectance of the middle-infrared band (Band 6 for Landsat OLI and MODIS, Band 5 for Landsat TM/ETM) and SWIR stands for the reflectance of the SWIR1 band, which corresponds to Band 6 for Landsat OLI and MODIS data, and Band 5 for Landsat TM/ETM+ datasets.

The CART models lie on the supervised branch of the machine learning (ML) algorithms and may be used for both classification and regression problems. They find homogeneous subsets by recursively partitioning the input data, based on independent variable splitting criteria employing variance-minimizing algorithms. Being a non-parametric algorithm, intuitive and with a significant ability to characterize complex interactions among variables, the CART models have been widely used in different practical remote sensing applications [32,33]. For this study, we employed the CART algorithm provided by the GEE, with the seasonal averaged values of the RS indices (NDVI, NDWI, MNDWI, WNDWI and WRI) for each year as the input data. The surface water results for each season of the period investigated were then predicted based on the trained classification model.

To evaluate the performance of the RSI indices and the trained classification algorithm, three Landsat images for each period studied were randomly selected (Table 3).

Table 3. Selected scenes of Landsat TM/ETM+/OLI surface reflectance imagery used for accuracy assessment.

Data	Satellite	Date	Cloud Cover (%)
D1	Landsat 5 TM	16 July 1988	2.00
D2	Landsat 5 TM	3 September1994	4.00
D3	Landsat 5 TM	28 May 1999	14.00
D4	Landsat 7 ETM+	7 June 2000	0.00
D5	Landsat 5 TM	22 March 2004	0.00
D6	Landsat 5 TM	18 September 2011	3.00
D7	Landsat 8 OLI	10 September 2014	10.30
D8	Landsat 8 OLI	08 May 2015	0.83
D9	Landsat 8 OLI	29 July 2016	1.16

In order to represent the water and the land features in the study area, we used the stratified random sampling method to select 240 test samples (120 water and 120 non-water samples). The water samples consist of river, lake and sea water pixels, and the non-water samples are composed of soil, vegetation and built-up area pixels. The validation of the accuracy assessment has been achieved by referring to the original satellite images and Google Earth images.

3. Results and Discussion

3.1. Human Intervention

A review of the principal scientific literature [27,34–40] concerning human intervention in the Nuntasi-Tuzla Lake region demonstrates huge landscape modification that began more than a century ago. Two series of works were carried out in the region: the first began in the period 1903–1905 and the second started in 1970. The first series was carried out for fisheries and finalized in four stages [37]: (i) 1903–1916, when a link between the Sf Gheorghe arm and the Razim-Sinoe Lake was made in order to improve the fresh water circulation [41]; work began in 1905 with the Dunavat canal and was continued with the Dranov (1912) and Crasnicol canals; (ii) in 1930–1948 and (iii) 1950–1963, a great number of canals were excavated in order to connect the Danube with the lakes (Lipoveni, Fundea, Mustaca canals) and these lakes with each other; (iv) in 1963–1985, the existing canals were broadened and deepened [37] and a water gate at Portita mouth was built [39]. In 1963, a study was conducted concerning the territorial planning of the Dranov-Razelm-Sinoie Lacuster Complex [40]. This document recommended the closure of the Razim Sinoe Lagoon complex and the transformation of this system into an irrigation reservoir. In this context, in the period 1971–1975 the Portita mouth was closed by a dam [28] and a series of hydraulic works were made to control the communication between Razim Lake and the Black Sea through Sinoe Lake. From 1971 to 1976, the irrigation system was built and consisted of six subsystems. The Sinoe irrigation subsystem, of which the study area is a part, came into operation in 1976. The Sinoe SPA (Supply Pumping Station) covered a surface of 57,162 ha and had a flow rate of 46.1 mc/s. The distribution network had a length of 157.3 km [36]. The irrigation system was in operation from 1976 to 1990. After this year, water demand for irrigation decreased. Under the new property law, a conservation process for the irrigation system began [41,42], resulting in their destruction (over 75% of these arrangements were inoperable in 2016 [42,43]. According with the National Agency for Land Improvement, in 2021 only 10,930 ha were contracted (19%) in the study area and only 1250 ha have been irrigated. All this human intervention modified the hydrological condition of the Nuntasi-Tuzla ecosystem. The former Research and Design Institute for Water Management (the current Aquaproiect Company) shows a high demineralization of lake water from 57.8 g/L (1934) to 1.6 g/L (1982), a decrease in production of therapeutic mud, which was 0 in 1980, and an increase in the fresh water supply through the two tributaries, from 5 mil/mc in 1977 to 15 mil/mc in 1980 [44]. Braier [20] specified in her study that the salinity of this lake decreased due to the freshwater input and to the fact that the connection between this lake and the Istria lake was interrupted.

3.2. Hydrological Drought Analysis

Unfortunately, there are no meteorological or hydrological drought studies for this region. The idea to investigate this area appeared when, in 2020, the Istria town hall informed the authorities that Lake Nuntasi-Tuzla had dried up.

The Nuntasi Tuzla Lake is supplied by the Nuntasi and Sacele Rivers and we assume that there is a breakpoint in the time series data based on the variation of annual discharge presented in Figure 5, which shows a descending trend. As we showed in Section 2.2.1, in order to detect changes in the time series data, we used the annual discharge series for the 1965 to 2020 period. The multiannual average discharge for Nuntasi River is 0.348 m^3/s, and is 0.081 m^3/s for the Sacele River. It should also be specified that, for the Nuntasi River, the values for 2015 and 2017 are missing.

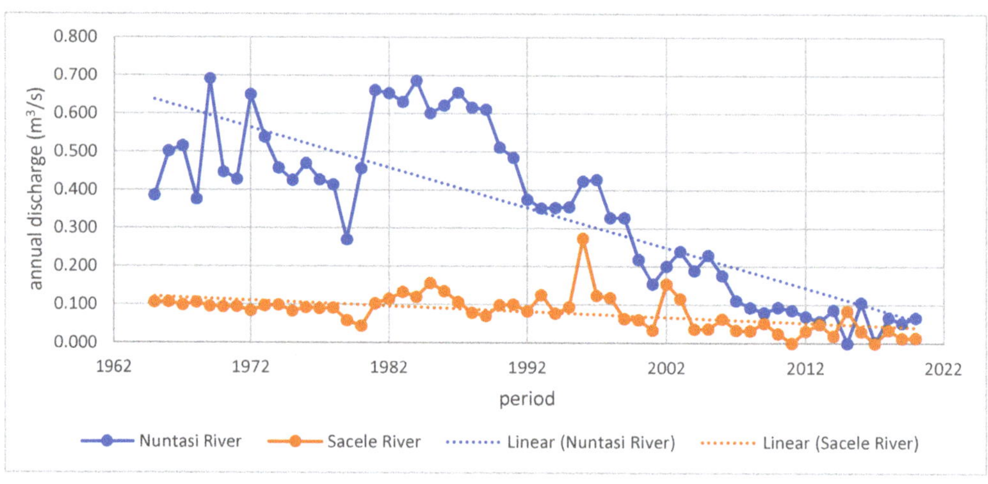

Figure 5. Variation of annual average discharge of the Nuntasi and Sacele Rivers.

The hydrological data are not normally distributed, even after several transformations. For the Nuntasi and Sacele annual discharge series, Buishand and Lee and Heghinian tests were not performed. The Pettitt test rejected the null hypothesis H_0 = no break for all confidence levels (90%, 95% and 99%) and admitted a break in 1997. The Hubert segmentation procedure detected four break points (1980, 1989, 1997 and 2006). For the Sacele discharge series, the Pettitt test rejected the null hypothesis for all confidence levels and admitted a break point in 2003, whilst the Hubert segmentation detected two break points, one in 1996 and one in 2003. For both Nuntasi and Sacele annual discharge, the breakpoints detected by the Pettitt and Hubert tests corresponded (1997 for the Nuntasi time series and 2003 for Sacele, respectively). As shown in the following table (Table 4), the average calculated for each subseries period detected by the Hubert procedure corresponded to the different stages of hydraulic work implemented in the study region and described in the previous Section 3.1. For the Nuntasi River, during the maximum operating period (1981–1989) of the irrigation system, the average annual discharge is 1.4 times greater than that from the previous period (1965–1980). This increase is due to the infiltration resulting from the irrigation water. The drill installed in the region by the former Land Improvement Office provided an increasing in the aquifers level [36]. Moreover, after 1989 the multiannual discharge for the period 1990–1997 returned to the baseline (1965–1980) and continued to decrease, reaching a dangerous level (6.75 times less than the average annual discharge for the 1965–1980 period). The same behavior was seen in the Sacele River, but the decrease was only by 3.1 times. Based on this situation, in order to detect hydrological drought using the two procedure described in the Section 2.3 (TLM and SPA, respectively) we used the daily discharge time series data for the 2007–2020 period.

Table 4. Average of the subseries detected via the Hubert segmentation procedure.

Hydrometric Station	Subseries Period	Average (m^3/s)	Observation
	1965–1980	0.466	hydraulic work construction and
	1981–1989	0.637	irrigation system operation
Nuntasi	1990–1997	0.410	
	1998–2006	0.229	gradual disrupting of
	2007–2020	0.069	irrigation system
	1965–1996	0.103	
Sacele	1997–2003	0.095	id.
	2004–2020	0.033	

Firstly, we construct the FDC by plotting the empirical cumulative frequency (EFQ) of the river daily discharge against exceedance frequency. Figure 6 shows the FDC for the Nuntasi River.

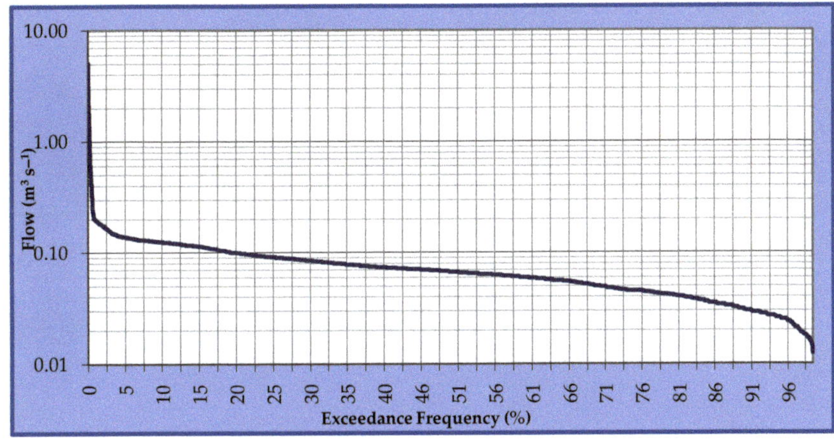

Figure 6. Flow Duration Curve for the Nuntasi River.

The difference between Q95%, Q90%, Q80% and Q75% is small (Table 5). In this context we chose as threshold (Qo) the value corresponding to Q90% for both rivers.

Table 5. Threshold values determined by FDC.

Hydrometric Station	EFQ (%)	Qo (m³ s⁻¹)
Nuntasi	95	0.03
	90	0.03
	80	0.04
	75	0.05
Sacele	95	0.01
	90	0.01
	80	0.02
	75	0.02

For a threshold value of 0.03 (Q90%), the TLM identified not only major droughts but also minor droughts (Figure 7). There are 11 periods of hydrological drought in total (Table 6) and the driest years are 2013 and 2020 with 16% and 20% deficit, respectively, of the period investigated. For the Sacele river, only two hydrological drought periods have been detected, in 2019 (4 days) and 2020 (4 days).

Table 6. Drought characteristics obtained with TLM and SPA for the Nuntasi station.

Date Start	Date End	Length Period (Days)
7/20/2008	7/22/2008	3
9/28/2011	9/30/2011	3
9/27/2012	9/29/2012	3
6/13/2013	6/30/2013	18
7/2/2013	7/2/2013	1
7/23/2013	8/30/2013	39
8/15/2019	8/16/2019	2
8/29/2019	9/1/2019	4
8/3/2020	9/14/2020	43
9/16/2020	9/30/2020	15
10/7/2020	10/20/2020	14

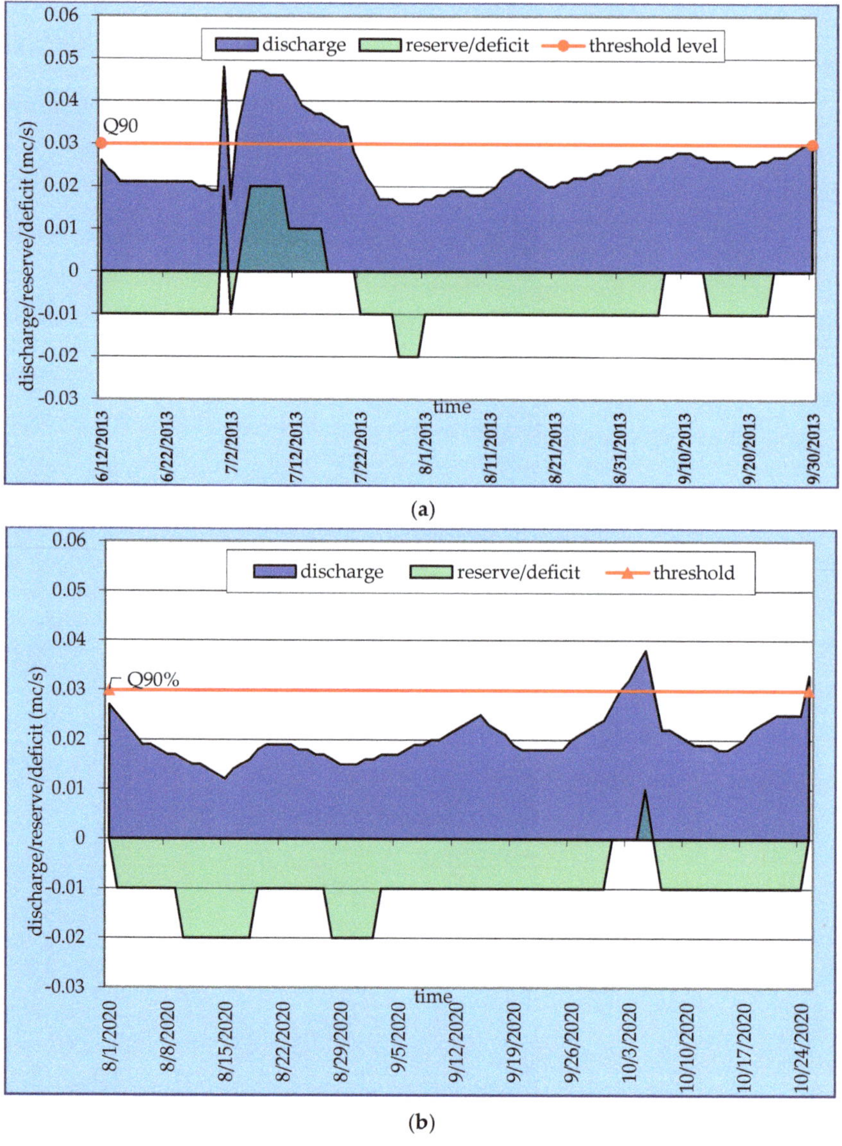

Figure 7. The TLM results for the Nuntasi station (2013): (**a**) zoom for 13 June 2013–30 August 2013 period; (**b**) zoom for 3 August 2020–20 October 2020 period.

3.3. Remote Sensing Analysis

3.3.1. Accuracy Assessment

Table 7 presents the confusion matrix of seasonal surface water detection results for the selected imagery introduced in Table 3. The overall accuracy (OA), Kappa coefficient, producer accuracy (PA) and user accuracy (UA) [45] of the surface water mapping are above 90%. On that account, the approach presented in this paper can be considered one of the optimal methods for surface water extraction for the study area (Figure 8).

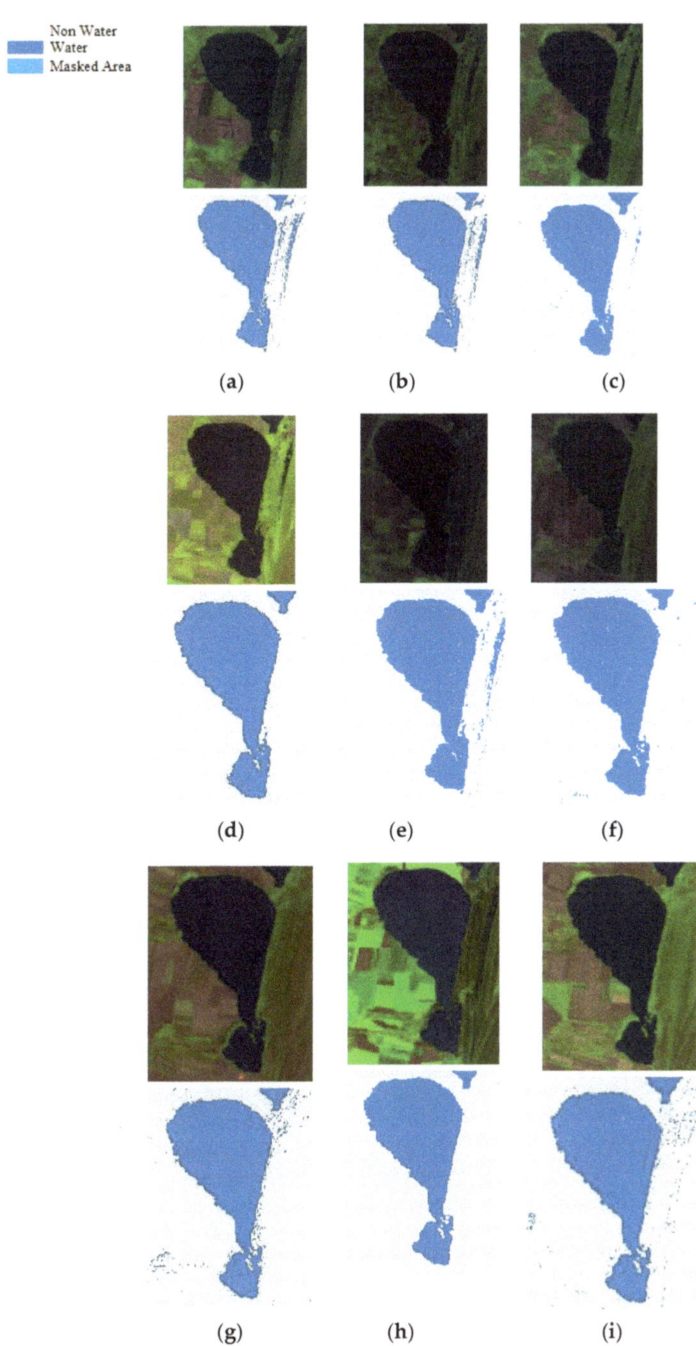

Figure 8. Earth Explorer Images (Source: https://earthexplorer.usgs.gov/ (accessed on 05 December 2021)) and seasonal surface water extraction maps for image (**a**) D1 (**b**) D2 (**c**) D3 (**d**) D4 (**e**) D5 (**f**) D6 (**g**) D7 (**h**) D8 (**i**) D9.

Table 7. Summary of accuracy assessments for Landsat dataset.

Data No.	Class	UA (%)	PA (%)	OA (%)	Kappa
D1	Water	95.74	100.00	98.21	0.96
	Non-Water	100.00	97.31		
D2	Water	100.00	98.11	99.09	0.98
	Non-Water	98.27	100.00		
D3	Water	98.36	100.00	99.01	0.98
	Non-Water	100.00	97.61		
D4	Water	96.77	100.00	98.03	0.96
	Non-Water	100.00	95.23		
D5	Water	100.00	98.36	99.09	0.98
	Non-Water	98.03	100.00		
D6	Water	98.50	100.00	99.06	0.98
	Non-Water	100.00	98.00		
D7	Water	98.07	100.00	99.03	0.98
	Non-Water	100.00	98.11		
D8	Water	98.21	100.00	99.03	0.98
	Non-Water	100.00	97.95		
D9	Water	97.95	100.00	99.09	0.98
	Non-Water	100.00	98.38		

Similarly, an accuracy verification procedure was applied on MODIS extracted surface water results (Table 8). The confusion matrix provided accuracies of higher than 90%, proving that the trained classification algorithm could effectively detect surface water from the MODIS dataset.

Table 8. Summary of accuracy assessments for MODIS dataset.

Date	Class	UA (%)	PA (%)	OA (%)	Kappa
20 August 2012	Water	95.75	95.55	95.74	0.91
	Non-Water	95.94	95.91		
23 September 2012	Water	96.07	98.00	96.66	0.94
	Non-Water	97.91	95.91		

Further, in order to evaluate the effectiveness of the surface water detection results from the MODIS dataset for 2012, a correlation analysis was performed for the period 2000–2011 between the MODIS and Landsat results. The multiple correlation coefficient of surface water between Landsat and MODIS was 0.97 and the coefficient of determination (R-squared) was 0.95 (Figure 9), showing that the surface water results from the MODIS dataset closely correspond to those from Landsat. On that account, the surface water detected from MODIS can effectively be used for seasonal change analysis of the surface water cover in the study area.

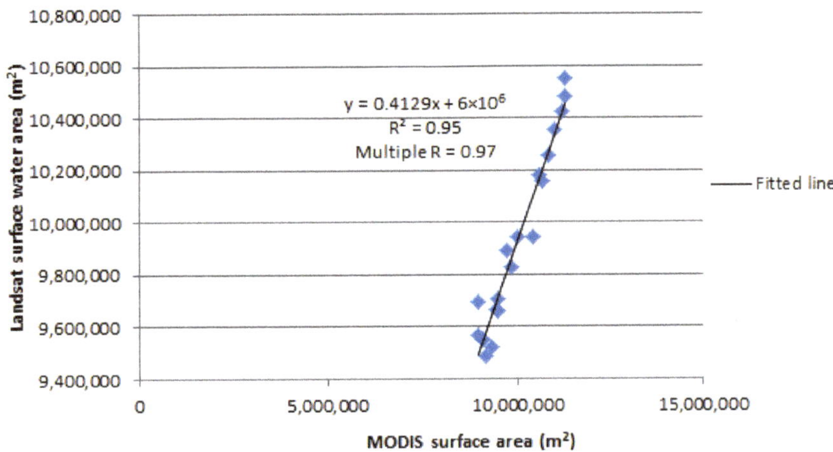

Figure 9. Correlation of surface water between Landsat and MODIS.

3.3.2. Seasonal Water Surface Variation

Figure 10 shows the variation of the seasonal average surface water from 1984 to 2021 of Nuntasi-Tuzla Lake.

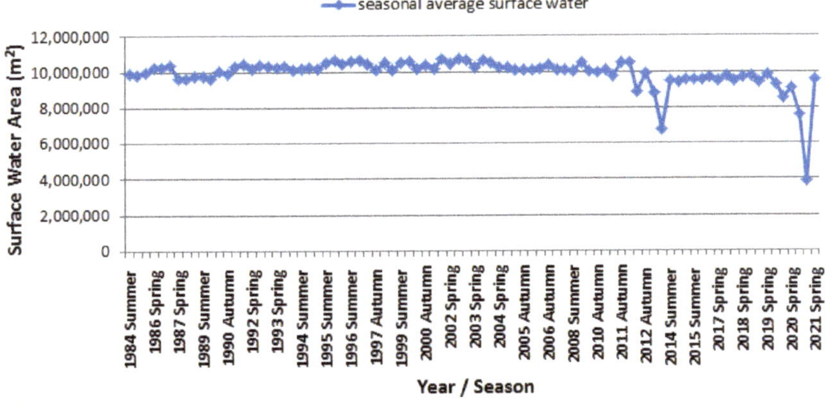

Figure 10. Variation in the seasonal average surface water area for the investigated period.

The largest seasonal average value from 1984–1999 period was recorded in autumn 1995 (Figure 11a), and the minimum was determined in autumn 1989 (Figure 11b). For the 2000–2012 period, the maximum average surface water was estimated in summer 2001 (Figure 11c), and the minimum in summer 2012 (Figure 11d). For the last period considered, 2013–2021, the largest seasonal average surface water was determined in spring 2019 (Figure 11e), and the smallest in autumn 2020 (Figure 11f). Figure 12a,b show the lake surface maps overlaid to produce lake surface area changes maps for the periods investigated.

The Hubert test detected five subseries in the seasonal water surface data series (in summer 1991, spring 2012, spring 2013, summer 2019 and summer 2020) which coincide with the detection in the flow data series of the Nuntasi and Sacele Rivers and full irrigation system operation.

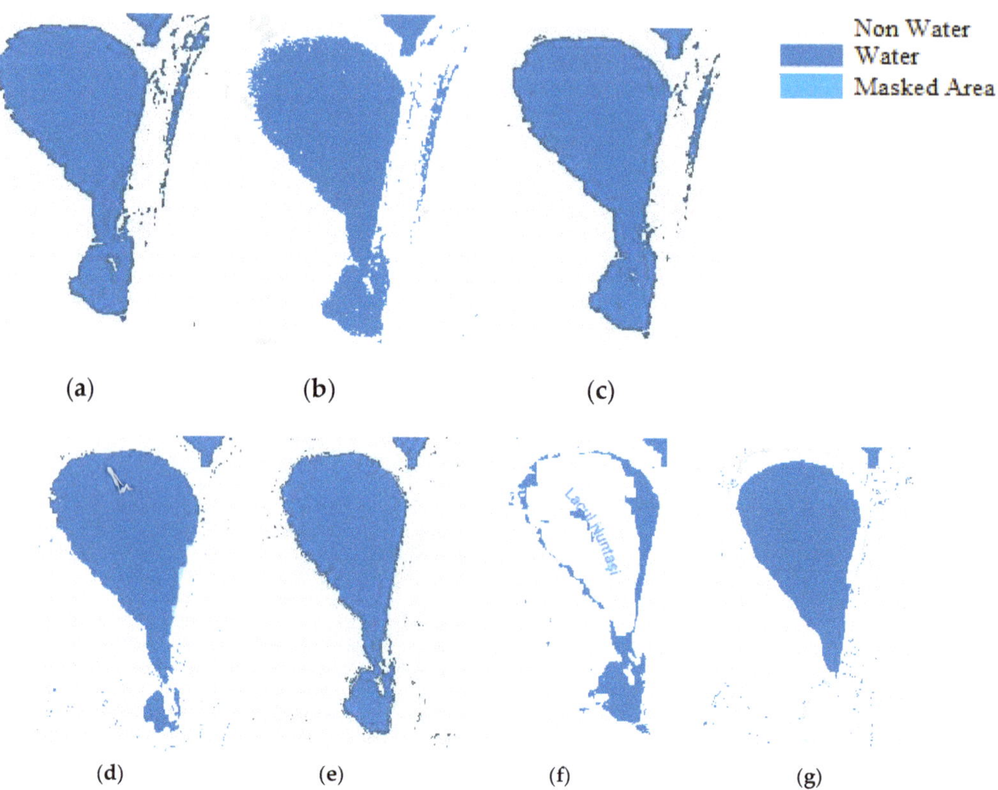

Figure 11. Seasonal surface water extraction maps (**a**) autumn 1995 (**b**) autumn 1989 (**c**) summer 2001 (**d**) summer 2012 (**e**) spring 2019 (**f**) autumn 2020 (**g**) summer 2013.

The seasonal average surface water was approximately 992 ha from 1984 to summer 1991, and increased by 4%, on average, after 1991. It did not then suffer significant changes until spring 2012 (Figure 10). Figure 12c shows the maximum growth of surface water area, 11.34%, from the smallest area in autumn 1989 to the largest in summer 2001. In summer 2012, the extracted surface water area decreased by 16.57% compared to the previous season, and it further reduced considerably till summer 2013 (Figure 11g), when the lake surface reached 670 ha, approximately. Figure 12d shows the decrease in surface water area from spring 2013 to summer 2013. Following 2014, the seasonal average surface water area increased steadily until spring 2019 (Figure 11e). Figure 12e shows the decrease in water area from spring 2019 to autumn 2019.

On average, the surface water from 2013 to spring 2020 decreased by about 9.26% compared to values from 2000 to 2012 and by 8.86% compared to the 1984–1999 period. In summer and autumn of 2020, it dramatically decreased by 18.76% and 57.83%, respectively, compared to the average surface water of previous years.

It is noted that the periods with the lowest values for lake surface coincide with the prolonged drought periods determined according to the TLM method (Table 6).

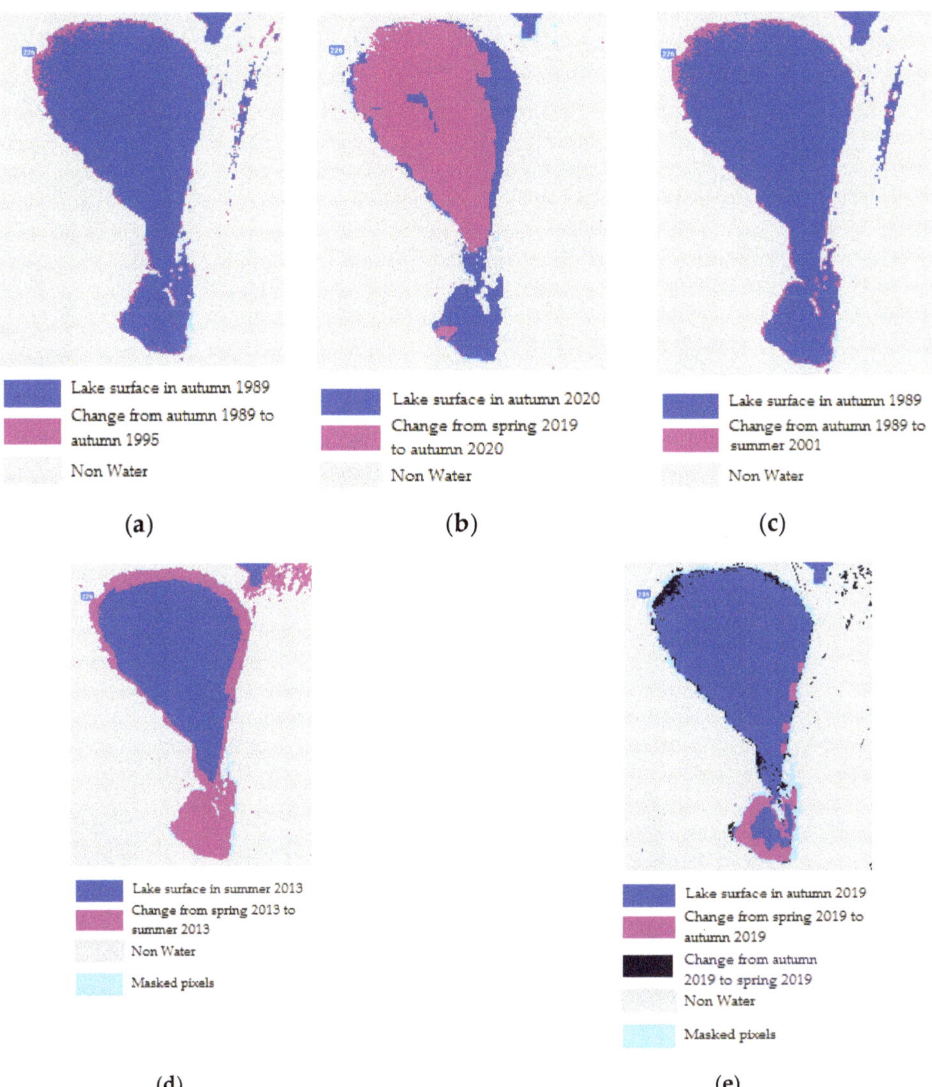

Figure 12. Lake Nuntasi-Tuzla average surface area changes map in the period (**a**) autumn 1989–autumn 1995 (**b**) spring 2019–autumn 2020 (**c**) autumn 1989–summer 2001 (**d**) spring 2013–summer 2013 (**e**) spring 2019–autumn 2019.

4. Conclusions

The main purpose of this study was to detect via remote sensing techniques the long-term variation of the Nuntasi-Tuzla Lake surface water during the 1984–2021 period in correlation with human intervention and climate change. We have reached several conclusions:

(i)   In natural conditions, the lake system is connected to the Black Sea via the Portita, Periboina and Edighiol outlets (Figure 2) and with the St George Arm of the Danube River via a system of canals and marshes. The nine lakes which compose the Razim Sinoe System are interconnected by the above mentioned system of canals. Human intervention has led to a deterioration of the ecosystem of the entire Razim Sinoe system, which has been isolated from the Black Sea, and the connections between the lakes have been cut by sluices. In this context, the water became freshwater. The cannal between Nuntasi-Tuzla Lake and Istria Lake has been silted since 1976. In view of the fact that the Dobrogea region is the driest region of Romania [35] and in order to ensure agricultural development, the state authorities from the period 1968–1975 built an important irrigation system in this region. In the irrigation period, the two tributaries (Nuntasi and Sacele rivers) have supplied the lake constantly, but after 1990, when the irrigation was stopped, the river flow decreased over time reaching its lowest level in the 2004 (2007)–2020 period. In a recent publication, the authors [21] showed that "the precipitation increased starting with 2012 but the evapotranspiration losses are much larger than the precipitation increase". We can conclude that the budget is negative. It is apparent that the surface lake may be a subject of irreversible changes.

(ii) The analysis of the daily flows of the two rivers during the 2007–2020 period detected several important drought events. Among these, two drought periods with long duration were determined, in 2013 and 2020. In this context, we further investigated if there was any influence of flow decreasing on the lake water surface and if there have been any other similar situations in the past.

(iii) Using the CART model with the seasonal averaged values of RS indices (NDVI, NDWI, MNDWI, WNDWI and WRI) as the input data, we assessed the seasonal water lake surface variation during the period 1984–2021 with over 90% mapping accuracy, user accuracy and overall accuracy. The results of the proposed classification method revealed that the evolution of surface lake water is correlated with human intervention and the hydrological drought identified with the TLM method. Significant decrease during the 2003–2020 period was identified in the surface water lake's evolution, thus the hydrological drought identified in 2011, 2012, 2013 and 2020 corresponds with the lowest values for water lake surface. In our opinion, the method based on remote sensing data and the CART model is calibrated, due to the results obtained. Unfortunately, we have only a short series of daily records, which limits this study.

Overall, the findings of this study provide some insights into the evolution of the Nuntasi Tuzla Lake and the factors driving it through a long period of time, which may be further used to formulate scientific measures to prevent other drought disasters similar to that of August 2020. Future work may imply improvement of the surface water changes' detection method, by investigating other spectral indices and comparing with state-of-the-art classification methods. Although the CART model is a very intuitive and powerful classifier, it often involves longer time to train and fails to meet the best performance if the problem has many uncorrelated variables. Other classification methods, such as support vector machines (SVM) or multilayer perceptron (MLP) neural network, which work well with large input data, can be adopted to identify surface water in Landsat satellite images.

According to the Romanian Water National Administration—Littoral Branch (ABADL), an extensive restoration campaign started in September 2020 and the canal between Istria and Nuntasi-Tuzla lakes was restored (Figure 13).

A fragile equilibrium was reestablished and in spring of 2021 a group of flamingo birds was spotted.

An Integrated Drought Management Programme (IDMP) was launched in 2013 by GWP (Global Water Partnership) for Central and Eastern Europe. The main objective [21] is to support the responsible authorities to prepare and integrate the Drought Management plan (DMP) within the future Basin Management Plan (BMP). In this context, more efforts are needed to validate the method proposed in this study.

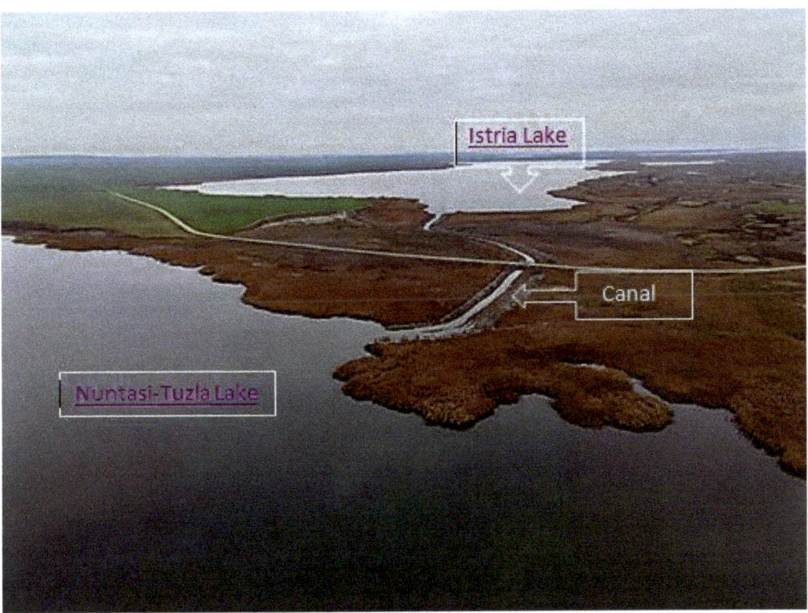

Figure 13. Nuntasi Tuzla Lake restauration (ABADL source) in September 2020.

Author Contributions: Conceptualization, C.Ș. and C.M.; Methodology, C.Ș. and C.M.; Software, C.Ș.; Validation, C.M. and G.D.; Formal analysis, C.M.; Investigation, C.M. and C.Ș.; Resources, G.D.; Data curation, G.D.; Writing—original draft preparation, C.M.; Writing—review and editing, C.Ș.; Visualization, G.D.; Supervision, C.M.; All authors have read and agreed to the published version of the manuscript.

Funding: This research received no external funding.

Institutional Review Board Statement: Not applicable.

Informed Consent Statement: Not applicable.

Data Availability Statement: Not applicable.

Conflicts of Interest: The authors declare no conflict of interest.

References

1. Peng, Y.; He, G.; Wang, G.; Cao, H. Surface Water Changes in Dongting Lake from 1975 to 2019 Based on Multisource Remote-Sensing Images. *Remote Sens.* **2021**, *13*, 1827. [CrossRef]
2. Bărbulescu, A.; Dumitriu, C.Ș.; Maftei, C. On the Probable Maximum Precipitation method. *Rom. J. Phys.* **2022**, *67*, 801.
3. Bărbulescu, A.; Postolache, F.; Dumitriu, C.S. Estimating the Precipitation Amount at Regional Scale Using a New Tool, Climate Analyzer. *Hydrology* **2021**, *8*, 125. [CrossRef]
4. Cui, B.-L.; Xiao, B.; Li, X.-Y.; Wang, Q.; Zhang, Z.-H.; Zhan, C. Exploring the geomorphological processes of Qinghai Lake and surrounding lakes in the northeastern Tibetan Plateau, using Multitemporal Landsat Imagery (1973–2015). *Glob. Planet. Chang.* **2017**, *152*, 167–175. [CrossRef]
5. Adrian, R.; O'Reilly, C.M.; Zagarese, H.; Baines, S.B.; Hessen, D.O.; Keller, W.; Livingstone, D.M.; Sommaruga, R.; Straile, D.; van Donk, E.; et al. Lakes as sentinels of climate change. *Limnol. Oceanogr.* **2009**, *54*, 2283–2297. [CrossRef]
6. Maftei, C.; Buta, C.; Carazeanu Popovici, I. The Impact of Human Interventions and Changes in Climate on the Hydro-Chemical Composition of Techirghiol Lake (Romania). *Water* **2020**, *12*, 2261. [CrossRef]
7. Conserving Iran and Iraq's Wetlands, UNEP. 2018. Available online: http://www.unep.org/news-and-stories/story/conserving-iran-and-iraqs-wetlands (accessed on 16 June 2021).
8. Benson, L.V.; Paillet, F.L. The Use of Total Lake-Surface Area as an Indicator of Climatic Change: Examples from the Lahontan Basin. *Quat. Res.* **1989**, *32*, 262–275. [CrossRef]

9. Rouse, W.; Haas, R.H.; Schell, J.A.; Deering, D. *Monitoring Vegetation Systems in the Great Plains with ERTs*; NASA: Washington, DC, USA, 1974; p. 309.

10. Peters, A.J.; Walter-Shea, E.A.; Ji, L.; Vliia, A.; Hayes, M.; Svoboda, M.D. Drought Monitoring with NDVI-Based Standardized Vegetation Index. *Photogram. Eng. Remote Sens.* **2002**, *68*, 71–75.

11. Mishra, A.K.; Singh, V.P. A review of drought concepts. *J. Hydrol.* **2010**, *391*, 202–216. [CrossRef]

12. Rokni, K.; Ahmad, A.; Selamat, A.; Hazini, S. Water Feature Extraction and Change Detection Using Multitemporal Landsat Imagery. *Remote Sens.* **2014**, *6*, 4173–4189. [CrossRef]

13. McFeeters, S.K. The use of the Normalized Difference Water Index (NDWI) in the delineation of open water features. *Int. J. Remote Sens.* **1996**, *17*, 1425–1432. [CrossRef]

14. Xu, H. Modification of normalised difference water index (NDWI) to enhance open water features in remotely sensed imagery. *Int. J. Remote Sens.* **2006**, *27*, 3025–3033. [CrossRef]

15. Shen, L.; Li, C. Water body extraction from Landsat ETM+ imagery using adaboost algorithm. In Proceedings of the 2010 18th International Conference on Geoinformatics, Beijing, China, 18–20 June 2010; pp. 1–4.

16. Guo, Q.; Pu, R.; Li, J.; Cheng, J. A weighted normalized difference water index for water extraction using Landsat imagery. *Int. J. Remote Sens.* **2017**, *38*, 5430–5445. [CrossRef]

17. Acharya, T.D.; Subedi, A.; Yang, I.T.; Lee, D.H. Combining Water Indices for Water and Background Threshold in Landsat Image. *Proceedings* **2017**, *2*, 143. [CrossRef]

18. Acharya, T.D.; Subedi, A.; Lee, D.H. Evaluation of Water Indices for Surface Water Extraction in a Landsat 8 Scene of Nepal. *Sensors* **2018**, *18*, 2580. [CrossRef]

19. Preoteasa, L.; Vespremeanu-Stroe, A.; Hanganu, D.; Katona, A.; Timar-Gabor, A. Coastal changes from open coast to present lagoon system in Histria region (Danube delta). *J. Coast. Res.* **2013**, *65*, 564–569. [CrossRef]

20. Breier, A. *Lacurile de pe Litoralul Romanesc al Marii Negre. Studiu Hidrogeografic*; Editura Academiei Republicii Socialiste România: București, Romania, 1976.

21. Maftei, C.; Dobrica, G.; Cerneaga, C.; Buzgaru, N. Drought Land Degradation and Desertification—Case Study of Nuntasi-Tuzla Lake in Romania. In *Water Safety, Security and Sustainability: Threat Detection and Mitigation*; Vaseashta, A., Maftei, C., Eds.; Springer International Publishing: Cham, Switzerland, 2021; pp. 583–597. [CrossRef]

22. Breiman, L.; Friedman, J.H.; Olshen, R.A.; Stone, C.J. *Classification and Regression Trees*; Routledge: Boca Raton, FL, USA, 2017. [CrossRef]

23. Maftei, C.; Barbulescu, A. Statistical analysis of precipitation time series in the Dobrudja region. *Mausam* **2012**, *63*, 553–564. [CrossRef]

24. Dumitriu, C.S.; Dragomir, F.-L. Modeling the Signals Collected in Cavitation Field by Stochastic and Artificial Intelligence Methods. In Proceedings of the 2021 13th International Conference on Electronics, Computers and Artificial Intelligence (ECAI), Pitesti, Romania, 1–3 July 2021; pp. 1–4. [CrossRef]

25. Vogt, J.V.; Somma, F. *Drought and Drought Mitigation in Europe*; Springer: Dordrecht, Netherlands, 2000. [CrossRef]

26. Fatulová, E.; Majerčáková, O.; Houšková, B.; Bardarska, G.; Alexandrov, V.; Kulířová, P.; Gayer, J.; Molnár, P.; Fiala, K.; Tamás, J.; et al. *Guidelines for Preparation of the Drought Management Plans: Development and Implementation of Risk-Based Drought Management Plans in the Context of the EU Water Framework Directive—As Part of the River Basin Management Plans*; Global Water Partnership Central and Eastern Europe: Bratislava, Slovakia, 2015.

27. Smakhtin, V.; Hughes, D.A.; International Water Management Institute. *Review, Automated Estimation and Analyses of Drought Indices in South Asia*; International Water Management Institute: Colombo, Sri Lanka, 2004.

28. Tallaksen, L.; van Lanen, H.A.J.; *Hydrological Drought. Processes and Estimation Methods for Streamflow and Groundwater.* Elsevier. 2004. Available online: https://research.wur.nl/en/publications/hydrological-drought-processes-and-estimation-methods-for-streamf (accessed on 13 August 2021).

29. Yevjevich, V.M. An objective approach to definitions and investigations of continental hydrologic droughts. *J. Hydrol.* **1969**, *7*, 353. [CrossRef]

30. Mateo-García, G.; Gómez-Chova, L.; Amorós-López, J.; Muñoz-Marí, J.; Camps-Valls, G. Multitemporal Cloud Masking in the Google Earth Engine. *Remote Sens.* **2018**, *10*, 1079. [CrossRef]

31. Bărbulescu, A.; Dumitriu, C.Ș.; Dragomir, F.L. Detecting Aberrant Values and Their Influence on the Time Series Forecast. In Proceedings of the International Conference on Electrical, Computer, Communications and Mechatronics Engineering (ICECCME), Mauritius, Mauritius, 7–8 October 2021. [CrossRef]

32. Pencue-Fierro, E.L.; Solano-Correa, Y.T.; Corrales-Muñoz, J.C.; Figueroa-Casas, A. A Semi-Supervised Hybrid Approach for Multitemporal Multi-Region Multisensor Landsat Data Classification. *IEEE J. Sel. Top. Appl. Earth Obs. Remote Sens.* **2016**, *9*, 5424–5435. [CrossRef]

33. Duan, H.; Deng, Z.; Deng, F.; Wang, D. Assessment of Groundwater Potential Based on Multicriteria Decision Making Model and Decision Tree Algorithms. *Math. Probl. Eng.* **2016**, *1*, 2064575. [CrossRef]

34. Gastescu, P. The Danube Delta: Geographical characteristics and ecological recovery. *GeoJournal* **1993**, *29*, 57–67. [CrossRef]

35. Aurel, L.; Liliana, M. Drought Management in the Agriculture of Dobrogea Province, 2013, 65–70. Available online: https://mpra.ub.uni-muenchen.de/53403/1/MPRA_paper_53403.pdf (accessed on 24 December 2021).

36. Grumezea, N.; Kleps, C.; Tusa, C. *Evolutia Nivelului si Chimismului Apei Freatice din Amenajarile de Irigatii in Interrelatie cu Mediul Inconjurator*; Intreprinderea Poligrafica Oltenia: București, Romania, 1990.

37. Pons, L.J. Natural resources. In *Conservation Status of the Danube Delta*; IUCN: Newbury, Berkshire, UK, 1992; Volume 4, pp. 23–36.

38. Bretcan, P.; Murărescu, O.; Samoilă, E.; Popescu, O. The Modification of the Ecological Conditions in the Razim-Sinoaie Lacuster Complex as an Effect of the Anthropic Intervention. Presented at the XXIVth Conference of the Danubian Countries on the Hydrological Forecasting and Hydrological Bases of Water Management, Bled, Slovenia, 2–4 June 2008.

39. Vadineanu, A.; Cristofor, S.; Ignat, G.; Romanca, G.; Ciubuc, C.; Florescu, C. Changes and opportunities for integrated management of the Razim-Sinoe Lagoon System. *Int. J. Salt Lake Res.* **1997**, *6*, 135–144. [CrossRef]

40. Bretcan, P.; Murărescu, O.; Samoilă, E.; Popescu, O. Water Management in the Razim—Sinoie Lacustrine Complex. Presented at the International Symposium on Water Management and Hydraulic Engineering, Ohrid/Macedonia, 1–5 September 2009; pp. 791–801.

41. Turnock, D. Water resource management problems in Romania. *GeoJournal* **1979**, *3*, 609–622. [CrossRef]

42. Proiect de Deabilitare si Reforma a Irigatiilor. Available online: http://old.madr.ro/pages/strategie/proiect-de-reabilitare-si-reforma-a-irigatiilor.pdf (accessed on 13 August 2021).

43. Ministerul Agriculturii și Dezvoltării Rurale. Strategia Națională de Reabilitare și Extindere a Infrastructurii de Irigații din România. 2019. Available online: https://www.madr.ro/download (accessed on 13 August 2021).

44. Rusu, G. *Sinteza Cercetarilor in Perioada 1975–1982 Pentru Lacurile Techirghiol, Nuntasi si Istria*; ICPGA: Bucuresti, Romania, 1982.

45. Foody, G.M. Status of Land Cover Classification Accuracy Assessment. *Remote Sens. Environ.* **2002**, *80*, 185–201. [CrossRef]

Article

Changing Causes of Drought in the Urmia Lake Basin—Increasing Influence of Evaporation and Disappearing Snow Cover

Maral Habibi [1,*], Iman Babaeian [2] and Wolfgang Schöner [1]

[1] Department of Geography and Regional Science, University of Graz, 8010 Graz, Austria; wolfgang.schoener@uni-graz.at

[2] Climate Research Institute, Atmospheric Science and Meteorological Research Center, Mashhad 91735-676, Iran; i.babaeian@gmail.com

* Correspondence: maral.habibi@uni-graz.at; Tel.: +43-(0)316-380-5678

Abstract: The water level of the Urmia Lake Basin (ULB), located in the northwest of Iran, started to decline dramatically about two decades ago. As a result, the area has become the focus of increasing scientific research. In order to improve understanding of the connections between declining lake level and changing local drought conditions, three common drought indices are employed to analyze the period 1981–2018: The Standard Precipitation Index (SPI), the Standard Precipitation-Evaporation Index (SPEI), and the Standardized Snow Melt and Rain Index (SMRI). Although rainfall is a significant indicator of water availability, temperature is also a key factor since it determines rates of evapotranspiration and snowmelt. These different processes are captured by the three drought indices mentioned above to describe drought in the catchment. Therefore, the main objective of this paper is to provide a comparative analysis of drought over the ULB by incorporating different drought indices. Since there is not enough long-term observational data of sufficiently high density for the ULB region, ECMWF Reanalysis data version 5(ERA5) has been used to estimate SPI, SPEI, and SMRI drought indicators. These are shown to work well, with AUC-ROC > 0.9, in capturing different classes of basin drought characteristics. The results show a downward trend for SPEI and SMRI (but not for SPI), suggesting that both evaporation and lack of snowmelt exacerbate droughts. Owing to the increasing temperatures in the basin and the decrease in snowfall, drought events have become particularly pronounced in the SPEI and SMRI time series since 1995. No significant SMRI drought was detected prior to 1995, thus indicating that sufficient snowfall was available at the beginning of the study period. The study results also reveal that the decrease in lake water level from 2010 to 2018 was not only caused by changes in the water balance components, but also by unsustainable water management.

Keywords: Urmia Lake Basin; drought; climate change; snow cover; evapotranspiration; ERA5; extremes

Citation: Habibi, M.; Babaeian, I.; Schöner, W. Changing Causes of Drought in the Urmia Lake Basin—Increasing Influence of Evaporation and Disappearing Snow Cover. *Water* **2021**, *13*, 3273. https://doi.org/10.3390/w13223273

Academic Editor: Alina Barbulescu

Received: 9 October 2021
Accepted: 15 November 2021
Published: 18 November 2021

Publisher's Note: MDPI stays neutral with regard to jurisdictional claims in published maps and institutional affiliations.

Copyright: © 2021 by the authors. Licensee MDPI, Basel, Switzerland. This article is an open access article distributed under the terms and conditions of the Creative Commons Attribution (CC BY) license (https://creativecommons.org/licenses/by/4.0/).

1. Introduction

Droughts are significant natural hazards worldwide, with widespread impacts on humans and ecosystems. While the frequency and magnitude of droughts are expected to change in the coming decades due to climate change, the regional evolution of future droughts remains highly uncertain [1,2]. The Middle East and southwest Asia are water-stressed regions, societally vulnerable, and prone to severe droughts. Since the 1940s, there have been two particularly severe drought periods in these regions, 1999–2001 and 2007–2008, respectively [3–5]. Different drought indices have confirmed that 2001 was one of the most severe periods of drought in Iran [6,7]. Studies regarding the periodic behavior of drought in Iran show that, in addition to the dominant short-term periods over the northwest of Iran, long-term periods of 10 and 30 years have also been observed. Although the average rainfall in southern Iran is much lower than in the north, short-term droughts in the north are more severe than in the south. Meanwhile, long-term droughts are more severe in Iran's west, east, southeast, south, and center [8–10].

Large-scale atmospheric circulation and teleconnection play an essential role in the development of drought in Iran. Dry periods are usually accompanied by westward displacement of a subtropical high, while wet periods are associated with eastward displacement of the high-pressure system toward the Arabian Sea. The Southern Oscillation Index (SOI) in its positive phase can decrease precipitation in the northwest of Iran [11]. Research has also confirmed that the late withdrawal of the Indian monsoon is associated with a delay in the onset of fall precipitation over Iran. This delay is accompanied by prolonged subtropical high pressure settling over Iran's plateau, with the latter preventing the formation of polar jet frontal systems [12–15].

Northwestern Iran has the highest risk of severe and long-lasting drought [16], and the shrinking of the Urmia Lake has been quite dramatic over the last two decades. Whether or not this is an effect of ongoing climate change or is the result of human activity, such as water management (e.g., water dams and unapproved wells) or agriculture [17,18], is the subject of considerable debate. The declining water level of Lake Urmia is now a major challenge in national water and environmental management. In addition, climate change and reduced precipitation [19–21], decreased water levels [22], and the expansion of saline areas around the lake have also all led to numerous environmental and economic impacts [23]. Over the Urmia Lake basin, a warming trend of 0.18 °C/decade has been detected, and precipitation has decreased by approximately 9 mm/decade [24]. This has resulted in an increase in evaporation from the lake of 6.2 mm/decade [24]. Consequently, the water level of Lake Urmia has been rapidly declining since 1995, with a 6.1 m decline for the period 1995–2009. The lake surface area has also decreased by about -188.3 km^2/yr, from 5503 km^2 in 1998 to 2323 km^2 in 2011. Mean precipitation in the ULB has decreased by 9.2%, and the average maximum temperature has increased by 0.8 °C over 1964–2005 [25]. The leading cause of the recession of Lake Urmia is the diminution of inflow from rivers [19,24,26–28]. Aziz et al. [29] found that compared with other impacts, the operation of dams (26%) and increasing water demand (16%) have played a clear role in reducing the water input to the lake. Changing climate contributed up to 16% of the lake level change within 1999–2014. The lake reached its highest level in 1995, but it has generally declined since then, reaching the lowest level in recent years [19,23,30]. The mean annual water inflow into the lake is 6900×10^6 m^3, of which 4900×10^6 m^3 is from rivers (with 2000×10^6 m^3 of this coming from the river Zarrineh Rood), 500×10^6 m^3 from floodwaters, and 1500×10^6 m^3 from precipitation onto the lake [31]. However, as water inflow into the lake has also recently declined [19,26], Lake Urmia's surface area and volume have continued to decrease [24].

The decline in the lake's water level and surface area [26] over the last two decades has caused an environmental disaster, led to increased salinity, and negatively affected agriculture, livelihoods, and health [32,33]. The strong natural climatic variability is being threatened and amplified by climate change, thus increasing the occurrence of hydrologically extreme drought events [34]. The impact of drought, particularly its socioeconomic impact, means that substantial improvements in water management practices are required in order to preserve or partially restore the lake [24]. Data from weather stations located in the ULB indicate that an increase in drought is to be expected in the future [35]. Salt storms around the former shoreline and lakebed have already begun to sweep across the region [36], negatively impacting local agriculture and adversely influencing human health [24].

Drought may be analyzed in terms of its duration, severity, spatial coverage, and water deficit characteristics. Several methods and indices have been developed based on climatic and hydrological variables to monitor and quantify drought intensity and impacts. These cover factors such as precipitation [37–39], soil humidity [40], evapotranspiration and vegetation conditions [41–43], or combine these in a number of specific indices [44–47]. Analyzing the impact of global climate change on drought has also become common [35].

Over the past few decades, new indices have been developed for drought quantification, and apart from data on precipitation, they have added variables such as temperature, snowpack, evapotranspiration, soil moisture, runoff, streamflow, groundwater, reservoir storage, and snowmelt [37–41,48–52].

Despite the many existing studies of the ULB, drought quantification remains an ongoing issue of high practical value. The present study focuses on quantifying drought for the ULB by using SPI, SPEI, and SMRI, and looks at precipitation, evapotranspiration, and snowmelt as the major components of water balance.

Snow droughts remain relatively unexplored compared to other drought types, and only a few studies have included snow information in drought characterization [41,52–57]. Snow droughts have become more prevalent, intensified, and prolonged across the Western United States. In addition, Eastern Russia, Europe, and the US have emerged as snow drought hotspots, with respective increases in drought duration of ~2%, 16%, and 28% in recent decades. While runoff from mountain snowmelt can support agricultural activities in downstream areas (e.g., California's Central Valley), snow also directly provides meltwater to croplands and protects winter wheat from frost and freezing (e.g., in Russia and Ukraine) [58]. The gradual increase in temperature in some regions in North America, such as the Athabasca River Basin (ARB), Central North America, and Alaska, resulting in snowpack melting in early spring, has already been described [59–61]. A snow drought or a deficit in snow water equivalent (SWE is the amount of water obtained if the snowpack is melted instantaneously) can have severe regional and global impacts on human activities and ecosystems, both in snow-covered and snow-free areas. Precipitation storage as snow in winter and spring can critically modulate hydrological droughts in summer. Concerning streamflow, droughts are often related to the presence or absence of snow in the preceding winter, whereas winter droughts can occur despite large amounts of precipitation falling as snow [62]. One extension of the SPI is the Standardized Snowmelt and Rain Index (SMRI). This accounts for the impact of rain and snowmelt deficits on streamflow. The SMRI has been found to be a valuable complementary index for characterizing streamflow droughts in catchments with a significant snowmelt component in runoff generation [52]. When the snow/precipitation ratio decreases, the SMRI approaches the SPEI. However, the difference between SPEI and SMRI increases with the increasing impact of snow cover on streamflow for the entire and dry periods only [52]. Moreover, meteorological drought indices that include evaporation or snowmelt can be better correlated with streamflow [63].

The water level of Lake Urmia also experiences monthly and seasonal variations, primarily determined by snowmelt [24]. Thus, depending on the snowmelt in the surrounding mountains, the Urmia water level peaks in May–June (precipitation peaks from October to May) and drops to a minimum in October to December due to the lack of snowmelt. Overall, monthly fluctuations in lake water level can be up to nearly 0.6 m [24]. Urmia Lake started drying out about twenty years ago, and the lake level has declined by more than eight meters during this period. Since reaching its highest level in 1995 (1278.48 m), the water level has decreased annually on average by 40 cm over the last two decades. In September 2015, it reached the lowest level, and southern parts of the lake dried out. Indeed, the whole lake is a moderately shallow water body (about 6 m deep on average) Figure 1 [64].

The main objective of the present study is to better understand drought events in the ULB by focusing on SMRI in addition to the SPI and SPEI indices. Use is made of the SPEI index in order to assess the impact of rising temperatures. The ERA5 dataset has been used for drought indices to provide for a higher density of meteorological data and allow for a more elongated data period. The paper is organized as follows: Section 2 introduces the materials and methods. Sections 3–5 contain results, discussion, and conclusions, respectively.

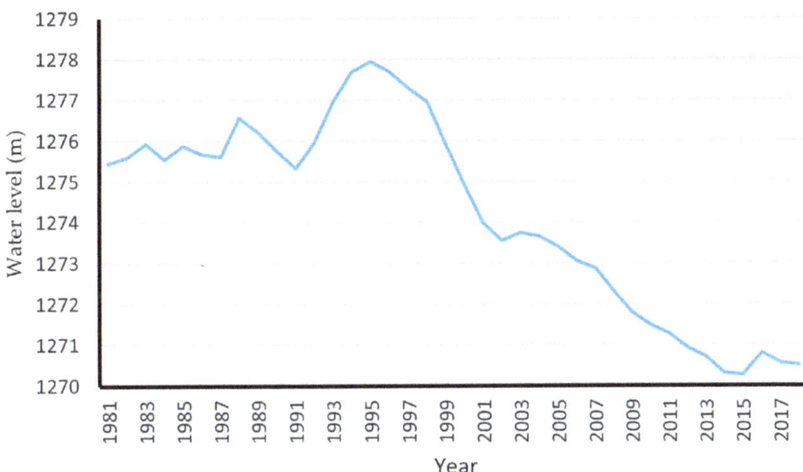

Figure 1. Measured water level (m) of Urmia Lake since 1981.

2. Materials and Methods

2.1. Lake Urmia Catchment

Located in the mountainous region of north-western Iran, Lake Urmia (Figure 2) and its catchment basin cover an area of approximately 52,000 km^2. Lake Urmia, the world's second-largest saline lake, covers about 5000 km^2 and belongs to Iran's Western Azerbaijan and Eastern Azerbaijan provinces.

The specific geology of this area, the high amount of evaporation, and the constant solute aggregation result in highly salty soils. The lake is surrounded by some freshwater wetlands, which are of ecological importance. The Mediterranean climate of the Urmia Basin is influenced by the surrounding high mountains and is characterized by cold winters and relatively temperate summers. The basin's long-term average temperatures are 0.2 °C in winter (December–January–February) and 23.9 °C in summer (June–July–August), while the average annual temperature is 12.3 °C. July and August are the warmest, and January and February are the coldest months of the year. Long-term average evaporation (for the 35-year record from 1966 to 2000) from Lake Urmia is 1373.7 mm yr^{-1}, and the highest evaporation has been observed in July and August. Precipitation in the Urmia basin is estimated to be 302.8 mm yr^{-1}, mainly falling in the period from October to March (with the highest amount in spring), during which the region is affected by the Rossby-forced advection of the Mediterranean, and sometimes, the Siberian air masses. In contrast, precipitation is comparatively low from June to September due to the dominance of upper-level ridges of high pressures. Annual evaporation is much higher than annual precipitation, suggesting that the lake is suffering from a water deficit [24].

2.2. ERA5 Reanalysis Data

The European Center for Medium-range Weather Forecast (ECMWF) has released the ECMWF Reanalysis v5 (ERA5) dataset as part of the Copernicus Climate Change Services. The dataset uses a spatial resolution of 31 km, covering the period from 1979 onwards [65]. We used ERA5 monthly precipitation and 2-m air temperature data with a 0.25° grid spacing for 1981–2018. ERA5 datasets were compared with gauge observations over the Karun basin in southwestern Iran and were found to be quite accurate [66]. In contrast, ERA-Interim, The Climate Forecast System Reanalysis (CFSR), and the Japanese 55-year Reanalysis (JRA-55) interpolated datasets show larger underestimations relative to observations [67]. In addition, the performance of ERA5 data is generally more consistent across different climate variables. In the absence of observational precipitation data, ERA5 and ERA-Interim are the best choices for data covering the Sistan and Baluchestan provinces, one of Iran's poorly gauged areas for rain [67]. The ERA5 data has also been validated against observations from the Ardabil province near the ULB [68]. It was found that, after correcting for bias, the ERA5 daily and monthly precipitation data was quite adequate, particularly given the data scarcity prevailing in the region. The ERA5 precipitation dataset was compared to observational datasets from meteorological stations in nine different precipitation zones of Iran for the period 2000–2018. After correcting for bias, ERA5 precipitation reanalysis datasets were found to be a very promising substitute for weather station data [69].

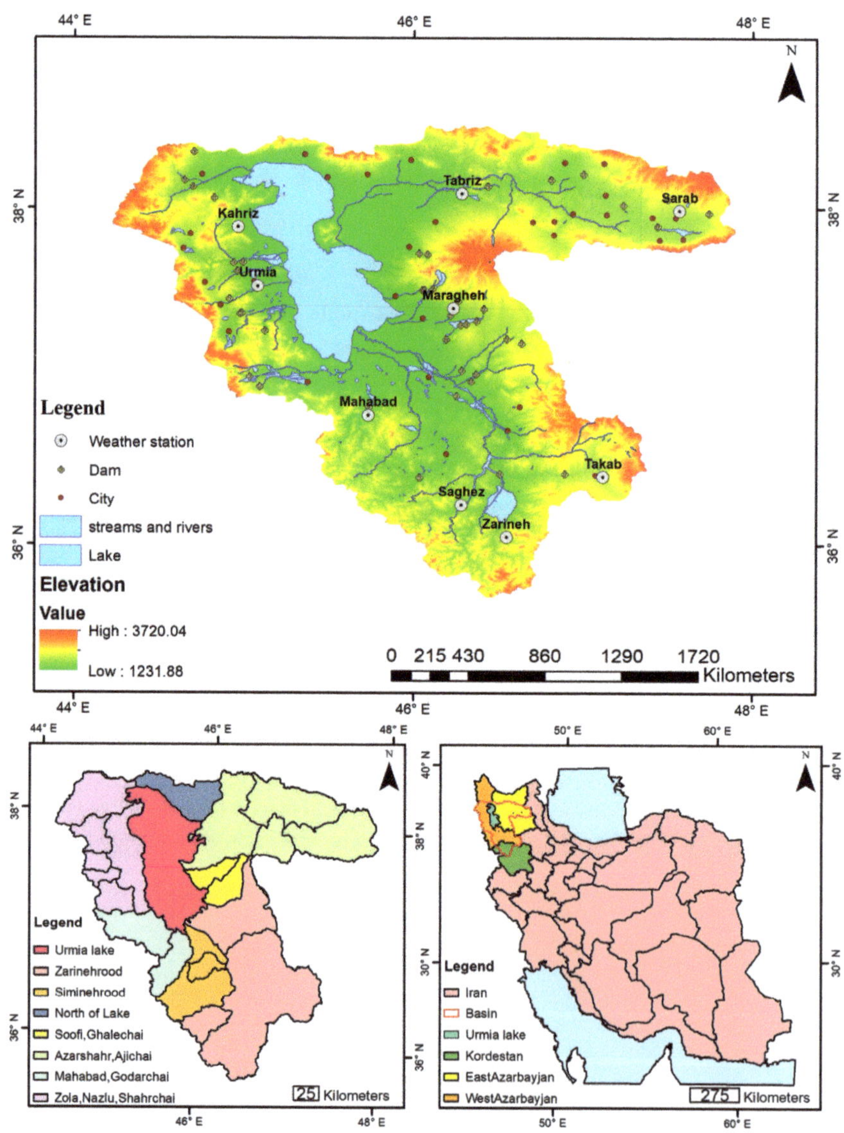

Figure 2. Geographical location of Lake Urmia.

2.3. Observational Data

Precipitation and temperature data was obtained from the Iran Meteorological Organization for nine synoptic weather stations from 1995–2018. Table 1 and Figure 2 show the geographical location of the weather stations in the basin. The synoptic stations, geographically dispersed around the Urmia Lake basin, were selected to represent the conditions prevailing among the mixture of plain and mountainous locations.

Table 1. Geographical characteristics of stations in the basin.

Station	Lon.	Lat.	Altitude(m)	Period of Data
Saghez	46.28	36.23	1522	1995–2018
Takab	47.12	36.40	1817	1995–2018
Mahabad	45.73	36.77	1352	1995–2018
Maragheh	46.23	37.40	1344	1995–2018
Urmia	45.08	37.53	1336	1995–2018
Sarab	47.57	37.98	1682	1995–2018
Kahriz	44.97	37.88	1336	1995–2018
Tabriz	46.28	38.08	1364	1995–2018
Zarineh	46.55	36.04	2142	1995–2018

2.4. Drought Indices

The SPI, SPEI, and SMRI drought indices were generated using the R software package [70]. This is a free software package for statistical computation and graphics.

2.4.1. Standardized Precipitation Index (SPI)

The SPI is designed to assess drought conditions based on the probability distribution of long-term precipitation using the gamma distribution [39]. Precipitation data is transformed into normalized values. The SPI is given as the number of standard deviations by which the observed precipitation deviates from the long-term mean for a normally distributed random variable. It can thus be used to define and compare drought conditions in different areas. The index gives a good and reliable estimate of drought magnitude, severity, and spatial extent. When precipitation is above the long-term mean value, the SPI is positive, and if precipitation falls below the long term, the SPI is negative. Unlike other drought indices, SPI is relatively easy to use because it only requires a single input data series of long-term precipitation [71]. As it is based on normalized data, the SPI is spatially invariant, and droughts can be assessed in different regions [72]. The index is calculated as follows:

$$\mathrm{SPI} = \frac{x_i - \overline{x}}{\sigma}$$

where x_i is the precipitation of the selected period during the year i, x is the long-term mean precipitation and σ is the standard deviation for the selected period.

2.4.2. Standardized Precipitation Evaporation Index (SPEI)

SPEI is calculated based on the non-exceedance probability of the differences between precipitation and potential evapotranspiration (PET), adjusted using a three-parameter logistic distribution which accounts for common negative values [51,73]. SPEI uses a three-parameter distribution to capture the deficit values since it is most likely that the moisture deficit can be damaging in arid and semi-arid areas. For two-parameter distributions as used in SPI, the variable x has a lower boundary of zero ($0 < x < \infty$), meaning that x can only take positive values. In contrast, for the three-parameter distributions used in SPEI, x can take values in the range ($\gamma < x < \infty$), implying that x can also take negative values; γ is the parameter of origin of the distribution [51]. Use of the log-logistic distribution is thus recommended for SPEI since it provides a better fit for extreme negative values [74]. The SPEI is obtained by normalizing the water balance into the log-logistic probability distribution. For the purposes of the present study, PET is estimated using the Thornthwaite method [75]. The difference (D_i) between precipitation (P) and PET for the month (i) is given by:

$$D_i = P_i - \text{PET}_i$$

The calculated D values are aggregated at different time scales as follows:

$$D_n^k = \sum_{n=0}^{k-1} P_{n-1} - (\text{PET})_{n-1}$$

where k is the timescale (months) of the aggregation and n is the calculation month. The probability density function of a log-logistic distribution is given as:

$$f(x) = \frac{\beta}{\alpha}\left(\frac{x-y}{\alpha}\right)^{\beta-1}\left(1+\left(\frac{x-y}{\alpha}\right)^{\beta}\right)^{-2}$$

where α, β and γ are scale, shape, and origin parameters respectively for $\gamma > D < \infty$. The probability distribution function for the D series is then given as:

$$D = \left[1 + (\alpha/x - y)^{\beta}\right]^{-1}$$

With $f(x)$ the SPEI can be obtained as the standardized values of F(x) according to the empirical method of [76]: where

$$\text{SPEI} = W - \frac{C_0 + C_1 W + C_2 W^2}{1 + d_1 W + d_2 W^2 + d_3 W^3}$$

And

$$W = \sqrt{-2ln(P)}$$

For

$$P \leq 0.5$$

P is the probability of exceeding a determined D_i value and is given as $P = 1 - f(x)$ while the constants are:

$C_0 = 2.515517, C_1 = 0.802853, C_2 = 0.010328, d_1 = 1.432788, d_2 = 0.189269, d_3 = 0.001308.$

The probability distribution function is given by; $F(x) = 1[1 + \exp(-y)]^{-1}$. The $F(x)$ values were then transformed to a normal variable by means of the above approximation by [77].

Hosking and Wallis [78] showed that the Di (P-ET0) distribution consistently produces the best goodness of fit to the generalized logistic (GLO) functions across all accumulation. The GLO is given by the above probability density function.

Since SPEI is a standardized variable, it can be used to compare droughts over different spatial and temporal scales. As with SPI values, negative SPEI values define drought conditions, and its accumulated values define the intensity, severity, magnitude, and duration of drought [79].

2.5. Standardized Snow Melt and Rain Index (SMRI)

The SMRI is based on precipitation and snowmelt minus snow accumulation. Snow accumulation, expressed as the amount of liquid water accumulated as snow, occurs when the mean temperature is smaller than a threshold temperature of 1 °C. In contrast, snowmelt, expressed as the amount of liquid water melted, is calculated with a simple temperature index model using a melt factor of 3 mm/°C-day (similar to [80]). The difference values, where P_i is precipitation, PET is potential evaporation, *SA* is snow accumulation, and *SM* is snowmelt [52,81], lead to the following SMRI equation:

$$D_i = P_i - \mathrm{PET}_i + \sum_{i=1}^{\infty} SM - \sum_{i=1}^{\infty} SA$$

2.6. Drought Characteristics—Duration, Severity, Frequency

Once a drought event was identified using the SPI, SPEI and SMRI indices, the drought starts and ends, drought duration (DD), and drought severity (DS) were derived from index data. DD is equal to the number of months between the starting month (included) and ending month (not included). DS is the absolute value of the integral area between line indices and the horizontal axis (SPI = 0, SPEI = 0, and SMRI= 0) from the beginning and ending month of drought. Drought frequency (DF) is represented by the number of events per 38 years (the whole period). Table 2 gives an overview of indices values for three drought categories. Drought begins when the index value is less than or equal to −1, and it ends when values become positive [39,82–84]. Figure 3 shows the drought characteristics using the run theory for a given threshold level. In runs theory, drought intensity is the average value of a drought parameter below the threshold level, which is measured as the drought severity divided by the duration [85]

Table 2. Drought classification scheme [39].

Value	Drought Category
−1 to −1.49	Moderate dryness
−1.5 to −1.99	Severe dryness
<−2	Extreme dryness

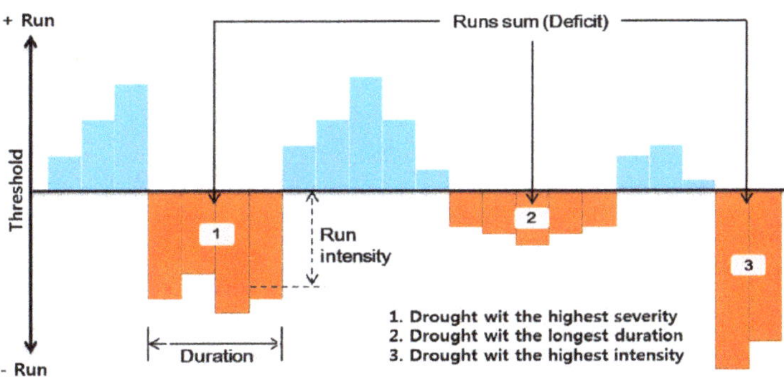

Figure 3. Drought characteristics using the run theory for a given threshold level [85].

The duration (*D*) of drought is the period in which the SPEI/SPI/SMRI value is continuously negative. It starts when the indices values are equal to −1 and ends when values become positive. The drought severity (*S*) is the cumulated index values within the drought duration, which is defined by:

$$S = -\sum_{i=1}^{D} Indexes_i$$

2.7. Relative Operative Characteristics (ROC) and Brier Skill Score (BSS)

The ROC curve is a performance measurement for classification problems at various threshold settings. The ROC method is a useful tool for assessing how well the ERA5-based drought classes can capture observational-based drought classes. It identifies how capable the model is of distinguishing between simulated classes (true positive, false positive, true negative, false negative). The ROC plots the True Positive Rate (*TPR*) against the False Positive Rate (*FPR*), where *TPR* is shown on the y-axis and FPR is shown on the x-axis. The rates are calculated as shown below [86]:

$$TPR = \frac{TP}{TP + FN}$$

$$FPR = \frac{FP}{FP + TN}$$

where,

TP: True Positives, are correct captures of drought classes.
FP: False Positives, are incorrect capture of drought classes.
TN: True Negatives, represent the capture of non-occurrence of specific drought classes
FN: False Negatives, represent the failure to capture the occurrence of specific drought classes.

The larger the Area Under Curve (AUC, which is the area under the ROC curve), the better the model simulates the classes (a perfect simulation is achieved when AUC = 1). For a model equivalent to random guessing, the AUC value is equal to 0.5. Table 3 shows a contingency table for distinguishing drought classes in terms of ERA5 and observational data.

Table 3. Contingency table for analysis of drought event detection by ERA5 data.

Drought Classes (Wet, Normal, and Dry)		ERA5	
		Yes	**No**
ERA5 Simulated	Yes	TP	FP
	No	FN	TN

3. Results

3.1. The Ability of ERA5 to Capture Drought Characterization over ULB

Figure 4 shows the basin's mean annual precipitation and temperature during 1995–2018 using ERA5 gridded (red line) and observational data (blue line). The basin's mean precipitation and temperature for the ERA5 and observational data are calculated using the mean gridded and Thissen methods, respectively. The correlation, RMSE, and bias of ERA5 precipitation and temperature with respect to observational data are shown in Tables 4 and 5. From Figure 4 and Table 4, it can be concluded that ERA5 data, after simple bias correction, can reasonably simulate the observed precipitation and temperature of the basin, specially over areas lacking adequate data. (Figure 4).

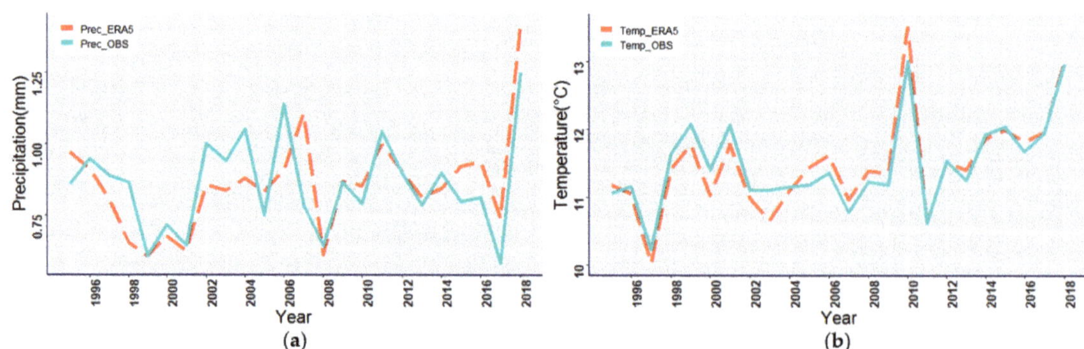

Figure 4. Comparison of observed (blue line) and ERA5 (red line) precipitation (**a**) and temperature (**b**) of the ULB.

Table 4. Correlation, bias and RMSE of ERA5 data comparing to observational data over Urmia Lake basin.

	Correlation	Bias	RMSE
Precipitation (mm)	0.389	0.112	0.205
Temperature (°C)	0.911	1.229	1.25

Table 5. Correlation and RMSE of ERA5 against observational indices over Urmia Lake basin.

Indices	Correlation	RMSE
SPI3	0.458	0.356
SPEI3	0.759	0.269
SMRI3	0.628	0.348
SPI6	0.638	0.438
SPEI6	0.815	0.3
SMRI6	0.632	0.445
SPI12	0.627	0.524
SPEI12	0.829	0.346
SMRI12	0.626	0.5772

As ULB weather stations data are often inadequate, we used ERA5 reanalysis data (thus providing us with full spatial coverage) to analyze the basin's drought conditions. In order to quantify systematic differences between the two data sets, the SPI, SPEI, and SMRI drought classes computed from station observations were compared with relevant ERA5- derived indices (for 3-, 6- and 12-monthly accumulated values). The ROC curves and Brier Skill Score (BSS) were then used to evaluate how well the ERA5 dataset captures different drought classes over the basin compared to measurements at meteorological stations (Table 6).

Table 6. Brier Skill Score (BSS) of ERA5 data representing drought over ULB using SPI, SPEI, SMRI.

Run	SPI			SPEI			SMRI		
	Dry	Normal	Wet	Dry	Normal	Wet	Dry	Normal	Wet
3-m	0.96	0.83	0.69	0.89	0.84	0.91	0.88	0.87	0.87
6-m	0.96	0.83	0.69	0.93	0.96	0.93	0.94	0.87	0.84
12-m	0.96	0.83	0.70	0.89	0.84	0.92	0.91	0.94	0.86

The BSS scores (Table 6) show high consistency between ERA5 and observational data for SPI, SPEI, and SMRI. To further illustrate the consistency of the drought indices from ERA5 data, we also analyzed ROC curves. The ROC curves are close to each other for the dry, wet, and normal SPEI category on 12-months running, with an average BSS score of 0.9–1 showing good to excellent consistency between ERA5 and observations over ULB. In SPI, the ROC curves for dry and normal events with AUC of 0.83–0.96 show excellent consistency of ERA5 data with observation, especially in the simulation of dry cases. The amount of AUC (BSS) for wet events is 0.70. Figure 5 shows that among the three indicators on a 12-month scale, the highest correlation exists between the SMRI drought index obtained from ERA5 and observation data. In addition, as can be seen for all drought indices, dry conditions are better identified by ERA5 data than are other conditions.

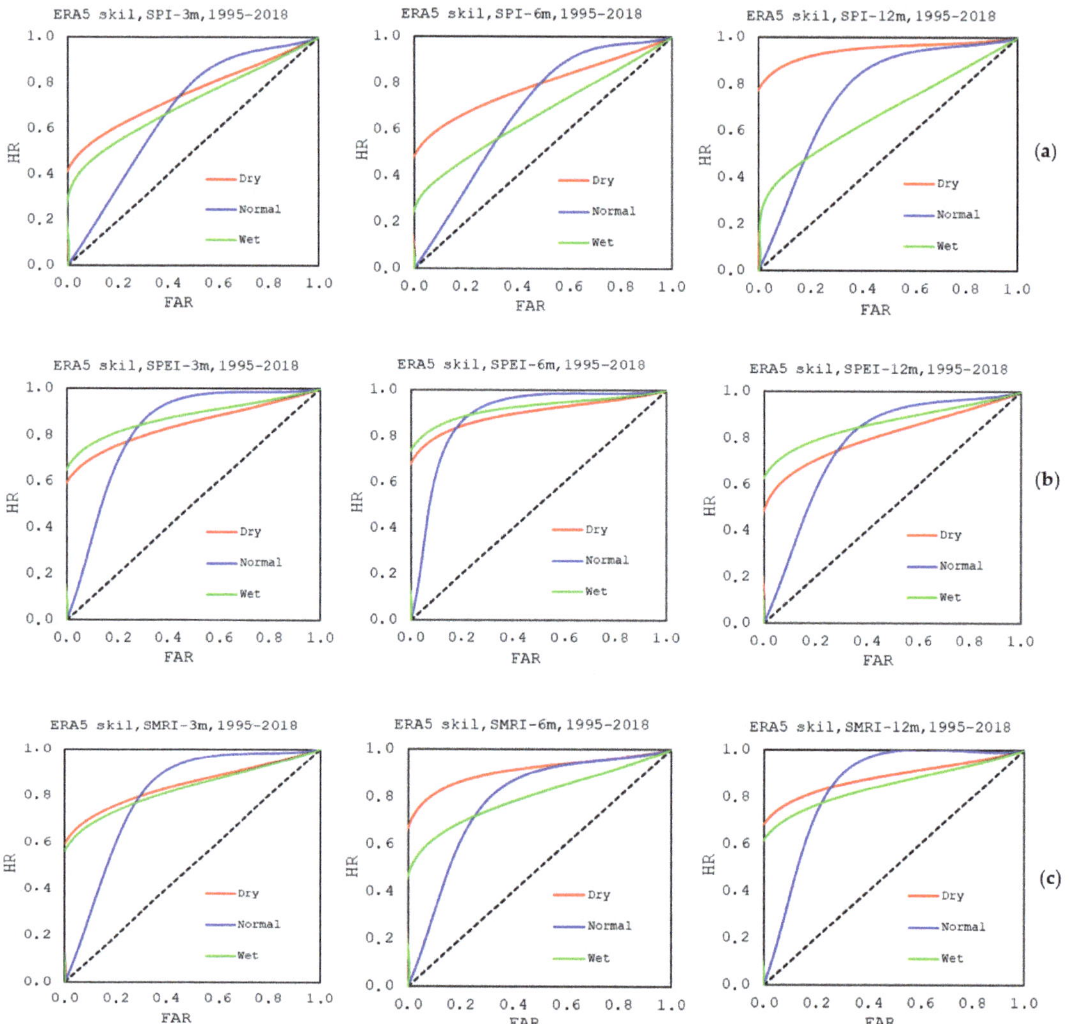

Figure 5. ROC curve for SPI (**a**), SPEI (**b**), SMRI (**c**) for three classes of dry, wet and normal drought conditions over ULB.

3.2. Temporal Evolution of the SPI, SPEI, and SMRI in ULB

To quantify the temporal evolution of SPI, SPEI, and SMRI for the ULB and to check the robustness of ERA5 results we compared the basin-wide indices for observational data with ERA5 over the period 1995–2018. This is shown in Figure 6. There is a general consistency between both input data sets for SPI (Figure 6a), SPEI (Figure 6b), and SMRI (Figure 6c). However, SPI and SPEI drought indices differ slightly depending on the time interval, especially at 3-, 6- and 12-months intervals. In recent decades, droughts indicated by SPEI have been more severe than those of SPI, mainly due to an increase in temperature and the related higher evapotranspiration rate. On the other hand, SMRI index values have also been rising, reflecting the conversion of solid precipitation to rain, the decline in snow storage in winter, and the relatively low amount of snowmelt during late spring to early summer. Figure 6 and Table 5 also imply that, as the time scale of the drought index increases from 3 months to 12 months, the correlation between ERA5 and observation data increases.

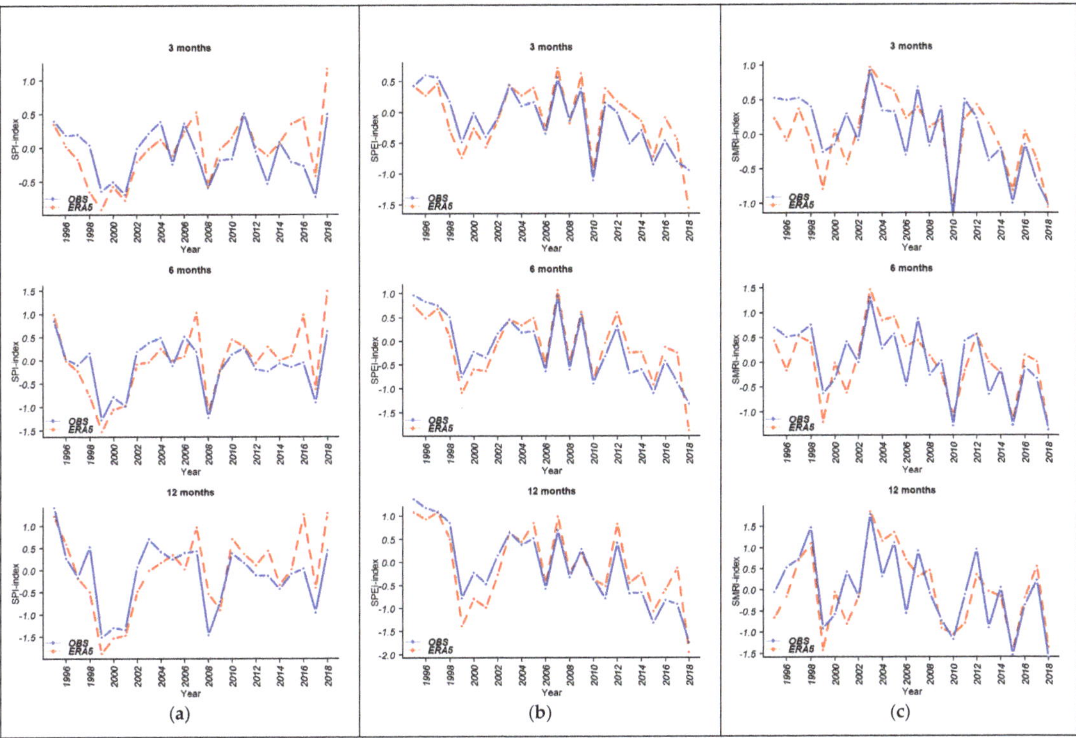

Figure 6. Time series showing 3-, 6- and 12- months accumulated values of SPI (**a**), SPEI (**b**) and SMRI (**c**) indices based on ERA5 (red) and observation (blue).

Regarding the seasonal distribution of the three drought indicators (averaged over the Urmia Lake Basin for the period 1995–2018) our analyses show that more severe droughts of the SPEI type (i.e., those induced by changes in precipitation and evapotranspiration) developed in most months, whereas for SPI-type droughts (precipitation driven) occurrence tends to be limited to late spring and summer (i.e., to the warmer months of the year). For SMRI-type droughts, severe droughts occurred from summer until to the end of fall) Figure 7.

In addition to seasonal variability, long-term variations and changes in SPI, SPEI, and SMRI data are also of interest, particularly with respect to the impact of climate change. On the one hand, drought frequencies and intensities differ, with differences decreasing with increasing drought timescale. The differences between the SPI and the other two indices have been increasing in recent years. This is particularly obvious from 2001 onwards. In fact, while there are a few moderate to extreme (index ≤ -1) droughts of SPI type from 2001 onwards, there is a clear increasing trend in extreme SPEI and SMRI type droughts (Figure 8). It is also clear that snowmelt-type droughts (SMRI) have become more common in recent years—especially since 2010—than other types of droughts (SPI and SPEI). This follows from the rising basin temperatures over the same period, which has led to a decrease in solid precipitation and a decreased snowmelt, and thus to an increase in the SMRI drought sequence. As a result of the associated rise in evapotranspiration, the SPEI type droughts are also affected by the basin temperature increase (Figure 8).

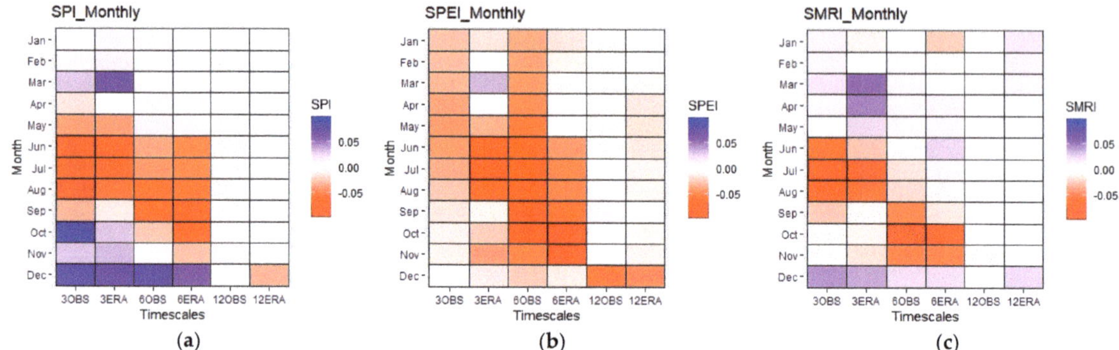

Figure 7. Monthly distribution of Indices (**a**) SPI, (**b**) SPEI, (**c**) SMRI.

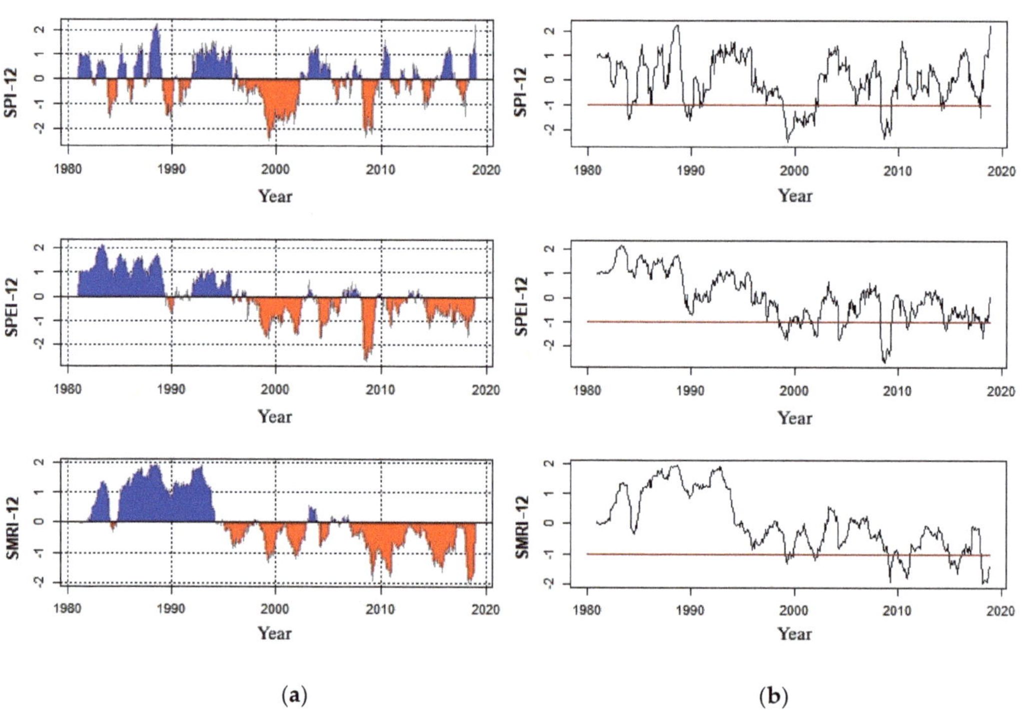

Figure 8. Time series of SPI, SPEI and SMRI indices of ULB over 1981–2018 (**a**) and their values under the threshold of −1 (**b**).

3.3. Drought Characteristics

The identified droughts were further characterized in terms of their severity, duration, and frequency at various timescales by analyzing their spatial distributions. For this purpose, we focused on specific dry and wet events derived from monthly SPI, SPEI, and SMRI data for the 38-year period from 1981 to 2018. In order to identify drought hotspots, we focused particularly on the years with moderate to extreme drought. Based on Table 2, threshold values related to moderate, severe, and extreme drought (index ≤ -1) were chosen to assess the drought characteristics using the three drought indices.

3.3.1. Drought Duration

Drought duration data in the ULB show a clear increase in snowmelt-driven droughts (SMRI) over the last years of the study period, particularly in 2010, 2015, and 2018. Similarly, drought duration for precipitation-driven droughts (SPI) were recorded in 1999, 2000, and 2001. Another noteworthy point on the SMRI droughts is that before 1995, no snowmelt drought occurred on any of the 3 to 12-month scales, indicating that there was not snowmelt in the water balance from 1981 to 1994. The decrease in snow precipitation in the basin then becomes noticeable and is clearly captured by the SMRI after 1995. The changes in the SPEI drought index are similar to those in the SMRI. This is no great surprise as the increase in ULB air temperatures and associated rise in evapotranspiration also began to play an ever-increasing role in drought events. Consequently, SMRI and SPEI indices are now more suitable for capturing drought conditions in the basin than the SPI index. Figure 9 also shows that based on SMRI data the most prolonged droughts occurred in the last decade in 2010, 2015, and 2018 (in particular for the 12-months timescale). This corresponds to other sources on drought information. The lake surface area diminished from 5650 square kilometers in 1998 to about 2005 square kilometers in 2010 [87]. The lowest annual surface discharge to the lake, recorded in 2015, was only 0.5 km^3 [88]. During their fieldwork in October 2018, previous research noticed that, especially on the eastern side of the lake, many people were complaining about the increasing occurrence of respiratory diseases as a result of a lake salt storm [89]. According to the SPI index, the 2015 and 2018 ULB droughts are classified as light drought and slightly wet, respectively. However, the SPEI index reveals large differences with respect to the impact of temperature, and the two droughts are classified as severe and moderate, respectively. According to the Ministry of Energy [90], the SPEI drought index results are consistent with the actual drought in the basin.

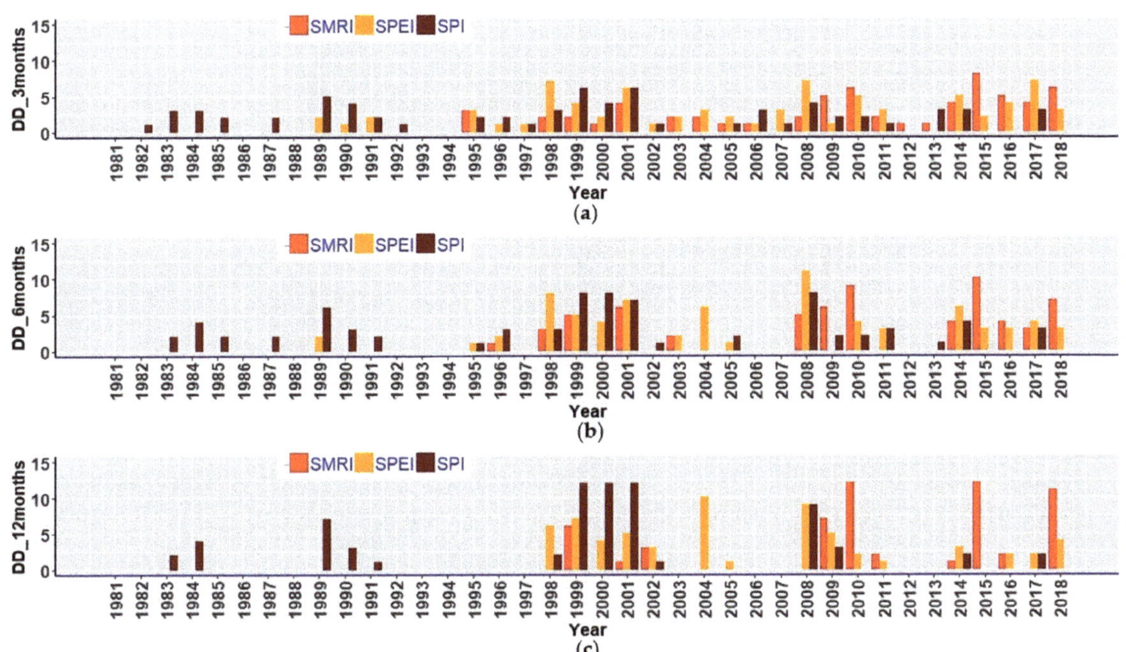

Figure 9. Drought Duration (DD) over ULB using SPI, SPEI and SMRI indices,1981 to 2018. (**a**) 3_months, (**b**) 6_months and (**c**) 12_ months.

3.3.2. Drought Severity

Developments in the time series of drought severity (Figure 10) are similar to those found for drought duration. From the beginning of the study period up to 1996, drought severity according to the SMRI index is almost zero. This indicates that before 1996, the Urmia Lake Basin did not face the challenge of snowmelt drought. The first severe SMRI drought event occurred around 1995/1996. From 1996 onward, however, the severity of snowmelt drought in the basin has increased. As was the case for drought duration, the most severe snowmelt droughts occurred in the more recent years, i.e., 2010, 2015, and 2018.The severity levels using SPEI and SPI indices were clearly less marked than those indicated by the SMRI data. Again, these results indicate that snowmelt drought is highly relevant for ULB, and droughts identified by SMRI give a more realistic picture of ULB droughts than the other two indicators. As can be seen in Figure 11, there is a clear correspondence between ULB drought severity and drought duration. The more severe the drought, the longer the drought duration. Use of the Palmer Drought Severity Index has also shown that while, on average, drought episodes have hit the Urmia Lake Basin every five years and most of them reached severe levels, the more recent droughts have also become more intense and of longer duration [25].

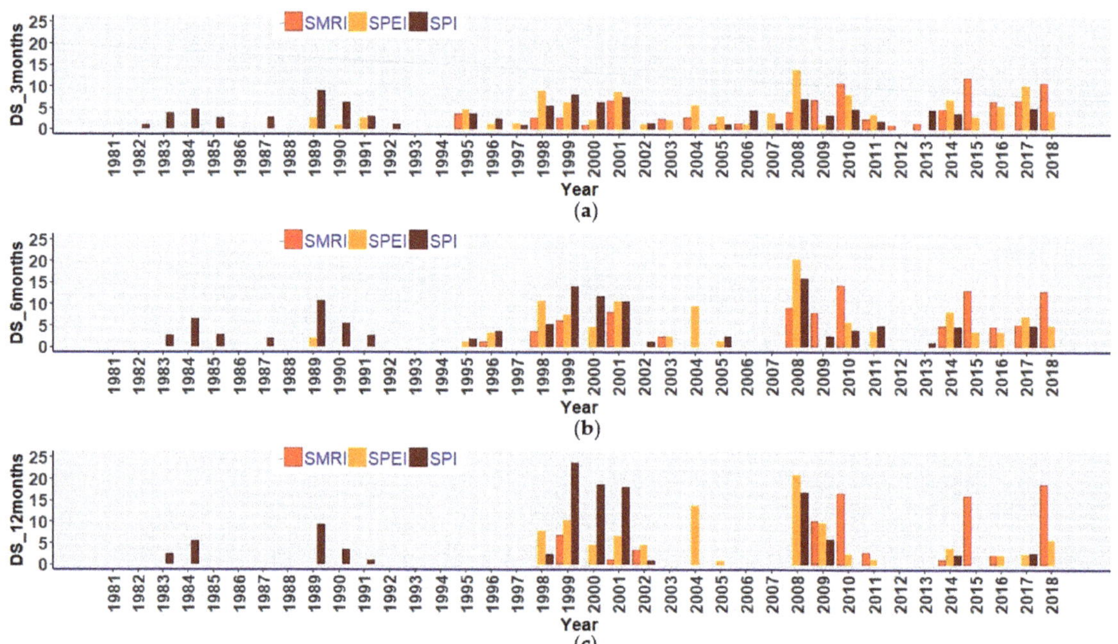

Figure 10. Drought Severity (DS) over ULB using SPI, SPEI and SMRI indices,1981 to 2018. (**a**) 3_months, (**b**) 6_months and (**c**) 12_ months.

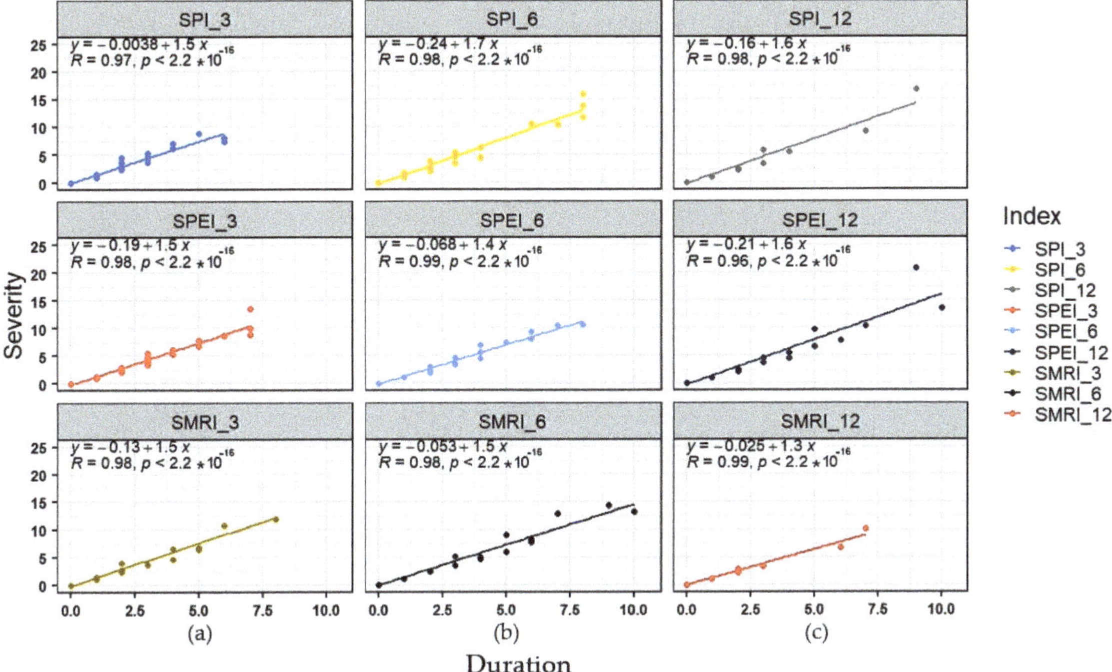

Figure 11. Aligning drought severity with its duration during period of 1981–2018: (**a**) 3_months, (**b**) 6_months and (**c**) 12_months.

3.3.3. Spatial Distribution of Drought Severity

Besides the temporal changes in drought events, the spatial patterns of drought events within the study region are also relevant for drought management. Figure 12 maps drought severity and shows some obvious drought hotspots. The spatial distribution of drought conditions in selected years with moderate and severe/extreme droughts were mapped for each index in order to evaluate spatial characteristics. The year 2001 was selected as the moderate year in the SPI index. The figure shows that the riverbed was most affected by severe droughts, but in 1999, which was considered the most severe year, not only the river but also the southern parts of the basin exhibit extreme drought. In evaluating SPEI spatial severity, 2009 and 2004 were selected as moderate and severe years, respectively. According to the SPI index, drought conditions were largely restricted to the riverbed. In 2009, the northeastern parts of the basin were subject to extreme drought. The SMRI index shows that the broadest range of drought severity includes parts of the east, north-east and south-eastern lake areas. This contrasts with spatial patterns as evidenced by the other two indicators. However, not much difference was found between the severe year 2015 and the extreme year 2018 in terms of snow drought patterns.

3.3.4. Spatial Distribution of Drought Frequency

Drought frequency (DF) is defined in terms of the number of drought events per 38 years. DF under −1 threshold maps indicate extreme conditions in the near lake region based on three drought indices. The drought frequency maps (Figure 13) show the difference between the drought frequency in the Urmia basin in terms of the three different indicators, SPI, SPEI, and SMRI. When considering SMRI data, the most extreme and prolonged events occurred in the Zarine river sub-catchment. Using SPI, all parts of the lake shore were affected by prolonged drought events, occurring in 30 out of 38 years, and drought patterns using the SPEI index show the prevalence of drought conditions ranging from the southern part of the lake to the southern parts of the basin for most years. It needs to be noted that in Figure 13, the SPI index only shows the occurrence of drought in the lake shore stations, while the SPEI index shows a relatively more expansive range of drought-affected areas. However, the SMRI index clearly identifies the drought area due to runoff deficiency in the south of the basin.

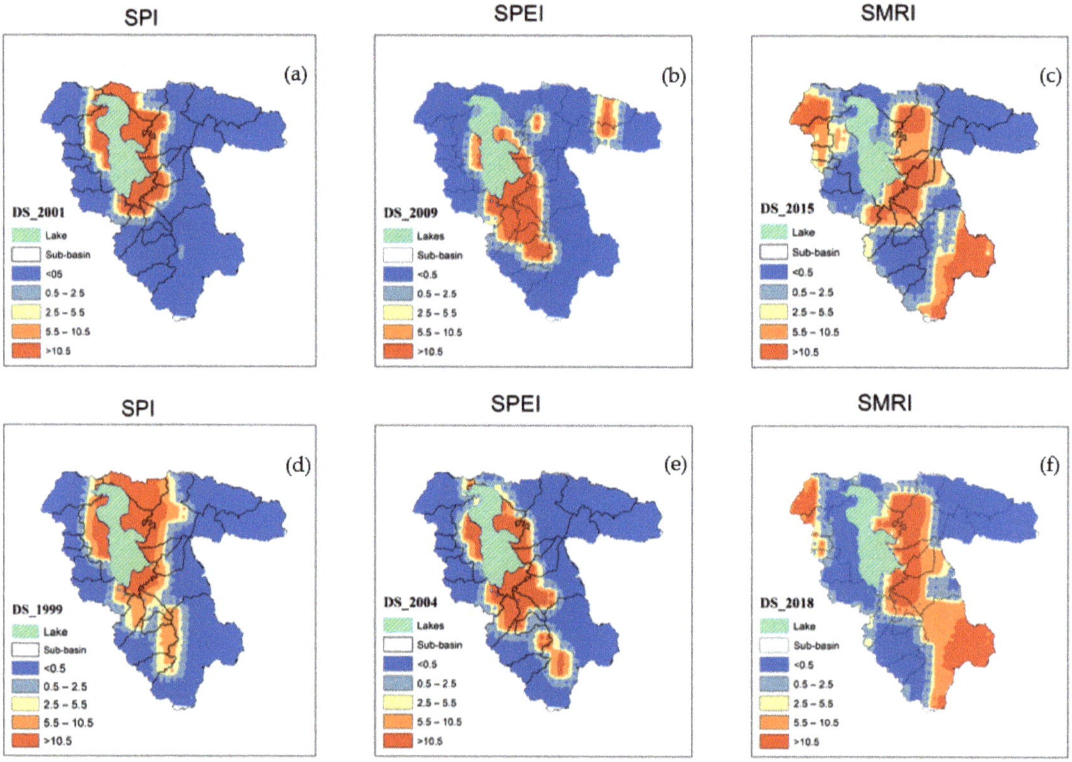

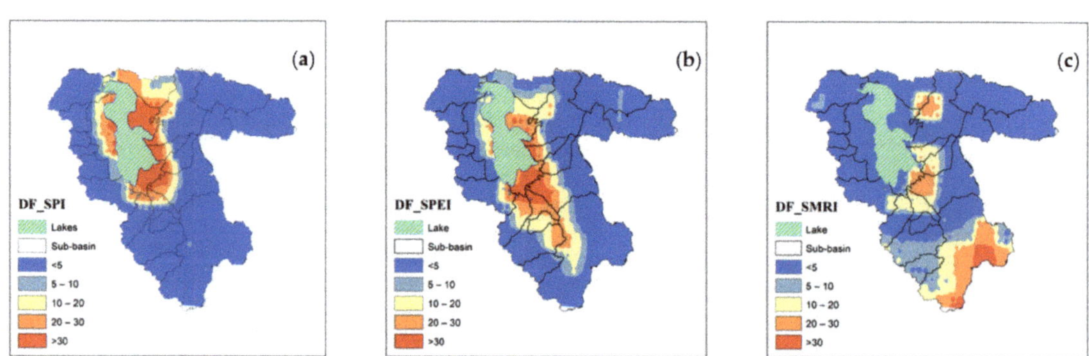

Figure 12. Drought Severity (DS) maps over ULB basin. (**a**) SPI dry year, (**b**) SPEI dry year, (**c**) SMRI dry year, (**d**) SPI drier year, (**e**) SPEI drier year and (**f**) SMRI drier year. Based on Table 2 threshold <-1 considered as drought hot spots [91].

Figure 13. Drought Frequency (DF) maps over ULB basin during 1981–2018. (**a**) SPI_DF, (**b**) SPEI_DF, (**c**) SMRI_DF. Based on Table 2 threshold <-1 considered as drought hot spots [91].

4. Discussion

The above comparison of drought events across the Urmia Lake Basin (ULB) has helped reveal specific drought behavior occurring over recent decades. The combined use of three common drought indices, i.e., SPI, SPEI, and SMRI, has provided clear insight into drought characteristics. Given the low density of meteorological stations, the inadequacies of existing statistics, the presence of geographical and topographical complexities (mountains, lowlands, and sea), an analysis of drought characteristics can only properly be carried out by making use of data with high spatial density. Therefore, the precipitation and temperature data of ERA5 were used to calculate drought indices. The suitability of the ERA5 dataset for purposes of drought monitoring was examined by making use of two series of data from observational and simulated ERA5. ROC and AUC (BSS) calculations were then used in order to check model performance. The results showed that the drought indices obtained by ERA5 are highly consistent with observational data and that they provide an excellent alternative to such data.

We also analyzed drought severity, duration, and frequency for various timescales, and additionally looked at spatial distribution, focusing on specific dry and wet events. There was a positive, significant, and robust correlation (p-value < 0.001) between the severity and duration of the drought, meaning that the more severe the drought occurrence, the longer its duration. The analysis revealed that both SMRI and SPEI are more suitable for capturing ULB drought characteristics than SPI. However, in recent decades, drought conditions as reflected by SPEI have been more severe than those indicated by SPI. This is mainly due to the fact that the former index is capable of capturing the impact of an increase in mean temperatures in the basin area. Furthermore, SMRI index data also exhibit a rising trend in drought conditions. The results above confirmed that the SMRI-type drought has become more abundant in recent years—especially since 2010—than other types of droughts (SPI and SPEI). This may be due to rising basin temperatures and global warming having led to a decrease in snowfall in the basin in the last decades, resulting in a decrease in snowmelt surface water and an increase in the SMRI drought sequence. In general, it can be confirmed that SMRI and SPEI indices are capable of capturing more real drought conditions in the basin than the SPI index. We found that the most extended periods of snowmelt-related drought, as measured by the SMRI, mainly occurred in the more recent years of the study period, i.e., in 2010, 2015, and 2018. While the SPI also indicates a few moderate to extreme (index ≤ -1) drought events, there is a clear and significant rising trend in extreme SPEI and SMRI drought. Our study has benefitted greatly from being able to make use of advances in SMRI data and from the availability of high-resolution ERA5 datasets.

Further tools are likely to become more widely available for investigating the possible relationships between climate data and water level or streamflow. For example, the use of robust wavelet analysis methods, such as the Least-Squares Cross-Wavelet Analysis (LSCWA) and the cross-wavelet transform, appears promising. Wavelet analysis can also show inter-annual and intra-annual variability within the climate and hydrological time series [92]. The scope for future research possibilities in the field thus remains undiminished.

5. Conclusions

The temporal evolution of drought indices was investigated to better understand the causes of drought in the ULB. This is the first time that the relevance of snow cover has been taken into account in analyzing drought in the ULB basin. In characterizing drought events, variables such as drought frequency, duration, and severity all need to be carefully examined. As the number of weather stations in the ULB is limited, and adequate data was lacking, ERA5 reanalysis data was drawn upon in order to capture drought behavior in the basin in a spatially consistent manner. ROC curves and BSS scores confirmed the ability of ERA5 to capture drought events in the catchment area. We found that the ERA5-based drought indices mainly capture dry events in the ULB well but are not as good at capturing wet events.

While drought events occurring before 1995 were mostly precipitation-driven (SPI), and reached a peak around 2000, ET (SPEI) and snowmelt-driven events (SMRI) were basically absent before 1995. Precipitation-driven droughts show no trend since 1980, while drought driven by evaporation and lack of snowmelt showed a marked increase since around 1995. In fact, the severe droughts found on the long timescale (12 months) in 2015 and 2018, were mainly driven by snowmelt. Other studies have also confirmed an increase in such drought events. Drought severity and duration in the ULB also appear to be highly interdependent (shown by means of correlation), i.e., there has been an increase in the severity of drought events, and their duration.

Although there has been no severe SPI drought in the last few decades, the water level in Lake Urmia has continued to decrease. This suggests that drought in the ULB is most likely due to non-sustainable water management, or to an increase in evaporation caused by global warming (and is not a direct result of rainfall variability). Most importantly, the results also show that snow droughts have not only been more frequent and severe in recent years but have also affected an increasing area over time. According to the snow drought index results, most extreme events were observed in the Zarinehrood and Siminehrood sub-basins, both of which play an important role in the revitalization of the lake.

Author Contributions: All authors contributed to the study's conception and design. M.H. undertook data preparation and processing and provided all maps. She also wrote the first draft of the manuscript. W.S. supervised the project and was the original source for the main idea. He also had a significant role in editing, reviewing, and finalizing the results. I.B. finalized the draft text and performed editing and reviewing. He was also active in gathering and reviewing data and was responsible for the statistical calculations. All authors have read and agreed to the published version of the manuscript.

Funding: This paper is academic research, and the authors declare that they have received funds from the University of Graz. Open Access Funding by the University of Graz.

Data Availability Statement: The datasets analyzed during the current study are available in the ERA5-ECMWF dataset repository (ERA5 | ECMWF), and stationary data are available in IRIMO.

Acknowledgments: The authors acknowledge the financial support provided by the University of Graz. We would like to thank all the Data providers. Data were provided by Iran Meteorological Organization, the European Centre for Medium-Range Weather Forecasts (ECMWF) and Integrated Climate Data Center—ICDC.

Conflicts of Interest: The authors declare that they have no conflict of interest.

References

1. Collins, M.; Knutti, R.; Arblaster, J.; Dufresne, J.-L.; Fichefet, T.; Friedlingstein, P.; Gao, X.; Gutowski, W.J.; Johns, T.; Krinner, G.; et al. Chapter 12—Long-Term Climate Change: Projections, Commitments and Irreversibility. In *Climate Change 2013: The Physical Science Basis. IPCC Working Group I Contribution to AR5*; IPCC Cambridge, Ed.; Cambridge University Press: Cambridge, UK, 2013.
2. Lu, J.; Carbone, G.J.; Grego, J.M. Uncertainty and hotspots in 21st century projections of agricultural drought from CMIP5 models. *Sci. Rep.* **2019**, *9*, 1–12. [CrossRef] [PubMed]
3. Barlow, M.; Zaitchik, B.; Paz, S.; Black, E.; Evans, J.; Hoell, A. A Review of Drought in the Middle East and Southwest Asia. *J. Clim.* **2016**, *29*, 8547–8574, Retrieved 11 February 2021. [CrossRef]
4. Agrawala, S.; Barlow, M.; Cullen, H.; Lyon, B. The drought and humanitarian crisis in central and southwest Asia: A climate perspective. *IRI Rep.* **2001**, 24. [CrossRef]
5. Lautze, S.; Stites, E.; Nojumi, N.; Najimi, F. *Qaht-e-Pool—A Cash Famine: Food Insecurity in Afghanistan 1999–2002*; Feinstein International Center, Tufts University: Boston, MA, USA, 2002.
6. Daryabari, S.J. Drought zoning of Iran in the past 50 years. *J. Geogr.* **2011**, *82*, 33–48.
7. Azadi, S.; Soltani, S.; Faramarzi, M.; Poumanafi, S. Calibration of Palmer drought index over Iran. *J. Iran's Water Res.* **2018**, *2*, 19–28.
8. Alijani, B.; Babaei-Fini, O. Spatial analysis of short-term drought in Iran. *J. Geogr. Reg. Plan.* **2009**, *1*, 109–121.
9. Babaei-Fini, O.; Alijani, B. Spatial analysis of long duration drought in Iran. *J. Phys. Geogr. Res. Q.* **2013**, *45*, 1–12. [CrossRef]

10. Daneshmand, H.; Mahmoudi, P. A spectral analysis of Iran's droughts. *Iran. J. Geophys.* **2017**, *10*, 28–47.

11. Nejad, M.T.; Hejazizadeh, Z.; Bosak, A.; Bazmi, N. The Effects of North Atlantic Oscillation on the atmospheric middle level Anomaly and precipitation of Iran. Case study: West of Iran. *Res. Geogr. Sci.* **2018**, *18*, 19–35. [CrossRef]

12. Azizi, G. Elnino and dry and wet period of Iran. *J. Geogr. Res.* **2000**, *32*, 71–84.

13. Khoshakhlagh, F.; Azixi, G.; Rahimi, M. Synoptic pattern of dry and wet winters over Southwest of Iran. *J. Appl. Res. Geogr. Sci.* **2012**, *12*, 57–77.

14. Babaeia-Fini, O.; Fattahi, E. Classification of synoptic patterns wet and dry condition over Iran. *J. Appl. Res. Geogr. Sci.* **2015**, *29*, 105–122.

15. Babaeian, I.; Rezazadeh, P. On the relationship between Indian monsoon withdrawal and Iran's fall precipitation onset. *Theor. Appl. Climatol.* **2018**, *134*, 95–105. [CrossRef]

16. Ghamghami, M.; Irannejad, P. An analysis of droughts in Iran during 1988–2017. *SN Appl. Sci.* **2019**, *1*, 1217. [CrossRef]

17. Farokhnia, A.; Morid, S. Assessment of the Effects of Temperature and Precipitation Variations on the Trend of River Flows in Urmia Lake Watershed. *J. Water Wastewater* **2014**, *25*, 86–97.

18. Fathian, F.; Morid, S.; Kahya, E. Identification of trends in hydrological and climatic variables in Urmia Lake basin, Iran. *Theor. Appl. Clim.* **2015**, *119*, 443–464. [CrossRef]

19. Zarghami, M. Effective watershed management; case study of Urmia Lake, Iran. *Lake Reserve Manag.* **2011**, *27*, 87–94. [CrossRef]

20. Delju, A.H.; Ceylan, A.; Piguet, E.; Rebetez, M. Observed climate variability and change in Urmia Lake Basin, Iran. *Theor. Appl. Climatol.* **2013**, *111*, 285–296. [CrossRef]

21. Azizzadeh, M.R.; Javan, K. Temporal and spatial distribution of extreme precipitation indices over the lake Urmia Basin, Iran. *Environ. Resour. Res.* **2018**, *6*, 25–40.

22. Tahroudi, M.N.; Ramezani, Y.; Ahmadi, F. Investigating the trend and time of precipitation and river flow rate changes in Lake Urmia basin, Iran. *Arab. J. Geosci.* **2019**, *12*, 219. [CrossRef]

23. AghaKouchak, A.; Norouzi, H.; Madani, K.; Mirchi, A.; Azarderakhsh, M.; Nazemi, A.; Nasrollahi, N.; Farahmand, A.; Mehran, A.; Hasanzadeh, E. Aral Sea syndrome desiccates Lake Urmia: Call for action. *J. Great Lakes Res.* **2015**, *41*, 307–311. [CrossRef]

24. Alizadeh-Choobari, O.; Ahmadi-Givi, F.; Mirzaei, N.; Owlad, E. Climate change and anthropogenic impacts on the rapid shrinkage of Lake Urmia. *Int. J. Climatol.* **2016**, *36*, 4276–4286. [CrossRef]

25. Eimanifar, A.; Mohebbi, F. Urmia Lake (Northwest Iran): A brief review. *Saline Syst.* **2007**, *3*, 5. [CrossRef]

26. Hassanzadeh, E.; Zarghami, M.; Hassanzadeh, Y. Determining them main factors in declining the Urmia Lake level by using system dynamics modeling. *Water Resour. Manag.* **2012**, *26*, 129–145. [CrossRef]

27. Jalili, S.; Kirchner, I.; Livingstone, D.M.; Morid, S. The influence of largescale atmospheric circulation weather types on variations in the water level of Lake Urmia, Iran. *Int. J. Climatol.* **2011**, *32*, 1990–1996. [CrossRef]

28. Javanshiri, Z.; Pakdaman, M.; Falamarzi, Y. Homogenization, and trend detection of temperature in Iran for the period 1960–2018. *Meteorol. Atmos. Phys.* **2021**, *133*, 1233–1250. [CrossRef]

29. Azizi, G.; Nazif, S.; Abbasi, F. Assessment of performance of Urmia basin dams using system dynamic approach. *J. Arid Reg. Geogr. Stud.* **2016**, *7*, 48–63.

30. Shadkam, S.; Ludwig, F.; van Oel, P.; Kirmit, Ç.; Kabat, P. Impacts of climate change and water resources development on the declining inflow into Iran's Urmia Lake. *J. Great Lakes Res.* **2016**, *42*, 942–952. [CrossRef]

31. Ghaheri, M.; Baghal-Vayjooee, M.H.; Naziri, J. Lake Urmia, Iran: A summary review. *Int. J. Salt Lake Res.* **1999**, *8*, 19–22. [CrossRef]

32. Kakahaji, H.; Banadaki, H.D.; Kakahaji, A. Prediction of Urmia Lake Water-Level Fluctuations by Using Analytical, Linear Statistic and Intelligent Methods. *Water Resour. Manag.* **2013**, *27*, 4469–4492. [CrossRef]

33. Karbassi, A.; Bidhendi, G.N.; Pejman, A.; Bidhendi, M.E. Environmental impacts of desalination on the ecology of Lake Urmia. *J. Great Lakes Res.* **2010**, *36*, 419–424. [CrossRef]

34. Lweendo, M.K.; Lu, B.; Wang, M.; Zhang, H.; Xu, W. Characterization of droughts in humid subtropical region, upper kafue river basin (Southern Africa). *Water* **2017**, *9*, 242. [CrossRef]

35. Davarpanah, S.; Erfanian, M.; Javan, K. Assessment of Climate Change Impacts on Drought and Wet Spells in Lake Urmia Basin. *Pure Appl. Geophys.* **2021**, *178*, 545–563. [CrossRef]

36. Ginoux, P.A.; Prospero, J.M.; Gill, T.E.; Hsu, C.; Zhao, M. Global-scale attribution of anthropogenic and natural dust sources and their emission rates based on MODIS Deep Blue aerosol products. *Rev. Geophys.* **2012**, *50*, RG3005. [CrossRef]

37. Palmer, W.C. Meteorological Drought, Office of Climatology. *US Weather Bureau, Research Paper.* 1965. No. 45. Available online: https://www.ncdc.noaa.gov/temp-and-precip/drought/docs/palmer.pdf (accessed on 9 October 2021).

38. Gibbs, W.J.; Maher, J.V. 1916–1993 & Australia. Bureau of Meteorology. In *Rainfall Deciles as Drought Indicators*; Bureau of Meteorology: Melbourne, Australia, 1967.

39. McKee, T.B.; Doesken, N.J.; Kleist, J. The relationship of drought frequency and duration to time scale. In Proceedings of the Eighth Conference on Applied Climatology, Anaheim, CA, USA, 17–22 January 1993; American Meteorological Society: Boston, MA, USA; pp. 179–184.

40. Palmer, W.C. Keeping track of crop moisture conditions, nationwide: The new Crop Moisture Index. *Weather* **1968**, *21*, 156–161. [CrossRef]

41. Shafer, B.A.; Dezman, L.E. Development of a surface water supply index (SWSI) to assess the severity of drought conditions in snowpack runoff areas. In *Proceedings of the Western Snow Conference*; Colorado State University: Fort Collins, CO, USA, 1982; pp. 164–175.

42. Kogan, F.N. Global droughts watch from space. *Bull. Am. Meteorol. Soc.* **1997**, *78*, 621–636. [CrossRef]

43. Kogan, F.N. World droughts in the new millennium from AVHRR-based vegetation health indices. *Eos Trans. Am. Geophys. Union* **2002**, *83*, 562–563. [CrossRef]

44. Keyantash, J.; Dracup, J.A. An aggregate drought index: Assessing drought severity based on fluctuations in the hydrologic cycle and surface water storage. *Water Resour. Res.* **2004**, *40*, W09304. [CrossRef]

45. Bhuiyan, C.; Singh, R.P.; Kogan, F.N. Monitoring drought dynamics in the Aravalli region (India) using different indices based on ground and remote sensing data. *Int. J. Appl. Earth Obs. Geoinf.* **2006**, *8*, 289–302. [CrossRef]

46. Abbas, S.; Nichol, J.E.; Qamer, F.M.; Xu, J. Characterization of drought development through remote sensing: A case study in Central Yunnan, China. *Remote Sens.* **2014**, *6*, 4998–5018. [CrossRef]

47. Cunha, A.P.M.; Alvalß, R.C.; Nobre, C.A.; Carvalho, M.A. Monitoring vegetative drought dynamics in the Brazilian semiarid region. *Agric. For. Meteorol.* **2015**, *214*, 494–505. [CrossRef]

48. Hollinger, S.E.; Isard, S.A.; Welford, M.R. A new soil moisture drought index for predicting crop yields. In *Preprints, Eighth Conference on Applied Climatology*; American Meteorological Society: Anaheim, CA, USA, 1993; pp. 187–190.

49. Meyer., S.J.; Hubbard, K.G.; Wilhite, D.A. A crop-specific drought index for corn: I. Model development and validation. *Agron. J.* **1993**, *86*, 388–395. [CrossRef]

50. Weghorst, K.M. *The Reclamation Drought Index: Guidelines and Practical Applications*; Bureau of Reclamation: Denver, CO, USA, 1996; p. 6.

51. Vicente-Serrano, S.M.; Beguería, S.; López-Moreno, J.I. A Multi-scalar drought index sensitive to global warming: The Standardized Precipitation Evapotranspiration Index—SPEI. *J. Clim.* **2010**, *23*, 1696–1718. [CrossRef]

52. Staudinger, M.; Stahl, K.; Seibert, J. A drought index accounting for snow. *Water Resour. Res.* **2014**, *50*, 7861–7872. [CrossRef]

53. Muhammad, A.; Kumar Jha, S.; Rasmussen, P.F. Drought characterization for a snowdominated region of Afghanistan. *J. Hydrol. Eng.* **2017**, *22*, 5017014. [CrossRef]

54. Hatchett, B.J.; McEvoy, D.J. Exploring the origins of snow drought in the northern Sierra Nevada, California. *Earth Interact.* **2018**, *22*, 1–13. [CrossRef]

55. Dierauer, J.R.; Allen, D.M.; Whitfield, P.H. Snow drought risk and susceptibility in the western United States and southwestern Canada. *Water Resour. Res.* **2019**, *55*, 3076–3091. [CrossRef]

56. Margulis, S.A. Characterizing the extreme 2015 snowpack deficit in the Sierra Nevada (USA) and the implications for drought recovery. *Geophys. Res. Lett.* **2016**, *43*, 6341–6349. [CrossRef]

57. Zhang, H.-M. Updated temperature data give a sharper view of climate trends. *Eos Trans. Am. Geophys. Union* **2019**, *100*, 1961–2018. [CrossRef]

58. Huning, S.L.; AghaKouchak, A. Global snow drought hot spots and characteristics. *Proc. Natl. Acad. Sci. USA* **2020**, *117*, 19753–19759. [CrossRef] [PubMed]

59. Ghaderpour, E.; Vujadinovic, T.; Hassan, Q.A. Application of the Least-Squares Wavelet software in hydrology: Athabasca River Basin. *J. Hydrol. Reg. Stud.* **2021**, *36*, 100847. [CrossRef]

60. Zhang, J.; Zheng, H.; Zhang, X.; VanLooy, J. Changes in Regional Snowfall in Central North America (1961–2017): Mountain Versus Plains. *Geosciences* **2020**, *10*, 157. [CrossRef]

61. Littell, J.S.; McAfee, S.A.; Hayward, G.D. Alaska.Snowpack Response to Climate Change: Statewide Snowfall Equivalent and Snowpack Water Scenarios. *Water* **2018**, *10*, 668. [CrossRef]

62. Van Loon, A.F.; Van Lanen, H.A.J. A process-based typology of hydrological drought. *Hydrol. Earth Syst. Sci.* **2012**, *16*, 1915–1946. [CrossRef]

63. Tijdeman, E. Natural and Human Influences on the Link Between Meteorological and Hydrological Drought Indices for a Large Set of Catchments in the Contiguous United States. *Water Resour. Res.* **2018**, *54*, 6005–6023. [CrossRef]

64. Hydrological Data. 2015. Available online: https://www.ulrp.ir/en/hydrological-data/ (accessed on 1 January 2020).

65. Hersbach, H.; Bell, B.; Berrisford, P.; Horányi, A.; Sabater, J.M.; Nicolas, J.; Radu, R.; Schepers, D.; Simmons, A.; Soci, C.; et al. Global reanalysis: Goodbye ERA-Interim, hello ERA5. *ECMWF Newsl.* **2019**, *159*, 17–24. [CrossRef]

66. Fallah, A.; Rakhshandehroo, G.R.; Sungmin, P.; Orth, R. Evaluation of precipitation datasets against local observations in southwestern Iran. *Int. J. Climatol.* **2019**, *40*, 4102–4116. [CrossRef]

67. Yazdandoost, F.; Moradian, S.; Zakipour, M.; Izadi, A.; Bavandpour, M. Improving the precipitation forecasts of the North American multi model ensemble (NMME) over Sistan basin. *J. Hydrol.* **2020**, *590*, 125263. [CrossRef]

68. Azizi mobaser, J.; Rasoulzadeh, A.; Rahmati, A.; Shayeghi, A.; Bakhtar, A. Evaluating the Performance of Era-5 Re-Analysis Data in Estimating Daily and Monthly Precipitation, Case Study; Ardabil Province. *Iran. J. Soil Water Res.* **2021**, *51*, 2937–2951. [CrossRef]

69. Izadi, N.; Karakani, E.G.; Saadatabadi, A.R.; Shamsipour, A.; Fattahi, E.; Habibi, M. Evaluation of ERA5 Precipitation Accuracy Based on Various Time Scales over Iran during 2000–2018. *Water* **2021**, *13*, 2538. [CrossRef]

70. Beguería, S.; Vicente-Serrano, S.; Reig, F.; Latorre, B. Standardized Precipitation Evapotranspiration Index (SPEI) revisited: Parameter fitting, evapotranspiration models, tools, datasets and drought monitoring. *Int. J. Climatol.* **2014**, *34*, 3001–3023. [CrossRef]

71. Smakhtin, V.U.; Hughes, D.A. *Review, Automated Estimation and Analyses of Drought Indices in South Asia*; IWMI Working Paper 083: Drought Series Paper 1; International Water Management Institute (IWMI): Sri Lanka, CO, USA, 2004; 24p.

72. Guttman, N.B. Comparing the Palmer Drought Index and the Standardized Precipitation Index. *JAWRA J. Am. Water Resour. Assoc.* **1998**, *34*, 113–121. [CrossRef]

73. Středová, H.; Středa, T.; Chuchma, F. Climatic Factors of Soil Estimated System. In *Bioclimate: Source and Limit of Social Development*; SPU v Nitre: Nitra, Slovakia, 2011; pp. 137–138. ISBN 978-80-552-0640.

74. Hernandez, E.A.; Uddameri, V. Standardized Precipitation Evaporation Index (SPEI)-based drought assessment in semi-arid south Texas. *Environ. Earth Sci.* **2014**, *71*, 2491–2501. [CrossRef]

75. Thornthwaite, C.W. An approach toward a rational classification of climate. *Geogr. Rev.* **1948**, *38*, 55–94. [CrossRef]

76. Abramowitz, M.; Stegun, I.A. *Handbook of Mathematical, Functions*; National Bureau of Standards: Washington, DC, USA, 1965.

77. Abramowitz, M.; Stegun, I.A. *Handbook of Mathematical Functions: With Formulas, Graphs and Mathematical Tables*; Cambridge University Press: Cambridge, UK; Courier Corporation: Chelmsford, MA, USA, 1964.

78. Hosking, J.R.M.; Wallis, J.R. *Regional Frequency Analysis: An Approach Based on L-Moments*; Cambridge University Press: Cambridge, UK, 2005. [CrossRef]

79. Byakatonda, J.; Parida, B.P.; Kenabatho, P.K.; Moalafhi, D.B. Modeling dryness severity using artificial neural network at the Okavango Delta, Botswana. *Glob. Nest J.* **2016**, *18*, 463–481. [CrossRef]

80. Freudiger, D.; Kohn, I.; Stahl, K.; Weiler, M. Large-scale analysis of changing frequencies of rain-on-snow events with flood-generation potential. *Hydrol. Earth Syst. Sci.* **2014**, *18*, 2695–2709. [CrossRef]

81. Hogan, J.E.; Phillips, F.M.; Scanlon, B.R. *Groundwater Recharge in a Desert Environment: The Southwestern United States*; American Geophysical Union: Washington, DC, USA, 2004; Volume 9. [CrossRef]

82. Mishra, A.K.; Singh, V.P.; Desai, V.R. Drought characterization: A probabilistic approach. *Stoch. Environ. Res. Risk Assess.* **2009**, *23*, 41. [CrossRef]

83. Spinoni, J.; Naumann, G.; Carrπo, H.; Barbosa, P.; Vogt, J. World drought frequency, duration, and severity for 1951–2010. *Int. J. Climatol.* **2014**, *34*, 2792–2804. [CrossRef]

84. Spinoni, J.; Naumann, G.; Vogt, J.; Barbosa, P. European drought climatologies and trends based on a multi-indicator approach. *Glob. Planet. Change* **2015**, *127*, 50–57. [CrossRef]

85. Lee, S.H.; Yoo, S.H.; Choi, J.Y.; Bae, S. Assessment of the impact of climate change on drought characteristics in the Hwanghae plain, North Korea using time series SPI and SPEI: 1981–2100. *Water* **2017**, *9*, 579. [CrossRef]

86. Benjamin, B.M.; Michael, D.; Morphew, S. Developing Hydro-Meteorological Thresholds for Shallow Landslide Initiation and Early Warning. *J. Water* **2018**, *10*, 1274. [CrossRef]

87. Tisseuil, C.; Roshan, G.R.; Nasrabadi, T.; Asadpour, G. Statistical Modeling of Future Lake Level under Climatic Conditions, Case study of Urmia Lake (Iran). *Int. J. Environ. Res.* **2013**, *7*, 69–80. [CrossRef]

88. Schulz, S.; Darehshouri, S.; Hassanzadeh, E.; Tajrishy, M.; Schüth, C. Climate change or irrigated agriculture—What drives the water level decline of Lake Urmia. *Sci. Rep.* **2020**, *10*, 1–10. [CrossRef] [PubMed]

89. Schmidt, M.; Gonda, R.; Transiskus, S. Environmental degradation at Lake Urmia (Iran): Exploring the causes and their impacts on rural livelihoods. *GeoJournal* **2021**, *86*, 2149–2163. [CrossRef]

90. Iranian Students News Agency (ISNA). 96 Large Dams Have Less than 40% Water Storage. Available online: https://www.isna.ir/news/97091105033/ (accessed on 30 September 2021).

91. Brito, S.S.B.; Cunha, A.P.M.A.; Cunningham, C.C.; Alvalá, R.C.; Marengo, J.A.; Carvalho, M.A. Frequency, duration, and severity of drought in the Semiarid Northeast Brazil region. *Int. J. Climatol.* **2017**, *38*, 517–529. [CrossRef]

92. Canchala, T.; Cerón, W.L.; Francés, F.; Carvajal-Escobar, Y.; Andreoli, R.V.; Kayano, M.T.; Alfonso-Morales, W.; Caicedo-Bravo, E.; de Souza, R.A.F. Streamflow Variability in Colombian Pacific Basins and Their Teleconnections with Climate Indices. *Water* **2020**, *12*, 526. [CrossRef]

Article

Water Scarcity Risk Index: A Tool for Strategic Drought Risk Management

Fernanda Rocha Thomaz [1,*], **Marcelo Gomes Miguez** [1,2,*], **João Gabriel de Souza Ribeiro de Sá** [2], **Gabriel Windsor de Moura Alberto** [2] **and João Pedro Moreira Fontes** [2]

[1] Programa de Engenharia Civil—PEC, COPPE/Universidade Federal do Rio de Janeiro—UFRJ, Rio de Janeiro 21941-450, Brazil

[2] Escola Politécnica—POLI/Universidade Federal do Rio de Janeiro—UFRJ, Rio de Janeiro 21941-909, Brazil

* Correspondence: fer@hidro.ufrj.br (F.R.T.); marcelomiguez@poli.ufrj.br (M.G.M.); Tel.: +55-21-3938-7830 (F.R.T.); +55-21-3938-7833 (M.G.M.)

Abstract: Drought events have affected many regions of the world, having negative economic, environmental and social impacts. When accompanied by increasing water demands, these events can lead to water scarcity. Since droughts can significantly vary in each geographic area, several indices have been developed around the world. Hazard indexes are commonly used to predict meteorological, agricultural and hydrological droughts. These indexes intend to predict hazards, but they do not provide information on when and where deficits can have negative consequences. This study presents a new planning and decision-support tool for monitoring water scarcity situations in a given region. This tool, called the Water Scarcity Risk Index (*W-ScaRI*), is formed by two subindices, which are proposed to describe a hazard and its consequences. Each subindex was constructed using a group of indicators and indices selected from the technical literature or originally proposed in this work. The *W-ScaRI* was applied to the Rio de Janeiro Metropolitan Region (RJMR), supplied with water by the Guandu/Lajes/Acari system. The RJMR is one of the most densely populated regions in Brazil, located in an area that has no natural water bodies capable of meeting its supply needs. Therefore, the Guandu River, which, in fact, is formed by two discharge transpositions from the Paraíba do Sul River, is the main drinking water supply source for this region. The RJMR suffered the consequences of unexpected, prolonged droughts in the Southeast region in 2003 and 2014–2015, leading the local authorities to implement temporary emergency measures in the management system of Paraíba do Sul and Guandu Basins, avoiding water shortage but showing the urgent need for planning and management support tools to anticipate possible future problems. The results of the study show that the formulation of the *W-ScaRI* can represent the water scarcity risk in a relatively simple way and, at the same time, with adequate conceptual and methodological consistency.

Keywords: water scarcity risk; drought; hazard; consequence; vulnerability; water scarcity

check for updates

Citation: Thomaz, F.R.; Miguez, M.G.; de Souza Ribeiro de Sá, J.G.; de Moura Alberto, G.W.; Fontes, J.P.M. Water Scarcity Risk Index: A Tool for Strategic Drought Risk Management. *Water* **2023**, *15*, 255. https://doi.org/10.3390/w15020255

Academic Editor: Alina Barbulescu

Received: 18 November 2022
Revised: 23 December 2022
Accepted: 3 January 2023
Published: 7 January 2023

Copyright: © 2023 by the authors. Licensee MDPI, Basel, Switzerland. This article is an open access article distributed under the terms and conditions of the Creative Commons Attribution (CC BY) license (https://creativecommons.org/licenses/by/4.0/).

1. Introduction

Drought is characterized as a gradual natural hazard ([1], [2] cited in [3]) usually driven by climatic characteristics on a regional or even global scale [3]. For this reason, it is difficult to identify the beginning and ending of a drought period, as well as the affected area [4]. According to Wilhite [5] and De Stefano et al. [6], the main features that make droughts different from other natural hazards are as follows: (a) droughts are difficult to define—there is no universally accepted definition; (b) it is difficult to determine the beginning and the end of the event; (c) the impacts are mainly functional and nonstructural over socioeconomic systems, and they are spread over large geographical areas; (d) droughts are a normal, recurrent and cyclical aspect of climate in virtually all regions of the world (but varying in intensity and duration).

In recent decades, drought events have increased in intensity and frequency, affecting many regions of the world and having negative economic, environmental and social impacts [7]. Thus, several socioeconomic systems and sectors in the world have suffered heavy losses due to the consequences of droughts [8]. In Europe, from 1976 to 2006, drought costs were estimated at EUR 100 billion [4]. In 2015, California's agriculture losses during the drought period reached USD 1.84 billion, in addition to the loss of 10,000 seasonal jobs [9]. In Canada, an increase in tree mortality was observed across the Boreal Forest following a series of regional droughts from 1963 to 2008 [10].

While droughts are natural phenomena caused by abnormal precipitation deficits, water scarcity combined with human action results in a water availability insufficient to satisfy water demands for different socioeconomic uses [11]. According to Van Loon and Van Lanen [12], water scarcity represents the overexploitation of water resources when the water demand is greater than the water availability. Dolan et al. [13] said that water scarcity is dynamic and complex, emerging from the combined influences of climate change, basin-level water resources and managed systems' adaptive capacities.

The increasing water demand accompanied by a changing climate can lead to the unsustainable use of freshwater, consequently increasing water scarcity [14]. Regions with water scarcity may suffer from strong constraints in terms of social integrity and economic development [15]. This situation will be aggravated as rapidly growing urban areas place heavy pressure on local water resources [16,17]. Therefore, it is necessary to revise water management procedures, especially in areas with demographic changes and that are vulnerable to climatic conditions, in order to ensure a sustainable and safe water supply [15].

Understanding the evolution of and variation in drought events at different spatial and temporal scales is crucial in drought planning [8]. One way to monitor drought and water scarcity in a basin is to use indices and indicators. Indices are important decision-support tools, as they aggregate information from indicators of different types, forming a single representative value of a more complex situation. This integration of several indicators allows for fast and easy comparisons across time and space [18,19]. Well-constructed indicators can translate information about complex phenomena in a simple way by aggregating and quantifying information with diverse sources and scales so that their significance becomes more apparent [20].

Since drought events are significantly different around the world, several indices have been developed and published internationally. According to Wang et al. [21], more than 58 index types from different countries are listed in the World Meteorological Organization (WMO) technical reports. Many indices were developed to assess the characteristics of a given type of drought according to its meteorological, agricultural or hydrological origin [22]. Cuartas et al. [23], for example, analyzed different drought indicators to assess hydrological droughts in several regions of Brazil and their impact on hydropower generation.

Drought alert systems, usually called drought monitors, use indices to detect and predict drought hazard situations. For example, the US Drought Monitor uses five indices to classify droughts [24,25]: the Palmer Drought Severity index (*PDSI*), CPC Soil Moisture Model, USG Weekly Streamflow, Standardized Precipitation Index (*SPI*) and Objective Drought Indicator Blends. In Brazil, the Northeast Drought Monitor was recently developed based on the US Drought Monitor and experiences from Spain and Mexico [26]. The Northeast Drought Monitor uses three indices: SPI, the Standardized Precipitation-Evapotranspiration Index (*SPEI*) and the standardized runoff index (*SRI*). These drought warning systems, while able to predict hazards, do not provide information on when and where deficits can have negative consequences [27]. However, this is important information for drought risk planning and management strategies, which can be used to trigger and prioritize specific actions [28].

In recent years, the number of studies related to the consequences of droughts has increased due to the growing concern about the importance of changing the way of handling natural disasters, moving from a crisis management approach to a risk-management-based prevention approach. In addition, the rising severity of the impacts of droughts has also contributed to improving and optimizing management during these events [29]. One of the most common ways to assess vulnerability is the use of indicators, which may or may not be aggregated into indices [6], helping to jointly consider different system aspects.

Tsakiris [30], for example, assessed drought risk as a functional relation of hazard (H) and vulnerability (V). In the vulnerability assessment, the study presented several factors that can be used in formulations, such as exposure, the capacity of the system, social factors, the severity and destructive capacity of the event, conditions and interrelated factors. However, to show an application case, the study only used the *RDI* index to represent the drought severity classes, which was associated with crop production losses.

Dabanli [31] developed a framework to assess drought risk in Turkey using hazard components (*SPI*) and vulnerability (four socioeconomic indicators). The study did not use exposure in its formulation.

Meza et al. [32] presented an integrated drought risk assessment that considers hazard, exposure and vulnerability components to evaluate the impact of droughts on irrigated and rainfed systems (separately) at the national level. To assess vulnerability, they used more than 20 indicators. Tien Le et al. [33] also proposed a drought risk assessment using hazard, exposure and vulnerability components, and it was applied to 27 province areas in Vietnam. Both studies used specific indicators, indices and data to assess the drought risk in agricultural areas.

Carrão et al. [34] also used a combination of hazard, exposure and vulnerability indicators (most of them at the country scale). The drought hazard was derived from a non-parametric analysis of historical precipitation deficits; drought exposure was based on indicators of population and livestock densities, crop cover and water stress; and drought vulnerability was computed as the arithmetic composite of social, economic and infrastructural indicators, using 15 indicators. The study mapped the global distribution of drought risk, serving as a kind of first triage analysis to determine where local risk assessments should be carried out in detail. However, with significant intra-annual variations in water use and availability, it is important to understand when water is available [28].

The study by Dunne and Kuleshov [35] assessed the spatial–temporal distribution of agricultural drought risk across the Murray–Darling Basin. The developed drought risk index included nine indicators. One of the vulnerability component indicators was the same as that used by Carrão et al. [34], and it aggregates several socioeconomic factors.

Sayers et al. [3] described a new approach called "Strategic Drought Risk Management" (GERS). According to this approach, drought risk is defined as "an emerging property of natural and human systems that reflects the interaction between the hazard of meteorological drought, blue drought (hydrological) and green drought (agricultural) and the vulnerability of exposed people, ecosystems and economies".

Considering the presented aspects, this paper proposes a new index named the Water Scarcity Risk Index (*W-ScaRI*), which aims to assess the risk of water scarcity in a given region, especially focusing on urbanized watersheds and, particularly, in metropolitan regions. The drought hazard, in a broad way, is related to the meteorological drought, the blue drought and the green drought, as proposed by Sayers [3], while vulnerability is expressed to convey the environmental, social and economic consequences of the event. However, this new proposal attempts to maintain a simple vulnerability component, using a small number of representative indicators. In this context, the proposed index is built to be applied mainly in urban areas in order to assess the risk of water scarcity related to human and industry supplies. Thus, the indicators and indices chosen to compose the *W-ScaRI* were selected or created to represent this risk, within an adequate scale of analysis, in a relatively simple way, using fewer but representative parameters of the risk components. The main contribution of this work lies in the possibility of establishing a relatively simple index that can be integrated in daily management operations, allowing for the definition of both a spatial hierarchization of critical areas (according to the mapped vulnerabilities) and a set of threshold values for the *W-ScaRI* that can raise warning flags and implement specific actions to diminish drought risks (for example, limiting water supply to the least affected activities, rationalizing water uses and temporarily using alternative water source supplies).

The *W-ScaRI* was tested in the Rio de Janeiro Metropolitan Region (RJMR) for the period of 2014–2015, when a serious water crisis occurred in the Paraíba do Sul River Basin, the main water supply source for the region. The successful application of *W-ScaRI* can validate its use in the future in the preparation of a strategic drought risk management plan for the region, helping to understand the temporal evolution of drought risks and supporting actions that can promote reduced vulnerability of exposed systems and increased resilience. However, the *W-ScaRI* is not limited to the tested region—it can be applied to other basins, with the possible adaptation of weights and indicators.

2. Materials and Methods

The *W-ScaRI* index proposed in this paper is built based on the risk formulation described by [3], where the drought risk is determined through two main components: a hazard and its consequences. A hazard is a potentially threatening situation that causes damage, and it is composed of a combination of atmospheric processes and hydrological responses, reducing the available water in lakes, rivers, reservoirs and/or soil. The consequences reflect the exposure and vulnerability of a system to the environmental, social and economic impacts of droughts. The vulnerability component also includes resilience, which is a system's ability to adapt to or recover from damage. Thus, the risk caused by water scarcity comes from the drought itself, as well as from the aspects of water and land management.

The "hazard" and "consequence" subindices are constructed by combining the indicators or indices selected from the existing technical literature but also by using indicators originally proposed and developed in the current study. The *W-ScaRI* is based on a mixed formulation consisting of a weighted product of two weighted sums. Thus, each subindex is composed of weighted summations, and, subsequently, these subindices are weighted and multiplied to compose the *W-ScaRI*. The *W-ScaRI* is illustrated in Figure 1 and is represented by Equation (1):

$$W - ScaRI \ = \ HI^{wh} \times x \, CI^{wc} \tag{1}$$

where *HI* is the drought hazard subindex; *CI* is the consequence subindex; and *wh* and *wc* are the weights associated with the hazard (*HI*) and consequence (*CI*) subindexes, respectively.

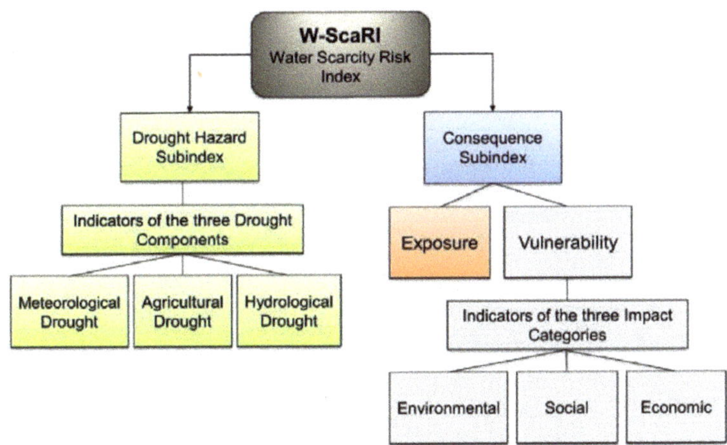

Figure 1. Water Scarcity Risk Index (*W-ScaRI*) representation.

Note that this formulation is chosen to represent the risk process conceptually. The product indicates that risk only exists if the hazard and possible negative consequences are both non-zero elements. However, the hazard and its consequences are subindices composed of a sum of a set of possible components that contribute to each one of these risk elements. Regarding the scale of the *W-ScaRI*, to clarify communication and maintain simplicity, its subindices and indicators are normalized within the range of 0 to 1, where the value "0" corresponds to the minimum risk (or the absence of risk, to be precise) and "1" corresponds to the maximum risk.

The drought hazard subindex consists of indicators representing the three components that characterize different aspects of droughts: meteorological, agricultural/green water and hydrological/blue water (Figure 1). The hazard subindex formulation is presented in Equation (2):

$$HI = MDI \times wm + ADI \times wa + HDI \times wh \tag{2}$$

where *MDI*, *ADI* and *HDI* are the hazard indicators of meteorological, agricultural/green water and hydrological/blue water droughts, respectively, and *wm*, *wa* and *wh* are the weights associated with each indicator.

The Standardized Precipitation Index (*SPI*) developed by McKee et al. [36] is used to represent the meteorological drought indicator (*MDI*). The *SPI* consists of the difference between the measured precipitation and the historic average rainfall (over a given period of time) divided by the standard deviation. Since precipitation does not typically follow a normal distribution, the gamma distribution is adjusted, and gamma transformation is applied to the normal distribution. To facilitate analyses, we propose a color scale for the drought classes and the *SPI* values, as shown in Table 1.

Table 1. Drought classification by *SPI* ranges [36,37].

Color Scale	*SPI*	Classification
	$SPI > 0$	Wet period
	0 to −0.99	Mild Drought
	−1.00 to −1.49	Moderate Drought
	−1.50 to −1.99	Severe Drought
	$SPI \leq -2.00$	Extreme Drought

To normalize the indicators, three equations are adjusted considering the typical probability values of the normal distribution and the *SPI* values between "0" and "−3". To normalize the indicator between 0 and 1, the cumulative probability values of a normal curve are adopted, assuming that the *SPI* values equal to −1, −2 and −3 are associated with the cumulative probabilities up to 1, 2 and 3 standard deviations at the right side of the curve, as shown in Table 2.

Table 2. *SPI* indicator normalization.

SPI Values	Cumulative Probability—Normal Distribution	Adjusted Equation
0	0.50	
−1	0.841	$y = -0.341x + 0.5$
−1	0.841	
−2	0.977	$y = -0.136x + 0.705$
−2	0.977	
−3	0.999	$y = -0.022x + 0.933$

The Reconnaissance Drought Index (*RDI*) developed by Tsakiris and Vangelis [38] is used to represent the agricultural drought indicator (*ADI*), which addresses water deficits as a kind of balance between the entry and exit of a water system. The *RDI* is calculated based on information about the accumulated rainfall (observed) and the potential evapotranspiration (calculated). The initial *RDI* values satisfactorily follow both the gamma and lognormal distributions [39,40].

The hydrological drought indicator (*HDI*) used to represent the hydrological drought hazard is the Streamflow Drought Index (*SDI*) [41], as shown in Equation (3):

$$SDI_{i,k} = \frac{V_{i,k} - \overline{V}_k}{S_k} \tag{3}$$

where $V_{i,k}$ is the cumulative flow volume of the hydrological year i in the reference period k, and $\overline{V}_k$ and S_k are the mean and standard deviation of the accumulated flow volumes, respectively.

The *SDI* is calculated by adjusting a two-parameter log-normal distribution [41].

For both *RDI* and *HDI*, the drought classification (Table 1) and the indicator normalization (Table 2) are the same as those used for the *SPI* index.

The consequence subindex (*CI*) is characterized by indicators that represent the system exposure and by three groups of vulnerability indicators that comprise the main drought impacts: environmental, social and economic impacts. The formulation of the consequence subindex is presented in Equations (4)–(6):

$$CI = ExI \times VI \tag{4}$$

$$ExI = ExI_1^{wex1} \times ExI_2^{wex2} \tag{5}$$

$$VI = \left(EnVI \times wen + SVI \times ws + \sum_{i=1}^{n} EVI_i \times we_i \right) \tag{6}$$

where ExI is the exposure indicator; VI is the vulnerability indicator; $EnVI$ is the environmental vulnerability indicator; SVI is the social vulnerability indicator; EVI is the economic vulnerability indicator; and wex, wen, ws and we are the weights of these indicators, respectively.

To represent the exposure (ExI), two indicators are used. The first indicator (ExI_1) is the water stress indicator, which is used in order to link the consumption demands of the various water uses in the basin to the water availability. Values close to zero indicate that there is a water surplus in the basin. Values close or equal to 1 denote that almost all the available water is used to supply the various water uses. The second exposure indicator (ExI_2) is the percentage of the reservoir equivalent storage ($VEqR$), which is used by the Brazilian National Water Agency (ANA) to monitor the evolution of reservoir useful storage in the main Brazilian basins [42]. This is calculated by dividing the sum of the accumulated storages of the existing reservoirs in the basin, at a given moment, by the sum of the total useful storages of these reservoirs. It is a strategic method to consider the joint effect of several in-line reservoirs. To apply this method to other places, the same logic can be used when one or more reservoirs are upstream of the interest catchment.

Equation (7) shows how to calculate ExI_2 from the $VEqR$ results:

$$ExI_2 = 1 - \frac{VEqR}{100} \tag{7}$$

Table 3 shows the drought classifications based on the $VEqR$ and ExI_2 value ranges and the color scales proposed in this study.

The ExI_2 indicator cannot be used if the basin does not have reservoirs. In this situation, the weight of the indicator is zero, and the weight of the ExI_1 indicator is 1.

Table 3. Drought classifications based on $VEqR$ and ExI_2 ranges.

Color Scale	$VEqR$ (%)	ExI_2	Classification
	$VEqR > 40$	$ExI_2 < 0.6$	Wet period
	$30 < VEqR \leq 40$	$0.70 > ExI_2 \geq 0.6$	Mild Drought
	$20 < VEqR \leq 30$	$0.80 > ExI_2 \geq 0.7$	Moderate Drought
	$10 < VEqR \leq 20$	$0.90 > ExI_2 \geq 0.8$	Severe Drought
	$0 \leq VEqR \leq 10$	$1 \geq ExI_2 \geq 0.9$	Extreme Drought

For the environmental vulnerability indicator ($EnVI$), the qualitative water balance indicator, $Bqual$ (%), is used according to [43] and as shown in Equation (8):

$$Bqual = \frac{Wdil + Wcons}{Wavail} \times 100 \qquad (8)$$

where $Wcons$ is the discharge representing the consumed water (m^3/s); $Wavail$ is the water availability (m^3/s); and $Wdil$ is the necessary flow to dilute a given effluent (m^3/s), which is calculated according to Equation (9):

$$Wdil = 0.001 \times \frac{BOD_L}{BOD_C} \qquad (9)$$

where BOD_L is the sum of the biochemical oxygen demand (BOD) load of the domestic sewage discharged in the basin (mg/s), and BOD_C is the maximum allowable BOD concentration in the water course (mg/L) that enables the following water uses: human consumption after conventional treatment; the protection of aquatic communities; primary contact recreation; and irrigation.

$EnVI$ is calculated by dividing $Bqual$ by 100. For values of $Bqual$ greater than 100, $EnVI$ is equal to 1.

Population density (PD) is used to represent the social vulnerability indicator (SVI). Thus, the higher the density, the greater the expected social impacts of water scarcity. The indicator is calculated for each municipality in the study area based on a linear equation that normalizes the population density. To adjust this equation, the country's average population density is associated with a vulnerability equal to 0.5, while the value of the city with the highest population density in the country is associated with a vulnerability of 1 (Table 4), creating a local scale (that can be adapted to other regions or countries) that can help with risk comparisons. This procedure allows the for the application and comparisons of the indicator in different urban or metropolitan regions of the reference country. The final integrated indicator for the study area is calculated using the weighted average of the population of each considered municipality.

Table 4. Social vulnerability indicator determination.

Population Density—PD (People/km^2)	Social Vulnerability Indicator—SVI (0–1)	Adjusted Equation
City with the highest population density	1	$SVI = a \times PD + b$, where PD is the population density; a and b are the adjusted parameters of the SVI equation.
Country's average population density	0.5	

The following indicators are used to represent economic vulnerability: gross domestic product (GDP) per capita (EVI_1) and the level of competition with human supply (EVI_2). The use of GDP per capita aims to characterize the exposure of economic activities to the risk of water scarcity (considering that water is an important input to most significant economic activities). Thus, the higher the GDP per capita of a region, the greater the impact in a situation of water scarcity. This indicator is calculated for each municipality in the study area based on a linear equation that normalizes the GDP per capita. To adjust this equation, the country's average *GDP* per capita is associated with a vulnerability equal to 0.5, and that of the city with the highest *GDP* per capita in the country is associated with a vulnerability of 1 (Table 5). This procedure allows for the application and relative comparisons of the indicator in different regions of the country. The final integrated indicator for the study area is calculated using the weighted average of each municipality's *GDP*.

Table 5. Economic vulnerability indicator determination.

GDP per Capita	Economic Vulnerability Indicator—EVI_1 (0–1)	Adjusted Equation
City with the highest *GDP* per capita	1	EVI_1 = c × *GDP* per capita + d where c and d are the adjusted parameters of the IVE_1 equation.
Country's average *GDP* per capita	0.5	

Usually, when water scarcity occurs, the use of water is prioritized for human consumption and animal watering. This principle is also stated in the Brazilian "water law" (Federal Act 9433/97). Therefore, any other type of water use can be firstly impacted in water scarcity cases. The second indicator of economic vulnerability (EVI_2) is the level of competition for the general use of human water supplies; this indicator is built with the purpose of characterizing the impacts on industrial activities located in basins where water is used for human supplies. Thus, in a basin where there are both industrial and human water supply abstractions, the lower the flows abstracted for industry supply compared to those abstracted for human supply, the greater the vulnerability. EVI_2 is expressed by Equation (10):

$$EVI_2 = \left(1 - \frac{Qi - Qa}{Qcap}\right) \tag{10}$$

where Qi is the flow abstracted by the industrial sector (m^3/s); Qa is the flow abstracted by the human supply sector (m^3/s); and $Qcap$ is the total flow abstracted (m^3/s).

$$EVI2 = 1,\ when \left(1 - \frac{Qi - Qa}{Qcap}\right) > 1$$

In the final assessment of each subindex, weights are applied to each group of hazard and vulnerability indicators. Similarly, weights are also used for the subindices in the *W-ScaRI's* final calculation. The weight sensitivity study, as well as the insertion of the resilience component in the *W-ScaRI*, which led to development of the proposed index, is carried out at a later stage of the research project.

It is important to highlight that the present work develops the structure of the index and evaluates its potential application. Thus, the definition of weights, although essential in practical applications to effectively use the index as a management tool, is of secondary relevance until the final index formulation proposal is achieved. Table 6 shows the weights used in this study, which were equally divided among the indicators.

The sum of the weights of the meteorological (*MDI*), agricultural (*ADI*) and hydrological (*HDI*) drought indicators are considered equal to 1 in the hazard subindex calculation. The weights of the exposure indicators, water stress and % *VEqR* are considered equal to 0.50.

Table 6. Weights used for *W-ScaRI* calculation.

Drought Hazard Subindex			Consequence Subindex				
Indicators		**Weight**	**Exposure**		**Vulnerability**		
			Indicators	**Weight**	**Indicators**		**Weight**
Meteorological Drought (*MDI*)	*SPI*	0.333	Water stress	0.5	Environmental vulnerability (*EnVI*)	*Bqual*	0.333
Agricultural Drought (*ADI*)	*RDI*	0.333	% *VEqR*	0.5	Social vulnerability (*SVI*)	*PD*	0.333
Hydrological Drought (*HDI*)	*SDI*	0.333			Economic vulnerability (*EVI*)	*GDP* per capita	0.1665
						% Competition with human supply	0.1665
Weight = 0.5					Weight = 0.5		

3. Study Area

The area where the *W-ScaRI* is tested is the Rio de Janeiro Metropolitan Region (RJMR) (Figure 2), the largest urban agglomeration of the Brazilian coastal zone, with a population of about 12.4 million in 2017 [44].

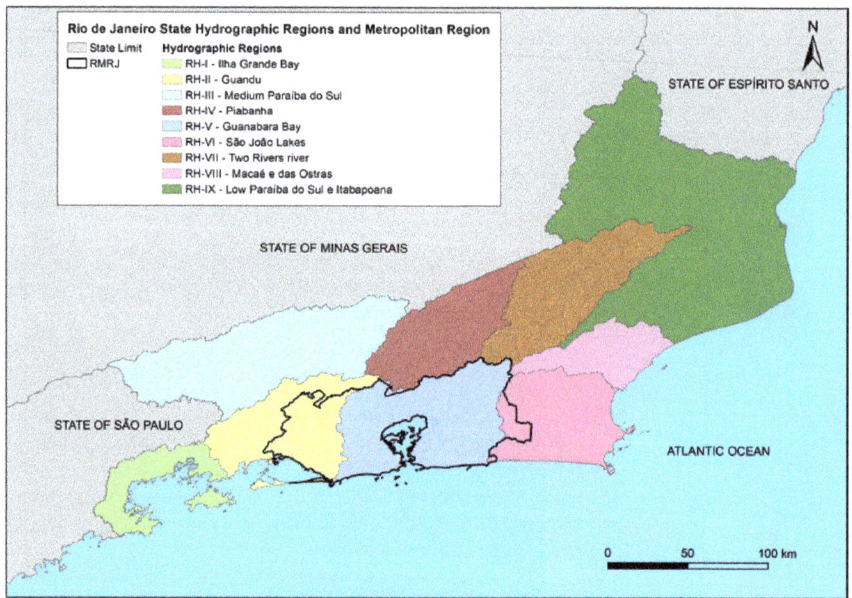

Figure 2. Rio de Janeiro State and Rio de Janeiro Metropolitan Region (RJMR).

A large part of the RJMR population, around 82%, representing almost 10 million inhabitants, is supplied water by the Guandu/Lajes/Acari system. This system currently produces a total discharge of 52.4 m^3/s, of which 45 m^3/s comes from the Guandu System, 5.5 m^3/s comes from the Lajes Reservoir (Lajes System) and 1.9 m^3/s comes from the Acari system [45].

The main source of water for this supply system is the Paraíba do Sul River Basin (Figure 3), which covers an area of 61,307 km^2, involving three of the most developed states in the country—São Paulo (13,934 km^2), Minas Gerais (20,699 km^2) and Rio de Janeiro (26,674 km^2) [46]. The Paraíba do Sul River is formed by the union of the Paraibuna and Paraitinga Rivers in the Bocaina Mountains, in the State of São Paulo, at an altitude of 1800 m.

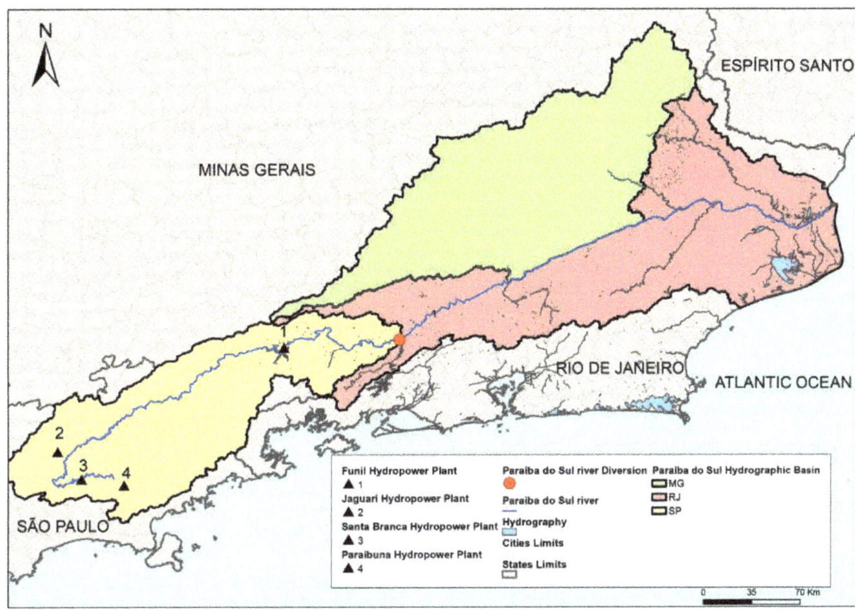

Figure 3. Paraíba do Sul River Basin.

The basin has a complex system, which includes accumulation reservoirs, power plants and pumping stations, as well as a water transposition system with the original purpose of generating electricity in the Lajes Hydroelectric Complex (RJ). However, today, this system supplies water to most of the RJMR.

The Paraíbuna, Santa Branca, Jaguari and Funil Hydroelectric Plants, located along the headwaters of the Paraíba do Sul River, regulate its flow, allowing water diversion from the main transposition of the basin. This diversion is accomplished by the Santa Cecília Dam, in the city of Barra do Piraí, offering a maximum pumping capacity of 160 m³/s. The water is carried to the Santana Reservoir, and it is pumped from there to the Vigário Reservoir. The accumulated water in the Vigário Reservoir is then diverted by gravity to the Serra do Mar Atlantic hill, and it is sent to the Nilo Peçanha and Fontes Nova Power Plants. The outflows from these plants and the Lajes Reservoir are sent to the Ponte Coberta Reservoir (Pereira Passos Power Plant), located on the Lajes Stream, which is the Guandu River's main source. The operation of the Pereira Passos Hydropower Plant (HPP) must ensure the continuity of the supply to the Guandu water treatment plant and other users of the Guandu River Basin [47]. Figure 4 illustrates the steps of this complex water transfer system from the Paraíba do Sul River to the Guandu River.

In the last 25 years, two critical droughts have occurred in the Paraíba do Sul Basin. In the first one, from 2003 to the beginning of 2004, the storage level of the flow regulating the reservoirs of the Paraíba do Sul River indicated the possibility of rationing, including for the RJMR. Faced with the possibility of a water crisis, the ANA issued resolutions that resulted in a flow reduction downstream of the Santa Cecília Dam, in the derivation to the Lajes Complex in Santa Cecília, and downstream of the Pereira Passos HPP on the Guandu River [43].

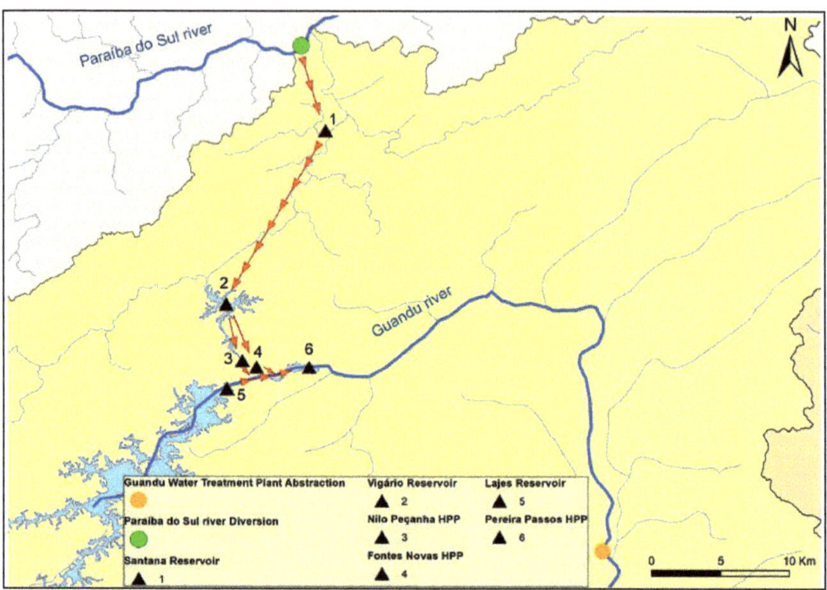

Figure 4. Schematic drawing of the water transfer from Paraíba do Sul to Guandu Rivers.

In the 2014–2015 period, the severe water scarcity in the Southeast region of Brazil affected the two most important metropolitan regions of the country: Rio de Janeiro (RJMR) and São Paulo (SPMR). Regarding the RJMR, the storage volume of the reservoirs of the Paraíba do Sul River Basin showed the need to use the dead volume of some reservoirs, which actually started to happen at the end of January 2015. In order to preserve reservoir stocks and, at the same time, ensure water uses, reductions in the minimum inflows to the Santa Cecilia Reservoir were gradually allowed, together with periodic assessments of the impacts on water uses downstream [48]. In addition, the flow reduction downstream of the Pereira Passos HPP was also authorized.

4. Results and Discussion

Since the Paraíba do Sul River is the main water supply source for the RJMR, the indicators of the hazard subindex were applied to this river basin, upstream of the Santa Cecília Dam, as follows:

- Meteorological and agricultural droughts: Several precipitation and temperature monitoring gauges.
- Hydrological drought: The *SDI* indicator was applied to the series of natural flows to the Santa Cecília Reservoir (transposition site), and for analysis purposes, it was applied to the series of other reservoirs in the basin.

The consequence subindex indicators were applied as follows:

- Exposure: The water stress indicator (ExI_1) was applied to the Guandu Basin between Pereira Passos HPP and the mouth. The equivalent reservoir volume percentage indicator (ExI_2) was applied to the reservoirs in the Paraíba do Sul River Basin, upstream of the Santa Cecília Dam (transposition site).
- Environmental vulnerability: The qualitative water balance indicator (IVA_1) was applied to the Guandu Basin and the rivers located in the municipalities supplied by the Guandu/Lajes/Acari system.
- Social vulnerability: The population density of the municipalities supplied by the Guandu/Lajes/Acari system.

- Economic vulnerability: The economic activities of the municipalities supplied by the Guandu/Lajes/Acari system and industries in the Guandu Basin.

Figure 5 shows the region where the *W-ScaRI* was applied. The *W-ScaRI* was applied to January 2015, when the equivalent reservoir of the Paraíba do Sul Basin reached one of its lowest values, and to October 2015, at the end of the water crisis. To apply the *SPI*, *RDI* and *SDI* indices, the DrinC computer program was used [37].

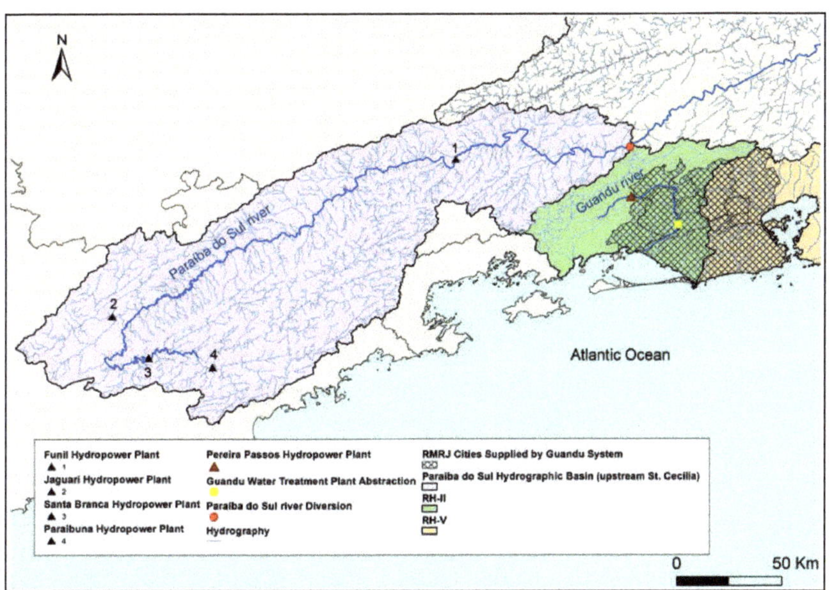

Figure 5. The *W-ScaRI* application area.

4.1. Data Collection and Analysis

The data used to calculate the *W-ScaRI* indicators/indices are presented in Table 7.

Table 7. Data used for *W-ScaRI* application.

Indicator	Type of Information	Period	Reference
MDI (*SPI*)	Monthly rainfall series. Gauges: Resende, Taubaté, Fazenda Santa Clara, S. Luiz do Paraitinga, Igaratá and Santa Isabel	1991–2016	[49–51]
ADI (*RDI*)	Monthly rainfall and monthly average and maximum and minimum temperature series. Gauges: Resende and Taubaté	1991–2016	[49,50]
HDI (*SDI*)	Average monthly natural flows at Paraibuna HPP, Santa Branca HPP, Funil HPP and Santa Cecília Dam	1931–2017	[52]
ExI_1 Water stress	Pereira Passos HPP outflows series, incremental 95% flow (Guandu mouth—Pereira Passos HPP), water consumption data (Guandu River users)	1994–2016 2014	[43,52]
ExI_2 % Equivalent volume of reservoirs	Daily series of accumulated useful volumes of Paraibuna, Santa Branca, Jaguari and Funil Reservoirs, and useful volume of reservoirs	1993–2017	[52]
EnVI Qualitative water balance	Consumed flows by the users of RJMR basins, dilution flows of RJMR basins and water availability in RJMR basins	2014	[43]
SVI Population density	Municipal population and area	2017 estimate	[44]
EVI_1 GDP per capita	*GDP* and population	2015	[53]
EVI_2 % Competition human supply	Industrial abstraction flows Human supply abstraction flows	2014	[43]

4.2. Hazard Subindex

4.2.1. Meteorological Drought Indicator (*MDI*)

The *SPI* index at each rainfall gauge was calculated for monthly durations (Figure 6) for a period of severe drought in the basin (2013–2015). Figure 6 shows that drought began to occur at the end of 2013, extending practically throughout 2014 until January 2015. Starting in February 2015, there was an increase in rainfall at all measuring gauges, reflecting the increase in *SPI*. However, even with a few wetter months, the water stress situation in the basin lasted for most of 2015. In November, the rainfall returned to normal at all gauges, and the basin's reservoirs began to recover.

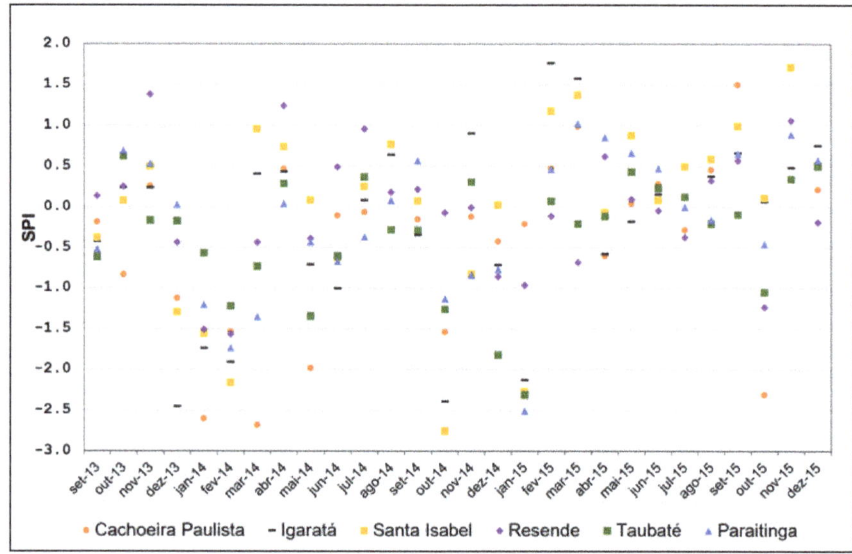

Figure 6. Monthly *SPI* of rainfall gauges in the region under study.

It is important to highlight that drought is more severe in the Paraíba do Sul Basin when critical *SPI* values occur in summer (rainy season). In this situation, the reservoir levels may not recover before the dry season begins.

Figure 7 shows the *SPI* calculated for January 2015 at all gauges used in the study, considering the color scale presented in Table 1. This figure also shows the area of influence of each gauge, as defined using the Thiessen method.

A duration of 12 months, comprising the reference month and the previous 11 months, was considered in the final calculation of the meteorological drought indicator (*MDI*), which integrates the hazard subindex. For example, for January 2015, the *SPI* was calculated from February 2014 to January 2015. The *SPI* of each rainfall gauge was normalized using the procedure shown in Table 2. The *MDI* (Table 8) was calculated by using the weighted summation of *SPI*, where the weight of each gauge was determined using the Thiessen method.

4.2.2. Agricultural Drought Indicator (*ADI*)

Figure 8 presents the *RDI* monthly duration results from September 2013 to December 2015.

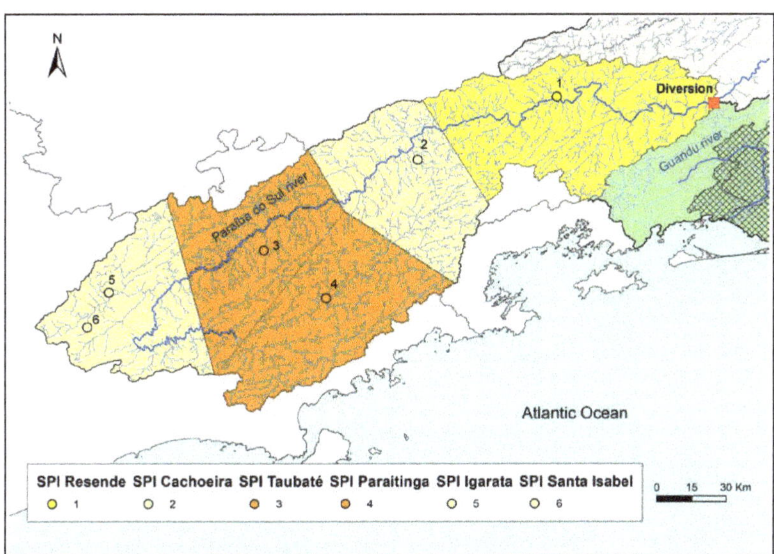

Figure 7. *SPI* index of rainfall gauges in the region (January 2015).

Table 8. Meteorological drought indicator *(MDI)* application.

Rainfall Gauges	Weight	January 2015		October 2015	
		SPI	*SPI* (0–1)	*SPI*	*SPI* (0–1)
Paraitinga	0.20	−2.74	0.99	−0.78	0.77
Taubaté	0.18	−2.36	0.98	−1.72	0.94
Resende	0.27	−1.35	0.89	−1.34	0.89
Cachoeira Paulista	0.18	−0.55	0.69	−0.08	0.53
Igaratá	0.11	−0.55	0.69	0.36	0.00
Santa Isabel	0.06	−0.94	0.82	0.37	0.00
Meteorological drought indicator—*MDI*		0.87		0.66	

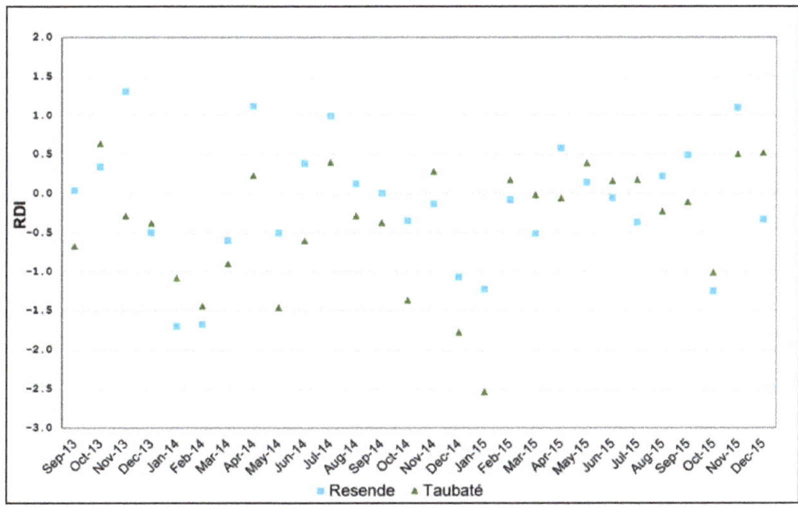

Figure 8. Monthly *RDI* of rainfall gauges in the region under study.

Figure 8 shows that 2014 was drier than 2015. When a drought occurs in the rainy months, as mentioned earlier, it is negatively reflected in the available discharges.

The normalization presented in Table 2 was considered in the final calculation of the agricultural drought indicator (*ADI*). Table 9 presents the application of the *ADI* for the Resende and Taubaté gauges.

Table 9. Agricultural drought indicator (*ADI*) application.

Gauges	Weight	January 2015		October 2015	
		RDI	*RDI* (0–1)	*RDI*	*RDI* (0–1)
Resende	0.50	−1.22	0.99	−1.25	0.84
Taubaté	0.50	−2.54	0.87	−1.01	0.87
Agricultural drought indicator—*ADI*		0.93		0.86	

4.2.3. Hydrological Drought Indicator (*HDI*)

To apply the *SDI* index, the color scale classification presented in Table 1 is used, along with the series of natural discharges inflowing to Santa Cecília (diversion place) and the series of HPP reservoirs located in the basin, namely, Paraibuna, Santa Branca, Jaguari and Funil.

Figure 9 presents the *SDI* results for a monthly duration during a severe drought in the basin (2013–2015). This figure shows that there was an extremely dry period between January 2014 and October 2015, especially in February 2014 and January 2015, usually wet months, as observed in the *SPI* values. However, the *SDI* values over the period analyzed are more severe than the *SPI* values. This is probably due to the duration of the drought event, which extended over a long period of time. It is therefore important to analyze the *SPI* for longer durations.

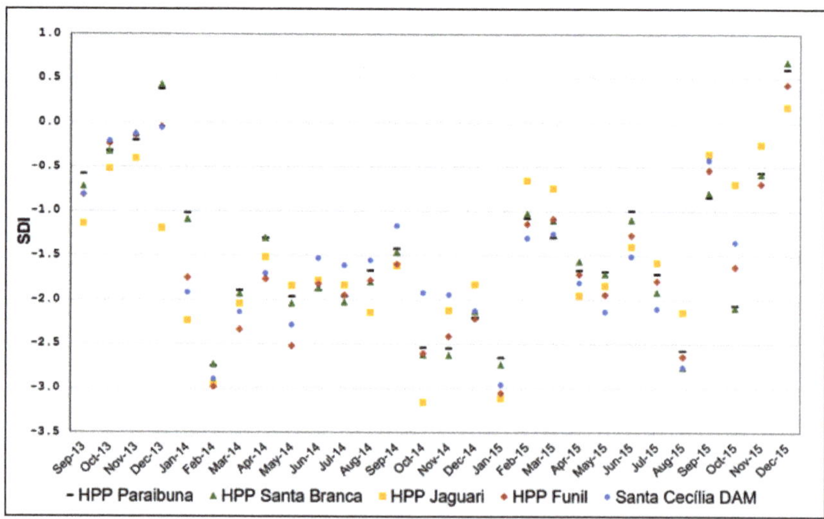

Figure 9. Monthly *SDI* in the basin's reservoirs and Santa Cecília Dam.

Figure 10 shows the application results for January 2015 at the Santa Cecília, Paraibuna, Santa Branca, Jaguari and Funil Reservoirs. The Paraibuna HPP Reservoir (Figure 10) has the largest useful volume among the reservoirs under discussion, and it is the most relevant due to its storage capacity. This reservoir presented very low *SDI* values throughout the period, indicating an extremely severe drought. In fact, the Paraibuna Reservoir reached a zero useful volume in January and February 2015.

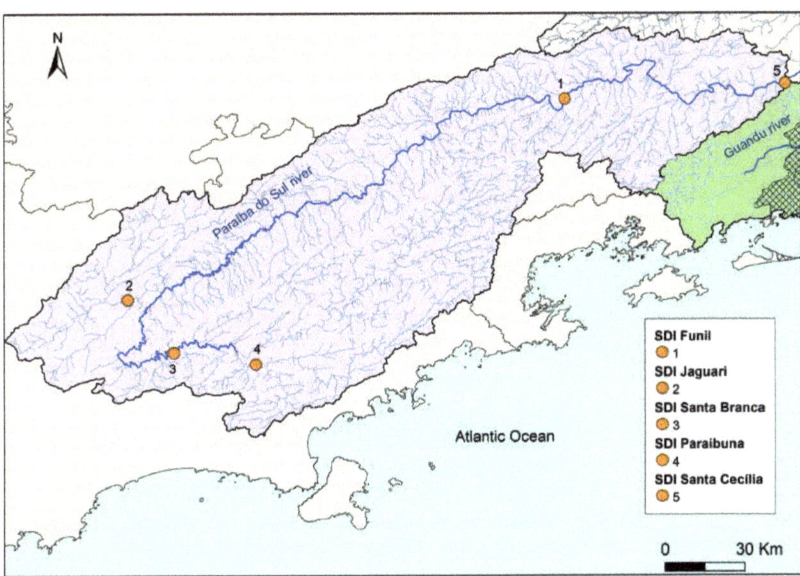

Figure 10. Monthly *SDI* in the basin's reservoirs and Santa Cecília Dam (January 2015).

To calculate the hydrological drought indicator (*HDI*) in order to compose the hazard subindex, the *SDI* index was only calculated for the Santa Cecília Dam, which reflects the flows from the entire upstream basin. The calculation used the monthly duration, and the normalization followed the procedure used for the meteorological drought indicator. Table 10 presents the final application results of the *HDI*, where an extremely severe drought situation is observed in January 2015.

Table 10. Hydrological drought indicator (*HDI*) application.

Place	January 2015		October 2015	
	SDI	*SDI* (0–1)	*SDI*	*SDI* (0–1)
Paraíba do Sul River in Santa Cecília	−2.96	1.0	−1.35	0.89
Hydrological Drought Indicator—*HDI*	1.0		0.89	

4.3. Consequence Subindex

The consequence subindex was determined using the method presented in the Section 2 and the data presented in Table 7.

4.3.1. Exposure Indicators (ExI)

For the application of the water stress indicator (ExI_1), the series of the monthly minimum outflows of the Pereira Passos HPP was used, plus an additional value of the 95% permanence flow of the incremental area between the Pereira Passos HPP and the mouth of the Guandu River.

Water consumption data from the various uses in the Guandu Basin were also employed: human supply, industrial use and environmental flow. From the total of 95.1 m^3/s, the 25 m^3/s portion corresponds to the environmental flow, 42.0 m^3/s corresponds to human supply, 29.1 m^3/s corresponds to industrial use and 0.95 m^3/s corresponds to the portion of returned sewage [43].

Table 11 shows that, in October 2015, the available flow was lower than the water uses in the basin, which means that the environmental flow was used to meet the needs of human and industrial supplies and, indeed, had a lower value than the established expected minimum. This could be observed for several days in 2015, probably as a way of avoiding impacts on human supply, as well as helping to preserve the reservoir stocks of the Paraíba do Sul Basin.

Table 11. Water stress indicator (ExI_1) application.

Local	Basin Water Uses (m³/s)	Available Flow (m³/s)	
		January 2015	October 2015
Guandu River Basin between mouth and Pereira Passos HPP	95.1	101.3	79.3
Water stress Indicator—ExI_1		0.94	1.0

The reservoir equivalent volume percentage indicator—$VEqR$—was calculated using the daily series of accumulated volumes of the reservoirs of the Paraibuna, Santa Branca, Jaguari and Funil HPPs, all located in the Paraíba do Sul Basin (Figure 7), upstream of the transposition, and the sum of the total useful volumes of each reservoir. Table 12 presents the minimum monthly values of $VEqR$ for the period between 2013 and 2015. The ExI_2 indicator is determined using Equation 7 and the lowest value of the equivalent volume percentage of the reference month (Table 13).

Table 12. Minimum monthly values of the indicator $VEqR$ (%).

Year	Jan	Feb	Mar	Apr	May	June	July	Aug	Sept	Oct	Nov	Dec
2013	42	54	60	68	69	65	62	57	51	48	45	48
2014	48	42	41	39	34	28	23	18	13	7	4	2
2015	0.4	0.3	8	16.3	17.1	15.3	11.6	7.1	6.6	5.4	5.8	9.9
2016	18	27	34	43	44	44	52	50	48	46	45	49

Table 13. Exposure indicator (ExI_2) application.

Year	Jan	Feb	Mar	Apr	May	June	July	Aug	Sept	Oct	Nov	Dec
2013	0.58	0.46	0.40	0.32	0.31	0.35	0.38	0.43	0.49	0.52	0.55	0.52
2014	0.52	0.58	0.59	0.61	0.66	0.72	0.77	0.82	0.87	0.93	0.96	0.98
2015	1.00	1.00	0.92	0.84	0.83	0.85	0.88	0.93	0.93	0.95	0.94	0.90
2016	0.82	0.73	0.66	0.57	0.56	0.56	0.48	0.50	0.52	0.54	0.55	0.51

The ExI_2 indicator was very high in January and February 2015 (Table 13), with the accumulated useful volume of the reservoirs close to zero (Table 12), showing that the basin ran without reserves.

4.3.2. Environmental Vulnerability Indicator ($EnVI$)

The environmental vulnerability was determined for each RJMR sub-basin supplied by the Guandu/Lajes/Acari system, using Equation (3). The final indicator was calculated using the weighted average of the discharge values needed for dilution purposes in each sub-basin, as shown in Table 14.

4.3.3. Social Vulnerability Indicator (SVI)

The social vulnerability indicator was determined for each RJMR municipality supplied by the Guandu/Lajes/Acari system, using the methodology described in Section 2. To normalize the population density, the linear equation shown in Table 4 was adjusted using the Brazilian average population density associated with a vulnerability equal to 0.5, while the density of the city of São Paulo (the city with the highest density in Brazil) was associated with 1, as shown in Table 15.

Table 14. Environmental vulnerability indicator (*EnVI*) application.

Sub-Basin	Dilution Flow—*Wdil* (m³/s)	Bqual (%)	Vulnerability (0–1)
Piraí River	3.3	50.3	0.5
Lajes Reservoir	-	33.4	0.33
Guandu River	7.3	81.9	0.82
da Guarda River	16.9	478.3	1
Guandu-Mirim and Litorâneos Rivers	105.6	1195.1	1
Iguaçu and Saracuruna Rivers	226.6	1305.4	1
Jacarepaguá and Marapendi Lake Rivers	156.0	1184.8	1
Pavuna-Meriti, Faria-Timbó and Maracanã Rivers; Governador and Fundão Island Rivers; Rodrigo de Freitas Lake Rivers	171.3	1248.7	1
Environmental Vulnerability Indicator—*EnVI*			=0.99

Table 15. Social vulnerability indicator determination.

City/Country	Population Density—*PD* (People/km²)	Social Vulnerability Indicator—*SVI* (0–1)	Adjusted Equation
São Paulo	7959.27	1	$SVI = 6 \times 10^{-5} \times PD$
Brazil	24.40	0.5	$+ 0.4985$

The final social vulnerability indicator was calculated using the weighted average of the population of each municipality (Table 16).

Table 16. Social vulnerability indicator (*SVI*) application.

Municipality	Population Density (Inhabitants/km²)	Population (Inhabitants)	SVI (0–1)
Belford Roxo	6277	495,783	0.88
Duque de Caxias	1905	890,997	0.61
Itaguaí	446	122,369	0.53
Japeri	1237	101,237	0.57
Nilópolis	8164	158,329	0.99
Nova Iguaçu	1542	798,647	0.59
Mesquita	4130	171,280	0.75
Paracambi	264	50,447	0.51
Queimados	1921	145,386	0.61
Rio de Janeiro	5433	6,520,266	0.82
São João de Meriti	13,075	460,461	1.00
Seropédica	297	84,416	0.52
Social Vulnerability Indicator—*SVI*			=0.79

Figure 11 shows the results of the social vulnerability indicator application for each municipality. The vulnerability of Rio de Janeiro has a major influence on the final *SVI* value due to the significant amount of the population potentially affected in the municipality.

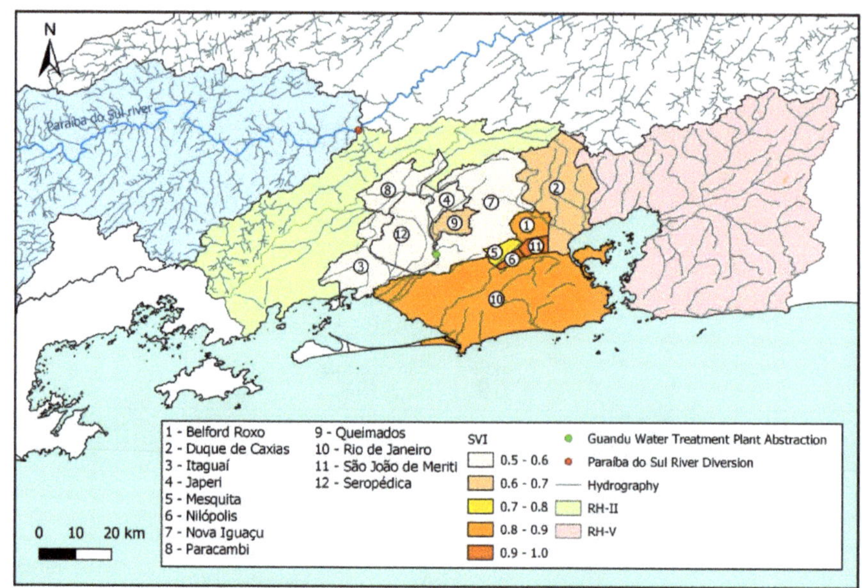

Figure 11. Social vulnerability indicator (*SVI*) application for each municipality supplied by the Guandu/Lajes/Acari system.

4.3.4. Economic Vulnerability Indicators (EVI_1 and EVI_2)

The economic vulnerability indicator (EVE_1) was calculated for each RJMR municipality supplied by the Guandu/Lajes/Acari system, according to the procedure previously discussed in the Section 2. To normalize the *GDP* per capita, the linear equation shown in Table 5 was adjusted considering, in our test case, the data of Brazil's *GDP* per capita, associated with a vulnerability equal to 0.5, and the data of the city of São Paulo's *GDP* per capita (the city with the highest *GDP* per capita in Brazil), associated with 1, as shown in Table 17.

Table 17. Economic vulnerability indicator (EVI_1) determination.

City/Country	*GDP* per Capita	Economic Vulnerability Indicator—EVE_1 (0–1)	Adjusted Equation
São Paulo	54.4	1	$EVI_1 = 0.02 \times GDP$ per capita—0.0856
Brazil	29.3	0.5	

The final EVI_1 value was calculated using the weighted average of each municipality's *GDP* (Table 18). Figure 12 shows the results of the economic vulnerability indicator (EVI_1) application for each municipality. The vulnerability of the Rio de Janeiro municipality has a major influence on the final EVI_1 value due to the significant value of the *GDP* per capita. Thus, this region could experience considerable impacts in situations of water scarcity.

Table 18. Economic vulnerability indicator (EVI_1) application.

Municipality	GDP (10^3 BRL)	GDP per Capita	IVE_1
Belford Roxo	7,479,539	16	0.23
Duque de Caxias	35,114,426	40	0.71
Itaguaí	7,404,493	62	1.00
Japeri	1,342,219	13	0.18
Nilópolis	2,525,559	16	0.23
Nova Iguaçu	15,948,718	20	0.31
Paracambi	2,084,163	12	0.16
Mesquita	843,386	17	0.26
Queimados	4,851,828	34	0.59
Rio de Janeiro	320,774,459	50	0.90
São João de Meriti	7,931,134	17	0.26
Seropédica	2,306,345	28	0.47
Economic Vulnerability Indicator—EVI_1			=0.82

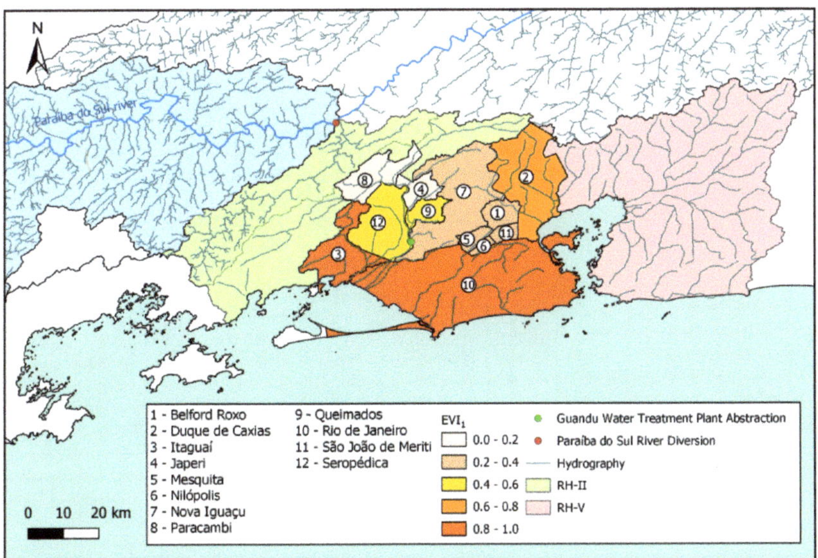

Figure 12. Economic vulnerability indicator (EVI_1) application for each municipality supplied by the Guandu/Lajes/Acari system.

The indicator of competition between industrial uses and human supply needs (EVI_2) was calculated for the Guandu Basin, where the Guandu water treatment system is located, and which also serves significant industrial uses. The results of the application are presented in Table 19, where it can be observed that the vulnerability is the maximum possible for this indicator. In the Guandu Basin, industrial abstractions are lower than abstractions for human supply, which can have a great impact on the industrial sector in situations of water scarcity. According to the Brazilian "water law" (Federal Act 9433/97), in these situations, the use of water is prioritized for human consumption and animal watering.

Table 19. Economic vulnerability indicator (EVI_2) application.

Water Use	Abstraction Flow (m³/s)	Percentage (%)
Industrial sector	29.1	41%
Human supply	42	59 %
Economic Vulnerability Indicator		$EVI_2 = 1.00$

4.4. W-ScaRI Calculation

The Water Scarcity Risk Index (*W-ScaRI*) was calculated for the entire available data series. Figure 13 shows a complete representation of the drought periods in the Paraíba do Sul River Basin (2003–2004 and 2014–2015) using the *W-ScaRI*, and it shows that the scarcity water was worse in the 2014–2015 period than in the previous one (which, in fact, is recognized as being true), with higher *W-ScaRI* values. It is interesting to note the importance of evaluating the risk index by combining hazards and consequences. For instance, for March 2012, the hazard subindex was 0.82, but the *W-ScaRI* was lower, with a value of about 0.60. In this case, if the planner uses an index that only represents the hazard, it could lead to the employment of unnecessary actions.

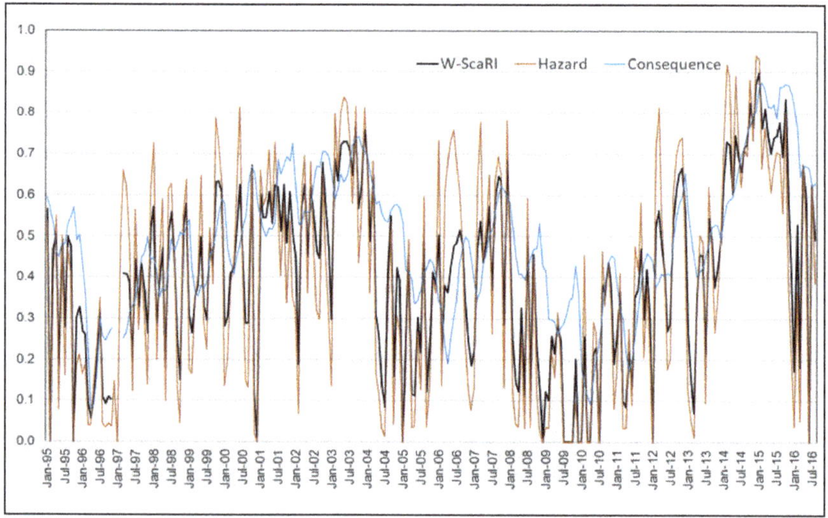

Figure 13. *W-ScaRI* application (1995–2015).

Tables 20 and 21 show the results of the *W-ScaRI* application for the RJMR in January and October 2015, where high index values can be observed in both months. In January, the basin reservoirs, represented by the equivalent volume indicator, had the lowest volume in all observed history, resulting in a maximum risk value (equal to 1). High values were also found for the *HDI* indicator, which, together with the *MDI* and *ADI*, increased the drought hazard subindex.

In October 2015, the hydrological drought indicators and the meteorological and agricultural indicators were lower than those in January, with the hazard subindex value being equal to 0.80, which is still high. However, the consequence subindex in October was higher than that obtained in January due to the exposure indicator. The results show that, in water stress situations, the RJMR is very vulnerable to environmental, social and economic impacts.

Table 20. *W-ScaRI* calculation for RJMR—January 2015.

Drought Hazard Subindex				Exposure			Consequence Subindex			
Indicators		Weight	Value	Indicator	Weight	Value	Indicators		Weight	Value
Meteorological drought indicator (*MDI*)	*SPI*	0.333	0.87	Water stress	0.5	0.94	Environmental vulnerability (*EnVI*)	*Bqual*	0.333	0.99
Agricultural drought indicator (*ADI*)	*RDI*	0.333	0.93	% *VEqR*	0.5	1.00	Social vulnerability (*SVI*)	*PD*	0.333	0.79
Hydrological drought indicator (*HDI*)	*SDI*	0.333	1.00				Economic vulnerability (*EVI*)	*GDP* per capita	0.167	0.82
								% Human supply competition	0.167	1.00
				Final Exposure		0.97	Final Vulnerability			0.90
Hazard Subindex		0.93					Consequence Subindex			0.87
Weight		0.50					Weight			0.50
Water Scarcity Risk Index—*W-ScaRI*										**0.90**

Table 21. W-ScaRI calculation for RJMR—October 2015.

Drought Hazard Subindex				Exposure			Consequence Subindex			
Indicators		Weight	Value	Indicator	Weight	Value	Indicators		Weight	Value
Meteorological drought indicator (*MDI*)	*SPI*	0.333	0.66	Water stress	0.5	1.00	Environmental vulnerability (*EnVI*)	*Bqual*	0.333	0.99
Agricultural drought indicator (*ADI*)	*RDI*	0.333	0.86	% *VEqR*	0.5	0.95	Social vulnerability (*SVI*)	*PD*	0.333	0.79
Hydrological drought indicator (*HDI*)	*SDI*	0.333	0.89				Economic vulnerability (*EVI*)	*GDP* per capita	0.167	0.82
								% Human supply competition	0.167	1.00
				Final Exposure		0.97	Final Vulnerability			0.90
Hazard Subindex		0.80					Consequence Subindex			0.87
Weight		0.50					Weight			0.50
Water Scarcity Risk Index—*W-ScaRI*										**0.84**

To summarize, in January 2015, all the drought hazard indicators increased the hazard subindex, and, thus, the *W-ScaRI* also increased, while in October 2015, the exposure (water stress) and volume equivalent percentage indicators of the reservoirs (% *VEqR*) caused the *W-ScaRI* to assume high values, practically the same as those observed in January 2015.

5. Conclusions

This study proposes a new index called the "Water Scarcity Risk Index (*W-ScaRI*), which aims to assess the risk of water scarcity in a given region, relating the drought hazard to the environmental, social and economic consequences of the event. Differently from other available indices, the *W-ScaRI* tries to maintain the simplicity of the consequence subindex, combining exposure with a set of few and representative vulnerability indicators. It also sheds light on urban/metropolitan issues and mainly focuses on human and industry supplies. To validate the index and test its representativeness, the *W-ScaRI* was applied to the Rio de Janeiro Metropolitan Region (RJMR) for the period of 2014–2015, when a serious water crisis occurred in the Paraíba do Sul River Basin, the main water supply source for the region.

The methods used to assess the drought hazard produced consistent results for the three components considered to be part of this representation: the meteorological drought, the agricultural drought and the hydrological drought. All indices used in the composition of the hazard subindex—*SPI*, *RDI* and *SDI*—clearly identified the periods of drought occurring in the case study proposed for the Paraíba do Sul Basin to validate the potential of the *W-ScaRI*. Although these three indices (used as indicators of the hazard subindex in the proposed formulation) are well-known, they are not usually integrated into the same formulation.

The procedures adopted to represent the consequence subindex were also considered adequate when applied to the RJMR supplied by the Guandu/Lajes/Acari system. All vulnerability indicators—environmental, social and economic—were found to be high, as expected based on the knowledge of the local reality. The value of 0.90 obtained for this component shows how vulnerable the region is and suggests that it may suffer severe impacts in the future during drought events. The "water stress" and "reservoir equivalent volume percentage" indicators also adequately represented the exposure of the system to possible impacts. These indicators directly access and affect the environmental, social and economic consequences in the risk assessment.

The results of this study also show the importance of evaluating the risk index by combining the hazard and its consequences. For instance, in some periods, the hazard subindex was high, but the *W-ScaRI* had a lower value. Only using the hazard subindex may lead to the employment of unnecessary actions, although we recognize that it is an important step in any case and that it can serve as a warning.

Therefore, the final formulation of the *W-ScaRI* represents the water scarcity risk in a relatively simple way, especially when considering the complexity of the real situation, and, at the same time, with adequate conceptual and methodological consistency. It is important to highlight that this study does not intend to present a complete representation of the physical process but rather to indicate the risks of drought in a region. This is one of the great challenges of building indices, since a rigorous representation of the physical process is not sought but rather indications of a given situation from the available information, accepting the limitations of its representation and preferably keeping its meaning simple. The results of the index's application were consistent and representative of the case studied. In this sense, W-ScaRI identified all periods of water scarcity risk in the basin, and it also identified periods when the hazard was high but the consequences were still low.

The *W-ScaRI* index showed to be a methodological tool capable of assessing the tendencies of water scarcity risks, and this is a significant feature to support the water resources management system of any basin. The system manager can define rules to effectively use this index according to each system's particularities. For example, a threshold could be defined to represent a safe operational range limit—in this sense, a given basin, for instance, could be considered safe if the *W-ScaRI* values are below 0.70. This arbitrary threshold can be defined in a different way depending on the system failure consequences or the absolute water availability. However, if the index is greater than the defined value, alerts can be provided so that actions can be proposed to control consumption, reduce distribution losses or reduce water abstractions. These measures may be introduced using a planned escalation logic, intending to prevent the index from achieving higher values and eventually approaching its maximum (which would indicate no available water reserves). In this way, the index can be used in a continuous manner to evaluate the current system safety status and to also map new trends after making management decisions, helping the water resources management systems to quantitatively assess the responses proposed to save water (depending on the physical interpretation of each basin, its consumption and its users).

6. Limitations and Recommendations

The *W-ScaRI* was developed mainly to assess the risk of water supply in urban/metropolitan areas. For this reason, some vulnerability indicators may be missing in a broad sense. For example, indicators such as farmland lost or sold due to droughts, as well as livestock lost or sold due to droughts, were not directly considered, although the *GDP* indicator does indirectly represent losses in agriculture and livestock. Future revisions of the index should assess the possibility of including these or other indicators, as well as indicators of environmental vulnerability, such as land fragmentation near water resources. However, the idea of maintaining the simplicity of the index should be preserved.

During the development of this study, we verified the need to apply the method to other regions in order to validate the obtained results in different situations. A sensitivity analysis of the weights used for the indicators and subindices will also be provided in a future study to guide their choice by managers and stakeholders. Another possible evolution is the inclusion of a resilience component in the formulation of the *W-ScaRI*, combining it in the consequence subindex or introducing a third component in the calculation. All these actions will make the *W-ScaRI* a more robust and reliable instrument for use by managers when facing possible water crises in the future.

Author Contributions: Conceptualization, F.R.T. and M.G.M.; Data curation, F.R.T., J.G.d.S.R.d.S., G.W.d.M.A. and J.P.M.F.; Formal analysis, F.R.T., J.G.d.S.R.d.S., G.W.d.M.A. and J.P.M.F.; Investigation, F.R.T., J.G.d.S.R.d.S., G.W.d.M.A. and J.P.M.F.; Methodology, F.R.T. and M.G.M.; Supervision, M.G.M.; Validation, F.R.T. and M.G.M.; Visualization, J.G.d.S.R.d.S., G.W.d.M.A. and J.P.M.F.; Writing—original draft, F.R.T.; Writing—review and editing, F.R.T. and M.G.M. All authors have read and agreed to the published version of the manuscript.

Funding: This research was funded by Coordenação de Aperfeiçoamento de Pessoal de Nível Superior (CAPES) (Finance Code 001); National Council for Scientific and Technological Development: edital GM/GD–Post-Graduation Programs share; Conselho Nacional de Desenvolvimento Científico e Tecnológico (CNPq) under Grant [303862/2020-3].

Data Availability Statement: Publicly available datasets were analyzed in this study. This data can be found here: Operation Data of the Paraíba Do Sul and Guandu River Basin Reservoirs (https://www.ana.gov.br/sar0/MedicaoSin, accessed on 17 November 2022); Monthly Rainfall Series (http://www.snirh.gov.br/hidroweb/publico/medicoes_historicas_abas.jsf, accessed on 17 November 2022); Population Estimates in 2017 (https://www.ibge.gov.br/estatisticas/sociais/populacao/9103-estimativas-de-populacao.html?=&t=downloads, accessed on 17 November 2022); Municipalities GDP—2015 (https://www.ibge.gov.br/estatisticas/economicas/contas-nacionais/9088-produto-interno-bruto-dos-municipios.html?=&t=resultados, accessed on 17 November 2022); Monthly Rainfall Series (http://www.hidrologia.daee.sp.gov.br/, accessed on 17 November 2022); Monthly Average Maximum and Minimum Temperature Series (https://bdmep.inmet.gov.br/, accessed on 17 November 2022).

Acknowledgments: The authors wish to acknowledge the administrative support provided by the Programa de Engenharia Civil of UFRJ (PEC-COPPE/UFRJ); the Coordenação de Aperfeiçoamento de Pessoal de Nível Superior—Brasil (CAPES); and the Conselho Nacional de Desenvolvimento Científico e Tecnológico—Brasil (CNPq) for supporting this research.

Conflicts of Interest: The authors declare no conflict of interest.

References

1. Tannehill Ray, I. *Drought Its Causes and Effects*, 1st ed.; Princeton University Press: Princeton, NJ, USA, 1947.
2. Gillette, H.P. A Creeping Drought Underway. *Water Sew. Work.* **1950**, 104–105. Available online: https://www.scirp.org/%28S%28czeh2tfqyw2orz553k1w0r45%29%29/reference/referencespapers.aspx?referenceid=2950033 (accessed on 17 November 2022).
3. Sayers, P.B.; Li, Y.; Moncrieff, C.; Li, J.; Tickner, D.; Xu, X.; Speed, R.; Li, A.; Lei, G.; Qiu, B.; et al. *Drought Risk Management*; UNESCO: Paris, France, 2016; ISBN 9789231000942.
4. Tsakiris, G. Drought Risk Assessment and Management. *Water Resour. Manag.* **2017**, *31*, 3083–3095. [CrossRef]
5. Wilhite, D.A. Drought as a Natural Hazard: Concepts and Definitions. In *Drought: A Global Assessment*; Routledge: London, UK, 2000; pp. 3–18.
6. De Stefano, L.; Tánago, I.G.; Ballesteros, M. Methodological Approach Considering Different Factors Influencing Vulnerability—Pan-European Scale. Technical Report. 2015. Available online: https://www.academia.edu/14388306/METHODOLOGICAL_APPROACH_CONSIDERING_DIFFERENT_FACTORS_INFLUENCING_VULNERABILITY_PAN_EUROPEAN_SCALE (accessed on 17 November 2022).

7. Mishra, A.K.; Singh, V.P. A Review of Drought Concepts. *J. Hydrol.* **2010**, *391*, 202–216. [CrossRef]
8. Xu, Y.; Zhang, X.; Wang, X.; Hao, Z.; Singh, V.P.; Hao, F. Propagation from Meteorological Drought to Hydrological Drought under the Impact of Human Activities: A Case Study in Northern China. *J. Hydrol.* **2019**, *579*, 124147. [CrossRef]
9. Howitt, R.; Macewan, D.; Medellín-Azuara, J. *Economic Analysis of the 2015 Drought for California Agriculture*; Center for Watershed Sciences, University of California—Davis: Davis, CA, USA, 2015; p. 16.
10. Peng, C.; Ma, Z.; Lei, X.; Zhu, Q.; Chen, H.; Wang, W.; Liu, S.; Li, W.; Fang, X.; Zhou, X. A Drought-Induced Pervasive Increase in Tree Mortality across Canada's Boreal Forests. *Nat. Clim. Chang.* **2011**, *1*, 467–471. [CrossRef]
11. Maia, R.; Costa, M.; Mendes, J. Improving Transboundary Drought and Scarcity Management in the Iberian Peninsula through the Definition of Common Indicators: The Case of the Minho-Lima River Basin District. *Water* **2022**, *14*, 425. [CrossRef]
12. Van Loon, A.F.; Van Lanen, H.A.J. Making the Distinction between Water Scarcity and Drought Using an Observation-Modeling Framework. *Water Resour. Res.* **2013**, *49*, 1483–1502. [CrossRef]
13. Dolan, F.; Lamontagne, J.; Link, R.; Hejazi, M.; Reed, P.; Edmonds, J. Evaluating the Economic Impact of Water Scarcity in a Changing World. *Nat. Commun.* **2021**, *12*, 1915. [CrossRef]
14. Joseph, N.; Ryu, D.; Malano, H.M.; George, B.; Sudheer, K.P. A Review of the Assessment of Sustainable Water Use at Continental-to-Global Scale. *Sustain. Water Resour. Manag.* **2020**, *6*, 18. [CrossRef]
15. Tzanakakis, V.A.; Paranychianakis, N.V.; Angelakis, A.N. Water Supply and Water Scarcity. *Water* **2020**, *12*, 2347. [CrossRef]
16. UN Water. Water Scarcity. Available online: http://www.unwater.org/water-facts/scarcity (accessed on 6 August 2022).
17. Procházka, P.; Hönig, V.; Maitah, M.; Pljučarská, I.; Kleindienst, J. Evaluation of Water Scarcity in Selected Countries of the Middle East. *Water* **2018**, *10*, 1482. [CrossRef]
18. Zonensein, J. Flood Risk Index as a Flood Management Tool. Master's Thesis, Universidade Federal do Rio de Janeiro—UFRJ, Rio de Janeiro, Brazil, 2007.
19. Miguez, M.G.; Gregório, L.T.; Veról, A.P. *Risk and Hydrological Disaster Management*, 1st ed.; Elsevier: Rio de Janeiro, Brasil, 2018.
20. Bellen, H.M. Van Sustainability Indicators: A Comparative Analysis. Ph.D. Thesis, Universidade Federal de Santa Catarina—UFSC, Florianópolis, Brazil, 2002.
21. Wang, L.; Yu, H.; Yang, M.; Yang, R.; Gao, R.; Wang, Y. A Drought Index: The Standardized Precipitation Evapotranspiration Runoff Index. *J. Hydrol.* **2019**, *571*, 651–668. [CrossRef]
22. Wu, J.; Liu, Z.; Yao, H.; Chen, X.; Chen, X.; Zheng, Y.; He, Y. Impacts of Reservoir Operations on Multi-Scale Correlations between Hydrological Drought and Meteorological Drought. *J. Hydrol.* **2018**, *563*, 726–736. [CrossRef]
23. Cuartas, L.A.; Cunha, A.P.M.; Alves, J.A.; Parra, L.M.P.; Deusdará-Leal, K.; Costa, L.C.O.; Molina, R.D.; Amore, D.; Broedel, E.; Seluchi, M.E.; et al. Recent Hydrological Droughts in Brazil and Their Impact on Hydropower Generation. *Water* **2022**, *14*, 601. [CrossRef]
24. NDMC—National Drought Mitigation Center Us Drought Monitor. Available online: https://droughtmonitor.unl.edu/data/docs/what_is_usdm.pdf (accessed on 30 September 2019).
25. Svoboda, M.D.; Fuchs, B.A.; Poulsen, C.C.; Nothwehr, J.R. The Drought Risk Atlas: Enhancing Decision Support for Drought Risk Management in the United States. *J. Hydrol.* **2015**, *526*, 274–286. [CrossRef]
26. Martins, E.S.Q.R.; From Nys, E.; Molejón, C.; Biazeto, B.; Silva, R.F.V.; Engle, N. Northeast Drought Monitor, Looking for a New Paradigm for Drought Management. In *Água Brasil Séries N° 10*; W.B.G.: Washington, DC, USA, 2015; ISBN 9788588192164.
27. Bachmair, S.; Svensson, C.; Prosdocimi, I.; Hannaford, J.; Stahl, K. Developing Drought Impact Functions for Drought Risk Management. *Nat. Hazards Earth Syst. Sci. Discuss.* **2017**, *17*, 1947–1960. [CrossRef]
28. Liu, J.; Yang, H.; Gosling, S.N.; Kummu, M.; Flörke, M.; Pfister, S.; Hanasaki, N.; Wada, Y.; Zhang, X.; Zheng, C.; et al. Water Scarcity Assessments in the Past, Present, and Future. *Earth's Future* **2017**, *5*, 545–559. [CrossRef]
29. González Tánago, I.; Urquijo, J.; Blauhut, V.; Villarroya, F.; De Stefano, L. Learning from Experience: A Systematic Review of Assessments of Vulnerability to Drought. *Nat. Hazards* **2016**, *80*, 951–973. [CrossRef]
30. Tsakiris, G. A Paradigm for Applying Risk and Hazard Concepts in Proactive Planning. In *Coping with Drought Risk in Agriculture and Water Supply Systems*; Iglesias, A., Garrote, L., Antonino, C., Francisco, C., Donals, W., Eds.; Springer: Berlin/Heidelberg, Germany, 2009; p. 320. ISBN 978-1-4020-9044-8.
31. Dabanli, I. Drought Risk Assessment by Using Drought Hazard and Vulnerability Indexes. *Nat. Hazards Earth Syst. Sci. Discuss.* **2018**, 1–15, *preprint*. [CrossRef]
32. Meza, I.; Eyshi Rezaei, E.; Siebert, S.; Ghazaryan, G.; Nouri, H.; Dubovyk, O.; Gerdener, H.; Herbert, C.; Kusche, J.; Popat, E.; et al. Drought Risk for Agricultural Systems in South Africa: Drivers, Spatial Patterns, and Implications for Drought Risk Management. *Sci. Total Environ.* **2021**, *799*, 149505. [CrossRef]
33. Le, T.; Sun, C.; Choy, S.; Kuleshov, Y. Regional Drought Risk Assessment in the Central Highlands and the South of Vietnam. *Geomatics Nat. Hazards Risk* **2021**, *12*, 3140–3159. [CrossRef]
34. Carrão, H.; Naumann, G.; Barbosa, P. Mapping Global Patterns of Drought Risk: An Empirical Framework Based on Sub-National Estimates of Hazard, Exposure and Vulnerability. *Glob. Environ. Chang.* **2016**, *39*, 108–124. [CrossRef]
35. Dunne, A.; Kuleshov, Y. Drought Risk Assessment and Mapping for the Murray–Darling Basin, Australia. *Nat. Hazards* **2022**, 1–25. [CrossRef]
36. Mckee, T.B.; Doesken, N.J.; Kleist, J. The Relationship of Drought Frequency and Duration to Time Scales. In Proceedings of the 8th Conference on Applied Climatology, Anaheim, CA, USA, 17–22 January 1993; pp. 17–22.

37. Tigkas, D.; Vangelis, H.; Tsakiris, G. DrinC: A Software for Drought Analysis Based on Drought Indices. *Earth Sci. Inform.* **2015**, *8*, 697–709. [CrossRef]
38. Tsakiris, G.; Vangelis, A. Establishing a Drought Index Incorporating Evapotranspiration. *Eur. Water* **2005**, *9*, 3–11.
39. Tigkas, D. Drought Characterisation and Monitoring in Regions of Greece. *Eur. Water* **2008**, *23*, 29–39.
40. Tsakiris, G.; Nalbantis, I.; Pangalou, D.; Tigkas, D.; Vangelis, H. Drought Meteorological Monitoring Network Design for the Reconnaissance Drought Index (RDI). In Proceedings of the 1st International Conference "Drought Management: Scientific and Technological Innovations", Zaragoza, Spain, 12–14 June 2008; Option Méditerranéennes, Series A. pp. 57–62.
41. Nalbantis, I.; Tsakiris, G. Assessment of Hydrological Drought Revisited. *Water Resour. Manag.* **2009**, *23*, 881–897. [CrossRef]
42. Agência Nacional de Águas—ANA. Situation Room of Paraíba Do Sul River Basin. Available online: http://www3.ana.gov.br/portal/ANA/sala-de-situacao/paraiba-do-sul/paraiba-do-sul-saiba-mais (accessed on 27 August 2018).
43. Fundação Coordenação de Projetos, Pesquisas e Estudos Tecnológicos—COPPETEC; Instituto Estadual do Ambiente—INEA. *Rio de Janeiro State Water Resources Plan, Diagnostic Report*; Rio de Janeiro, Brasil, 2014; Available online: https://www.worldbank.org/content/dam/Worldbank/Feature%20Story/SDN/Water/events/Rosa_Formiga_Johnson_Presentacion_Ingles-3.pdf (accessed on 17 November 2022).
44. Instituto Brasileiro de Geografia e Estatística—IBGE. Population Estimates in 2017. Available online: https://www.ibge.gov.br/estatisticas/sociais/populacao/9103-estimativas-de-populacao.html?=&t=downloads (accessed on 30 May 2018).
45. Integral Engenharia and FIRJAN. *Water Security Assessment of the Rio de Janeiro Metropolitan Region: Final Report*; FIRJAN: Rio de Janeiro, Brazil, 2015.
46. COHIDRO and Associação Pró-Gestão das Águas da Bacia Hidrográfica do Rio Paraíba do Sul—AGEVAP. *Paraíba Do Sul River Basin Integrated Water Resources Plan and Affluent Basins Water Resources Action Plans: Diagnostic Report, Tome 1*; COHIDRO/AGEVAP: Resende, Brazil, 2014.
47. Campos, J.D. Charging for Water Use in Transpositions of the Paraíba Do Sul River Basin Involving the Electric Sector. Master's Thesis, Universidade Federal do Rio de Janeiro, Rio de Janeiro, Brazil, 2001.
48. Agência Nacional de Águas—ANA. *Water Resources Conjuncture Report*; ANA: Brasília, Brazil, 2017.
49. Instituto Nacional de Meteorologia—INMET. Monthly Average Maximum and Minimum Temperature Series. *BDMEP-Teach. Res. Weather Database*; 2018. Available online: https://bdmep.inmet.gov.br/ (accessed on 20 March 2018).
50. Agência Nacional de Águas—ANA. Monthly Rainfall Series. Available online: http://www.snirh.gov.br/hidroweb/publico/medicoes_historicas_abas.jsf (accessed on 8 March 2018).
51. Departamento de Águas e Energia Elétrica—DAEE. Monthly Rainfall Series. Available online: http://www.hidrologia.daee.sp.gov.br/ (accessed on 8 March 2018).
52. Agência Nacional de Águas—ANA. Operation Data of the Paraíba Do Sul and Guandu River Basin Reservoirs. Available online: https://www.ana.gov.br/sar0/MedicaoSin (accessed on 3 April 2018).
53. Instituto Brasileiro de Geografia e Estatística—IBGE. Municipalities GDP—2015. Available online: https://www.ibge.gov.br/estatisticas/economicas/contas-nacionais/9088-produto-interno-bruto-dos-municipios.html?=&t=resultados (accessed on 30 May 2018).

Disclaimer/Publisher's Note: The statements, opinions and data contained in all publications are solely those of the individual author(s) and contributor(s) and not of MDPI and/or the editor(s). MDPI and/or the editor(s) disclaim responsibility for any injury to people or property resulting from any ideas, methods, instructions or products referred to in the content.

 *water*

Article

Characterization of Bias during Meteorological Drought Calculation in Time Series Out-of-Sample Validation

Konstantinos Mammas * and Demetris F. Lekkas

Department of Environment, University of the Aegean, 81100 Mytilene, Greece; dlekkas@env.aegean.gr
* Correspondence: mammas@env.aegean.gr

Abstract: The standardized precipitation index (SPI) is used for characterizing and predicting meteorological droughts on a range of time scales. However, in forecasting applications, when SPI is computed on the entire available dataset, prior to model-validation, significant biases are introduced, especially under changing climatic conditions. In this paper, we investigate the theoretical and numerical implications that arise when SPI is computed under stationary and non-stationary probability distributions. We demonstrate that both the stationary SPI and non-stationary SPI (NSPI) lead to increased information leakage to the training set with increased scales, which significantly affects the characterization of drought severity. The analysis is performed across about 36,500 basins in Sweden, and indicates that the stationary SPI is unable to capture the increased rainfall trend during the last decades and leads to systematic underestimation of wet events in the training set, affecting up to 22% of the drought events. NSPI captures the non-stationary characteristics of accumulated rainfall; however, it introduces biases to the training data affecting 19% of the drought events. The variability of NSPI bias has also been observed along the country's climatic gradient with regions in snow climates strongly being affected. The findings propose that drought assessments under changing climatic conditions can be significantly influenced by the potential misuse of both SPI and NSPI, inducing bias in the characterization of drought events in the training data.

Keywords: meteorological drought; SPI; bias; model-validation; drought class transitions

 check for updates

Citation: Mammas, K.; Lekkas, D.F. Characterization of Bias during Meteorological Drought Calculation in Time Series Out-of-Sample Validation. *Water* **2021**, *13*, 2531. https://doi.org/10.3390/w13182531

Academic Editor: Alina Barbulescu

Received: 3 August 2021
Accepted: 10 September 2021
Published: 15 September 2021

Publisher's Note: MDPI stays neutral with regard to jurisdictional claims in published maps and institutional affiliations.

Copyright: © 2021 by the authors. Licensee MDPI, Basel, Switzerland. This article is an open access article distributed under the terms and conditions of the Creative Commons Attribution (CC BY) license (https://creativecommons.org/licenses/by/4.0/).

1. Introduction

Droughts have significant environmental and socio-economic impacts comparable to other hazards, including floods, landslides, and earthquakes. In particular, the total economic damage of natural disasters during the 2003–2013 decade was estimated at USD 1.53 trillion [1], whilst the economic losses from droughts in Europe and the United Kingdom are expected to increase due to global warming between EUR 9.7 billion (under a +1.5 °C warming) and EUR 17.3 billion per year (under a +3 °C warming) [2]. Consequently, a variety of drought risk mitigation measures have been employed during the last decades in drought-sensitive sectors, including, for example, drought-resistant crops for agriculture, improved cooling techniques for health and early warning alerting systems for emergency management [2]. In addition, the Copernicus Emergency Management Service launched the European and Global Drought Observatories to improve preparedness by monitoring the occurrence and severity of droughts and forecasting the meteorological drought with a 3-month lead time [3].

The standardized precipitation index (SPI; [4]) has been widely used in early warning and climate services for the estimation of the onset, duration, and intensity of meteorological drought [5,6]. SPI expresses the accumulated precipitation over a specific period as a departure from the precipitation probability distribution and is very popular due to its simplicity; with only precipitation being required as input. It can be computed in different time scales (usually 3 to 24 months, indicated as SPI(3) to SPI(24), respectively) and can capture different aspects of the meteorological drought ranging from short- to long-term scales [7–10]. SPI(1) to SPI(3) address short accumulated periods and indicate relatively immediate precipitation responses such as reduction in soil moisture and snowpack. SPI(3) to SPI(12) address medium accumulation periods and indicate changes in seasonality of streamflow and reservoir storage [11]. For long accumulation periods, SPI(12) to SPI(48) are used to assess changes in slow responding fluxes, such as groundwater recharge.

SPI requires fitting a probability density function to the accumulated precipitation series (see Section in the Appendix A), with the Gamma distribution being the most popular as it is simple and can very well describe the accumulated precipitation at various scales [12,13]. However, the selection of a distribution may introduce bias to the index values by introducing over-/under-estimated drought events [14]. Alternatively, the Log-Normal, Normal [15], exponatiated Weibull [16], and the generalized extreme values (GEV) [17] distributions have been considered in many cases. In a changing climate where precipitation exhibits non-stationarity, traditional SPI calculation involves fitting the accumulated precipitation to a time-invariant probability density function, resulting to a trending SPI series that reflects the trend of accumulated precipitation [18]. To avoid this limitation, different versions of a non-stationary standardized precipitation index (NSPI) have been proposed using a time-varying probability density function that models precipitation under climate change. Russo et al. [19] modeled precipitation data with a linear trend in the scale parameter of the Gamma distribution, using generalized linear models on climate projections of global climate models. Wang et al. [20] developed a time-dependent SPI by fitting generalized additive models in location, scale, and shape (GAMLSS) to monitor regional droughts during the summer period in the Luanhe River basin in China. Results suggested that under non-stationarity in precipitation, the use of the traditional SPI does not lead to accurate drought classification.

Nevertheless, there are many studies where the stationary SPI has been used thoroughly to forecast meteorological drought at different time scales. Stochastic linear models, such as the autoregressive integrated moving average (ARIMA) and the seasonal autoregressive moving average (SARIMA) [21], have been used for SPI forecasting in different climatic domains [22–25]. Although these models address the non-stationary characteristics of drought, their ability to forecast non-linear components of the time series is limited. Methods such as the support vector regression (SVR) [26] and the artificial neural network (ANN) [27] have shown potential in drought forecasting applications due to their ability to capture non-linearities in the time series. The performance of these models is generally comparable, and hence both SVR and ANN have been recommended for forecasting applications [28,29].

Despite SPI's popularity in drought forecasting applications, the effect of the estimation of the index during model-validation has not been addressed until now. SPI is very sensitive to the temporal characteristics of precipitation and hence when computed on precipitation records at different accumulation periods, SPI leads to values with numerical differences. This is mainly due to the difference of the underlying probability distribution of precipitation from one period to another which increases in the long-term scales of the index [30]. In most studies that focus on drought forecasting applications, SPI has been computed on the entire dataset, omitting any model-validation for time-series, and further use of the validation and test datasets to estimate the respective distribution parameters [29,31–33]. In this case, the observations of SPI in the training dataset share the same parameter estimates with the validation and test datasets. This approach violates the fundamental principles of model-validation as accessing future information leads to different characterization of drought events in the training set, potentially enabling fitted models to access future information while it is not present.

Here, we investigate and quantify the complications related to incorrect computation of SPI, particularly with respect to climate change and the corresponding rainfall variability. Imprecise estimation of SPI, results to systematic biases leading to changes in the classification of drought events. The biases should not be ignored as they are propagated in the building process of any forecasting model. Most studies treat SPI as a traditional univariate sequence of observations, omitting the temporal dependence of the index on historical and future records [34,35]. Therefore, we note that drought forecasting is not the goal of this study, and instead, we address the violated fundamental principles of model-validation and pose the following scientific questions: (1) Are there any differences between the densities of accumulated precipitation when SPI is calculated using the entire data, prior to model-validation? (2) Are there deviations in drought events when stationary or non-stationary drought indices are computed? (3) Is the bias sensitive to the SPI scale? and (4) How does the bias vary in space depending on the underlying climate? To address these questions, we: (a) investigate the incorrect computation of SPI, (b) quantify the bias introduced to the index data when the SPI is computed prior to model-validation, and (c) assess the bias along a climatic gradient.

The paper is organized as follows. Section 2 presents a theoretical overview of model-validation for drought forecasting applications. Section 3 presents the proposed methodology. Section 4 presents the results, followed by a discussion in Section 5. Finally, in Section 6 we present the conclusions. The methodologies provided to calculate SPI and NSPI are provided in the Appendix A.

2. Theoretical Overview

2.1. Data Separation in Model-Validation for Time Series Forecasting

Two well-known model validation techniques are presented here; the out-of-sample (OOS) model-validation and the cross-validation (CV). OOS model-validation is the most common technique of data partitioning in traditional time series forecasting applications. The data are split into two different sets; the training set, which is used to train a model using a set of hyper-parameters and features, and the validation set, which is used to validate the model and it usually constitutes the last block of the series [36]. This approach preserves the temporal order of the series and copes with the dependency among observations (see Figure 1, OOS: train (blue), validation (orange)). Various extensions of the standard OOS validation approach have been introduced during the last years [37]; the fixed origin evaluation, within rolling-origin-recalibration evaluation, rolling-origin-update evaluation and rolling-window evaluation are some of the most important validation methods applied on individual series. However, using the last block of the series for model-validation does not always lead to a diverse validation error as the error reflects the characteristics of the series in the validation set, not present in the historical and future data [37].

CV is a statistical method that is used to evaluate the skill of regression and classification algorithms by measuring their performance on "unseen" datasets. It differs to the OOS validation method as the training and validation sets must cross-over in successive rounds, such that each data point has a chance of being validated. The *k-fold* cross validation is one of the most widely used approaches to assess the predictive performance of a model [38]. Here, the data are split into K roughly equal sized parts, while a model is fitted on the $K - 1$ parts of the data, and the prediction error is evaluated on the Kth part. The prediction error is a combination of K individual predictive errors and is used to select the best model across a series of models trained using different hyper-parameters and features (see Figure 1, CV: train (blue), validation (orange)). In traditional time series forecasting, *k-fold* CV receives little attention due to the theoretical and practical implications that violate the temporal dependency of the series [37]. To avoid violations on the temporal dependence, the validation set needs to be chronologically placed after the training set [36]. Blocked-CV (B-CV) is a variation of the standard *k-fold* CV, where the data are not partitioned randomly but sequentially into K sets, preserving the temporal order of the series [37] (see Figure 1, train (blue), validation (orange)).

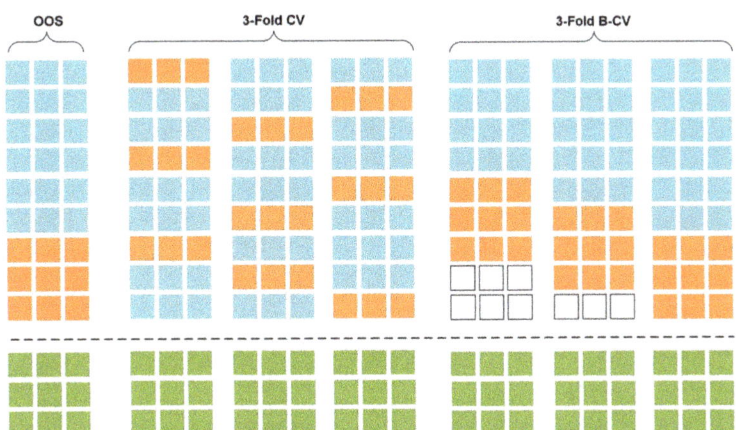

Figure 1. Conceptualization of data separation in model-validation. Training (blue), validation (orange) and test (green) datasets are shown for the OOS, 3-Fold CV and 3-Fold B-CV methods.

In both OOS and CV techniques, the test set should be kept in a *vault* (see Figure 1, test set is presented in yellow) and should be brought out at the end of the data analysis to perform model evaluation on unseen data (test error) as expected in real world applications [38].

In this study, we focus on the OOS model-validation method, since it is the key model-validation approach in drought forecasting applications [28], both for individual series analysis, regression and classification models. We denote the OOS model-validation function, $\mathrm{MV_{oos}}$, for time series forecasting tasks as follows:

Let $Y = [y_1, y_2, \ldots, y_M]$ be a vector of time series records with temporal dependence and length equal to M. Let also $k:\{1, \ldots, N\} \to \{1, \ldots, K\}$ be an indexing function that indicates the training set and $l:\{1, \ldots, N\} \to \{K+1, \ldots, N\}$ an indexing function that indicates the validation set such that $K < N < M$. Denote $\hat{f}^{-k}$ the fitted function with the kth part of the data removed, and $\hat{f}^{-k(i)}$ the function evaluated on the ith observation of the kth part of the data. Then the OOS model-validation estimate of the test error is defined as:

$$\mathrm{MV_{oos}}(\hat{f}) = \frac{1}{N-(K+1)} \sum_{i=K+1}^{N} L(y_i, \hat{f}^{-k(i)}(y_i)) \tag{1}$$

Given a set of models $\hat{f}(y, \beta)$ tuned by a parameter β, denote $\hat{f}^{-k}(y, \beta)$ the βth model evaluated with the kth part of the data. Then, for this set of models we define:

$$\mathrm{MV_{oos}}(\hat{f}, \beta) = \frac{1}{N-(K+1)} \sum_{i=K+1}^{N} L(y_i, \hat{f}^{-k(i)}(y_i, \beta)) \tag{2}$$

The function $\mathrm{MV_{oos}}(\hat{f}, \beta)$ provides an estimate of the test error and our goal is to find the tuning parameter $\hat{\beta}$ that minimizes it. Let $m:\{1, \ldots, M\} \to \{N+1, \ldots, M\}$ be an index that indicates the observations of the test set. The actual test error of the βth model is given by:

$$\mathrm{MV_{oos}}(\hat{f}, \beta) = \frac{1}{M-(N+1)} \sum_{i=N+1}^{M} L(y_i, \hat{f}^{-k(i)}(y_i, \beta)) \tag{3}$$

2.2. Addressing the Effect of Bias during Model-Validation in Drought Forecasting Applications

Even though OOS model-validation and *k-fold* B-CV deal with the temporal characteristics of the time series, there are additional sources of bias that violate the model validation process when predicting SPI. As presented in several studies (among others, [32,39]), SPI is computed using the entire available dataset; this means that the associated probability distribution parameters are estimated using information from historical and future records. The SPI series is often non-stationary and the probability distribution of monthly precipitation changes over time. When the SPI records of the training set are influenced by the properties of the precipitation distribution in the validation and test sets then *the model-validation estimate of the test error is biased as it accesses future information while it should not*. In this paper, we address the source and the magnitude of bias introduced during model-validation by demonstrating three different versions of the index calculated on different subsets of the data. To do so, we formulate the model-validation function for each one of these cases.

Let $X = [x_1, x_2, \ldots, x_M]$ be a time series of monthly precipitation records with length equal to M. Let also $k:\{1, \ldots, M\} \to \{1, \ldots, K\}$ be an indexing function that indicates the training set, $l:\{1, \ldots, M\} \to \{K+1, \ldots, N\}$ indicates the validation set and $m:\{1, \ldots, M\} \to \{N+1, \ldots, M\}$ indicates the test set, respectively, such that $K < N < M$. We denote by $Y^{(s)}$ the SPI at scale s that is computed for three different subsets of the data:

$$Y^{(s),(k)} = [y_{s-1}^{(s),(k)}, y_{s-1+1}^{(s),(k)}, \ldots, y_K^{(s),(k)}] \tag{4}$$

$$Y^{(s),(k,l)} = [y_{s-1}^{(s),(k,l)}, y_{s-1+1}^{(s),(k,l)}, \ldots, y_N^{(s),(k,l)}] \tag{5}$$

$$Y^{(s),(k,l,m)} = [y_{s-1}^{(s),(k,l,m)}, y_{s-1+1}^{(s),(k,l,m)}, \ldots, y_M^{(s),(k,l,m)}] \tag{6}$$

where $Y^{(s),(k)}, Y^{(s),(k,l)}, Y^{(s),(k,l,m)}$ being the different versions of SPI computed using the train (k), train and validation (k, l) and train, validation and test (k, l, m) sets, respectively (Figure 2). Consequently each version of the index is computed on data of different length. For each subset, a probability distribution function (e.g., Gamma, non-stationary Gamma) is fitted and its parameters are estimated to compute the SPI. Different lengths of data lead to different probability distribution functions [36] and subsequently to different SPI raw data, such that $\{\forall i \in R \to \{Y_{(i)}^{(s),(k)} \neq Y_{(i)}^{(s),(k,l)} \neq Y_{(i)}^{(s),(k,l,m)} : K < N < M\}\}$.

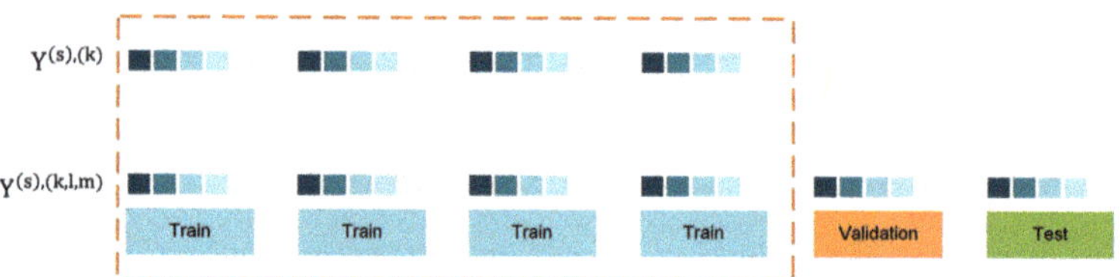

Figure 2. Calculation of the SPI using different subsets of the data within time series cross-validation.

The model-validation function for each version of the SPI index is defined as follows:

$$MV_{oos}(\hat{f}) = \frac{1}{N - (K+1)} \sum_{i=K+1}^{N} L(y_i^{(s),(k)}, \hat{f}^{-k(i)}(y_i^{(s),(k)})) \tag{7}$$

where $\hat{f}^{-k(i)}$ is the function evaluated on the validation set with index k and $y_i^{(s),(k)} \in Y^{(s),(k)}$ is the ith observation of the SPI index computed on the training set only (Equation (4)). The model-validation function of the SPI computed using the training and validation sets is given by:

$$MV_{oos}(\hat{f}) = \frac{1}{N - (K+1)} \sum_{i=K+1}^{N} L(y_i^{(s),(k,l)}, \hat{f}^{-k(i)}(y_i^{(s),(k,l)})) \tag{8}$$

where $y_i^{(s),(k,l)} \in Y^{(s),(k,l)}$ is the ith observation of the SPI. Equivalently, the model-validation function of the index computed on the training, validation, and test sets is as follows:

$$\mathrm{MV}_{\mathrm{oos}}(\hat{f}) = \frac{1}{N-(K+1)} \sum_{i=K+1}^{N} L(y_i^{(s),(k,l,m)}, \hat{f}^{-k(i)}(y_i^{(s),(k,l,m)})) \tag{9}$$

where $y_i^{(s),(k,l,m)} \in Y^{(s),(k,l,m)}$ is the ith observation of the index.

It is clear from the above definitions that different versions of SPI, with respect to data subsets, lead to different versions of the expected test error. The amount of bias introduced to the model-validation functions of Equations (8) and (9) is due to SPI's dependency on the parameter estimates of future and historical precipitation records. This leads to information leakage through $L(y_i^{(s),(k,l)}, \hat{f}^{-k(i)}(y_i^{(s),(k,l)}))$ and $L(y_i^{(s),(k,l,m)}, \hat{f}^{-k(i)}(y_i^{(s),(k,l,m)}))$. We call this *information leakage of SPI* introduced to the model-validation process in drought forecasting applications.

2.3. Estimating Bias during Model-Validation

The bias in SPI introduced from the data in the training set during model-validation is estimated by measuring the deviation between our baseline, $Y^{(s),(k)}$ (Equation (4)) and the SPI computed on the entire dataset, $Y^{(s),(k,l,m)}$ (Equation (6)). We employee different approaches that cover different aspects of the bias introduced to the distribution parameters in the training set: (1) the comparison between the distributions of accumulated precipitation that influence the estimation of drought in the training set, (2) the drought class transition approach that reflects potential changes in the characteristics of drought events, and (3) a set of statistical measures to quantify the deviation between the two versions of the index and generate insights when different scales are employed.

2.3.1. Comparison between the Distributions of Accumulated Precipitation

One of the main calculation steps of SPI, requires fitting a probability distribution function (e.g., Gamma, non-stationary Gamma) on the accumulated precipitation records for a given time scale. The parameters of the distribution are estimated using maximum likelihood and then the cumulative distribution is converted into a standard z-score (see Section in the Appendix A). When SPI is computed using the training, validation, and test data, then the probability distribution function is fitted on the entire data and this violates the fundamental principles of model-validation, described in Section 2.2. We compare the distribution parameter estimates between the two computational approaches of SPI, $Y^{(s),(k)}$ and $Y^{(s),(k,l,m)}$, to conclude whether potential change in the accumulated precipitation introduces bias in the training data. We perform further analysis by comparing the densities of accumulated precipitation for individual months and stations and demonstrate our findings in Section 4.

2.3.2. Drought Class Transition

A drought class transition is defined as the change in the characterization of a drought event when SPI is computed using the entire available dataset ($Y^{(s),(k,l,m)}$), instead of the training set ($Y^{(s),(k)}$). By measuring drought class transitions ($Y^{(s),(k)} \rightarrow Y^{(s),(k,l,m)}$), we aim at quantifying the magnitude of change in SPI classification with respect to the subsets used during model-validation (Figure 3). A transition from *Moderately Wet* when the $Y^{(s),(k)}$ is computed to *Moderately Dry* when the $Y^{(s),(k,l,m)}$ is instead computed suggests that the version of the SPI computed on the entire dataset incorporates bias to the training data by underestimating a wet event and classifying it as dry event.

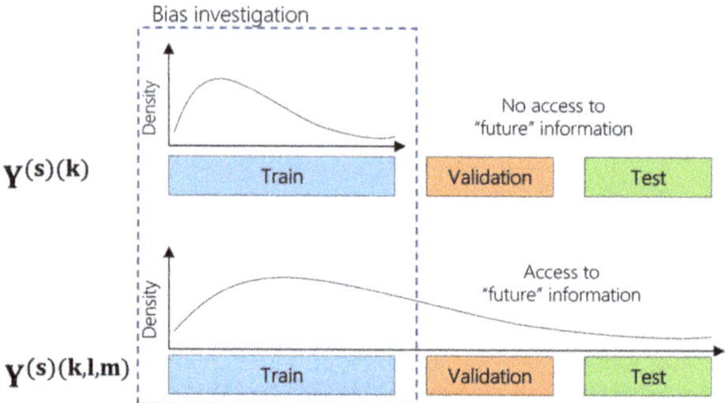

Figure 3. Two computational approaches of (N)SPI calculation. (**top**) (N)SPI calculation using the training set ($Y^{(s)(k)}$), (**bottom**) (N)SPI calculation using the training, validation and test sets ($Y^{(s)(k,l,m)}$).

2.3.3. Comparison between the Raw SPI Data

The mean absolute deviation (MAD) was used to quantify the amount of error introduced to the training data. MAD is a linear score that measures the average magnitude of the deviations from the mean:

$$\text{MAD} = \frac{\sum_{i=1}^{K} |y_i^{(s)(k)} - y_i^{(s)(k,l,s)}|}{K} \tag{10}$$

where K is the total number of records in the training set, s is the scale of the SPI and k, l, s are the index vectors for the training, validation and test sets, respectively.

3. Methodology

3.1. Data and Region of Interest

Sweden has a surface area of about 450,000 km^2, with its climate being characterized by a strong spatial gradient and a seasonal pattern. Precipitation is high in the west (mountainous areas) and is gradually reduced eastwards. The climatic patterns over the country can be clustered into three regions according to the Köppen-Geiger climate classification system [40]; snow, polar, and warm temperate (Figure 4). Sweden receives precipitation between 500 and 800 mm/year; however in the southwestern part of the country precipitation ranges between 1000 and 1200 mm, whilst in mountainous areas in the north precipitation can reach up to 2000 mm. An extensive 58-years record of observed daily precipitation (1 January 1961 till 30 September 2018) over Sweden at 36,662 basins has been provided by the Swedish Meteorological and Hydrological Institute (SMHI) (see Figure 4). The precipitation is spatially interpolated from SMHI's station network and elevation corrected to capture precipitation for hydrological applications at the basin scale [41]. This meteorological dataset is considered as state-of-the-art observation product in Sweden and has been used in various investigations, including seasonal forecasting and climate projections. These daily precipitation records were aggregated into monthly values, which were further used here.

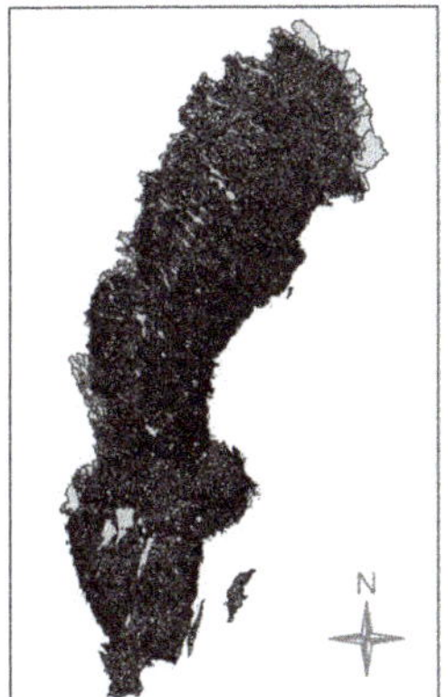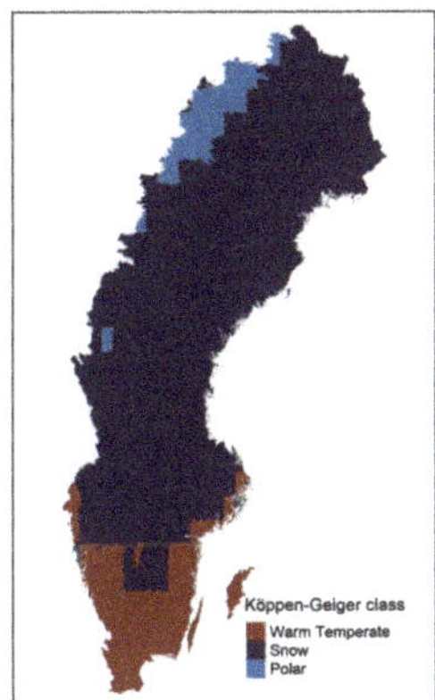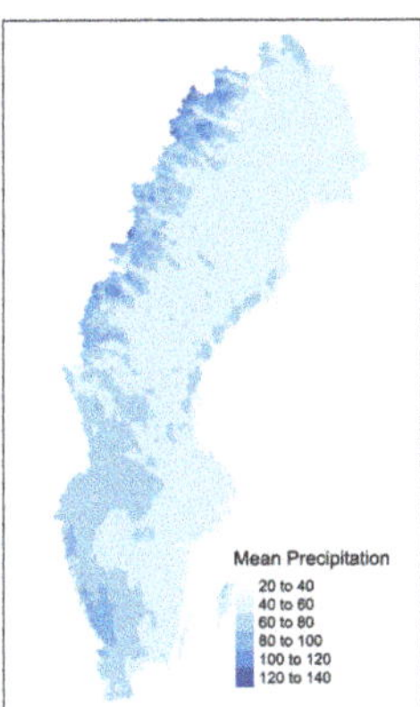

Figure 4. (**left**) Locations of precipitation stations used over Sweden. (**center**) Spatial distribution of the Köppen-Geiger climate classification system. (**right**) Mean monthly precipitation (mm) during the period 1961–2018.

3.2. Experimental Setup

A set of experiments were performed on the 36,662 basins using the available data to quantify the bias introduced to the training set. The analysis period is from January 1961 to September 2018 and each station consists of 693 monthly precipitation observations. At each location, the data were split into training (60%), validation (20%), and test (20%) sets using the OOS model-validation methodology described in Section 2.1. This ensures that the training set consists of at least 30 years of precipitation records to compute each version of SPI [4,42]. Two probability distribution functions were used: (1) the stationary Gamma, using maximum likelihood estimation, that leads to the computation of the traditional SPI (see Appendix A, Equations (A1)–(A7)), and (2) the non-stationary NS-Gamma (with time-varying *location* and *scale* parameters) that leads to the non-stationary SPI (NSPI) and is able to capture the increased precipitation trend during the last decades (see Appendix B). Additionally, different SPI scales were calculated (SPI(3), SPI(6), SPI(9), SPI(12), SPI(24)) to exploit potential relationships between the scale of the index and the level of bias introduced to the training set during model-validation. The two computational approaches of the SPI are employed; first, SPI is computed using only the training data, $Y^{(s)(k)}$, and second, SPI is computed using the entire dataset (training, validation, and test), $Y^{(s)(k,l,m)}$ (see Figure 3). In our experiments, we treat $Y^{(s)(k)}$ as the baseline since it does not violate the fundamental assumptions of the OOS model-validation process eliminating information leakage. The analyses are performed in the training set to:

1. Compare the densities of accumulated rainfall (see Section 2.3.1);
2. Count the number of drought class transitions (see Section 2.3.2);
3. Analyze the magnitude of the bias introduced at different SPI scales (see Section 2.3.3);
4. Assess the variation of bias along Sweden's climatic gradient. The error introduced to the model-validation is quantified based on one statistical metric; the mean absolute deviation (see Section 2.3.3).

Moreover, we showcase a set of experiments using data from a meteorological station the characteristics of which are presented in Table 1. It is clear that there is a change in the mean monthly precipitation between the two subsets.

Table 1. S-3357 meteorological station.

Station	Longitude	Latitude	Mean Monthly Rainfall (Train)	Mean Monthly Rainfall (Train, Valid, Test)
S-3357	67.37	22.28	77.6 mm	84.2 mm

4. Results

4.1. Comparison between the Distributions of Accumulated Precipitation

In Section 2.2, we provided the theoretical implications that arise during model-validation when SPI is computed using the entire data. Here, we identify the presence of bias by comparing the parameter estimates of Gamma and non-stationary Gamma distributions using two different subsets of the data and different scales of the index (SPI(3), SPI(6), SPI(9), SPI(12), and SPI(24)). Initially, we fitted the Gamma and non-stationary Gamma distributions on the accumulated precipitation records using the training data, and as a subsequent step, we fitted the same distributions using the training, validation, and test data. Figure 5, provides a graphical comparison of SPI(12) between the parameter estimates of $Y^{(12),(k)}$ (x-axis) and $Y^{(12),(k,l,m)}$ (y-axis) across the 36,662 basins.

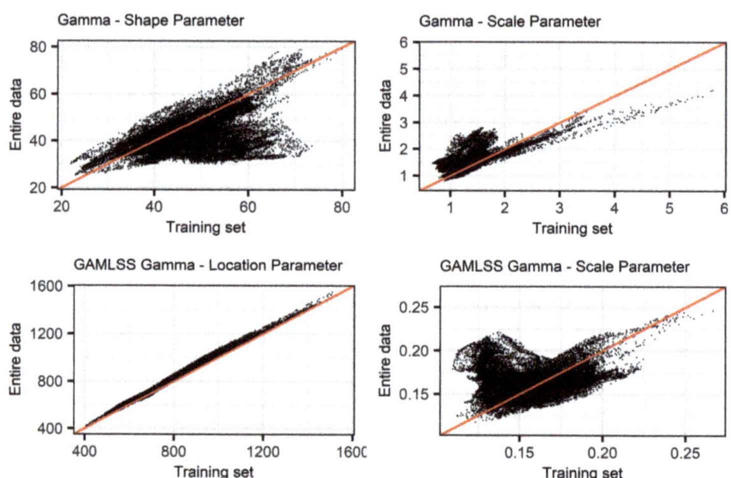

Figure 5. Comparison between the distribution parameter estimates of accumulated precipitation between $Y^{(12),(k,l,m)}$ (y-axis) and $Y^{(12),(k)}$ (x-axis) for 36,662 basins in Sweden during 1961–2018: (**top**) stationary SPI and (**bottom**) non-stationary SPI.

As presented in Figure 5 (top), results show the deviation of the *shape* and *scale* parameter estimates from the diagonal. This is an indication of systematic bias in the training data when $Y^{(12),(k,l,m)}$ is compared to $Y^{(12),(k)}$. The increased *shape* parameter in the training set, using $Y^{(12),(k)}$, leads to more symmetric shapes (71.1% of the basins have increased shape parameter in the training data). Additionally, the increased *scale* parameter in the entire data, using $Y^{(12),(k,l,m)}$, suggests that the distribution of accumulated precipitation has "heavier" tails and consequently higher variance compared to the Gamma distribution fitted on the training data, $Y^{(12),(k)}$ (76.2% of the basins have increased scale in the entire data). There are two main reasons behind the deviations observed in the parameter estimates: (1) the stationary SPI is time invariant; it uses a location specific Gamma distribution and results in a trending SPI that reflects the same trend as the increasing rainfall trend in Sweden during the last three decades; and (2) the rainfall trend is increasing even further in the validation and test data and this introduces even larger deviations between the parameter estimates of $Y^{(12),(k,l,m)}$ and $Y^{(12),(k)}$. In the small scales of the index, the deviations between the parameter estimates present smaller deviations; this is mainly attributed to the short "memory" property of SPI that shares the same parameters with series of shorter length in the validation and test set (Figure 5).

Furthermore, the results from the implementation of NSPI indicate that the *location* and *scale* parameter estimates deviate from the diagonal as well (Figure 5, bottom). GAMLSS are able to estimate time-varying *location* and *scale* distribution parameters as a function of the increasing trend of accumulated precipitation. However, they are not able to capture the change in distribution of accumulated precipitation in the validation and test sets.

4.2. Drought Class Transitions

Here, we address the question *"Are there drought class transitions in the training set?"*. Both the Gamma and NSGamma probability densities were used at different SPI scales (i.e., SPI(3), SPI(6), SPI(9), SPI(12), and SPI(24)) to measure the number of drought classes that change state in the training set when the two different versions of SPI are computed ($Y^{(12),(k)}$ versus $Y^{(12),(k,l,m)}$). Figure 6 presents the class-transitions in the training set, when either SPI(12) or NSPI(12) is computed using the two aforementioned approaches. The drought class transitions were computed for all 36,662 basins; each basin has available 693 monthly records, leading to 25,406,766 precipitation records in total that were analyzed for this experiment.

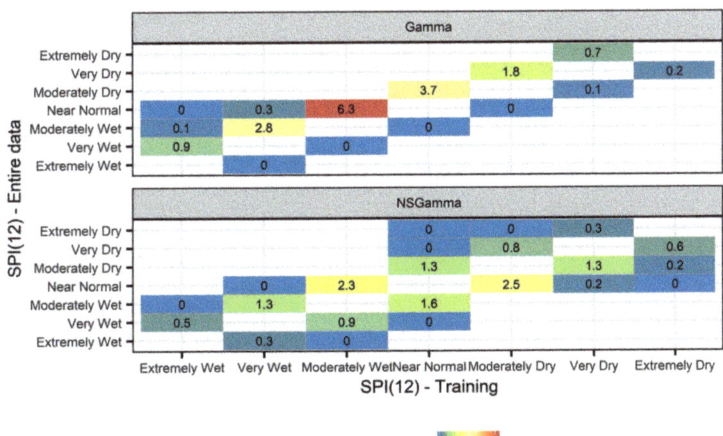

Figure 6. Transition of drought classes from $Y^{(12)(k)}$ (*x*-axis) to $Y^{(12)(k,l,m)}$ (*y*-axis) for 36,662 basins in Sweden during the period 1961–2018. Blue color leads to changes of lower magnitude while red color leads to changes of higher magnitude.

Results using both SPI and NSPI are subject to biases in the training data, leading to transitions of drought events when they are calculated using the entire dataset. The computation of a stationary SPI leads to a systematic underestimation of wet events when the index is computed using the entire dataset, i.e., $Y^{(12)(k,l,m)}$. The most frequent transitions are from *Moderately Wet* to *Near Normal* (change at 6.3% of the SPI values, which are 931,968 out of 25,406,766 values) and from *Near Normal* to *Moderately Dry* (change at 3.7% of the SPI values). In addition, outliers are observed, where *Very Dry* events switch to *Extremely Dry* (change at 0.7% of the SPI values). This systematic change is associated to the increase in the mean monthly precipitation amount in the country during the last three decades, that is not captured through the stationary SPI calculation. This observed increase in precipitation is probably attributed to climate change [43], which is consequently propagated in the SPI estimation and therefore drives the SPI performance. These findings reflect the difference between the monthly precipitation distribution in the training set and the training, validation, and test set. The mean monthly precipitation across all 36,662 basins is equal to 59 mm in the training set (1961–1995) and 64 mm in the validation (1995–2007) and test sets (2007–2018). The systematic trends in drought class transitions (upper triangular in Figure 6) is caused by the stationary nature of SPI and the incorrect computation of index that uses the entire dataset.

NSPI leads to a systematic overestimation and underestimation of drought events in the training data when the index is computed using the entire dataset, i.e., $Y^{(12)(k,l,m)}$. The most frequent transitions are from *Moderately Dry* to *Near Normal* (change at 2.5% of the NSPI values) and from *Moderately Wet* to *Normal* (change at 2.3% of the NSPI values) (see Figure 6 NSGamma). Although NSPI involves the calculation of time-varying *location* and *scale* distribution parameters in the training set, it is unable to capture the change in the distribution of accumulated precipitation in the validation and test sets, consequently leading to systematic change in the classification of drought events.

4.3. Comparison between the Raw SPI Data

Further analysis on the SPI data provides more in depth diagnostics regarding the effect of the *information leakage* (bias) in the training set. In Figure 7, SPI and NSPI were calculated using the two computational approaches presented in Section 2.3 for the station S-3357. The stationary SPI, $Y^{(12)(k,l,m)}$, shows systematically lower values in the training set compared to $Y^{(12)(k)}$, consequently leading to more dry events than could have been predicted with the latter approach. This finding is associated to the observed change in the mean monthly precipitation in the validation and test sets which is equal to 94.2 mm, as opposed to to the mean monthly precipitation in the training set which is equal to 77.6 mm. NSPI generates biases of lower magnitude when $Y^{(12)(k,l,m)}$ is compared to $Y^{(12)(k)}$ (see Figure 7 bottom). This is mainly due to the property of NSPI to capture the increasing precipitation trend and incorporate it in the NSPI calculation, resulting to a trend-free index. The observed deviations between the two NSPI computational approaches are mainly due to the change in the distribution of accumulated precipitation in the validation and test sets that is not captured in the NSPI calculation.

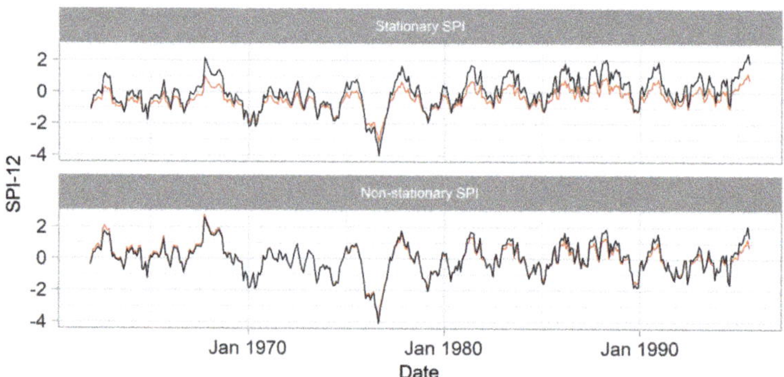

Figure 7. Comparison between $Y^{(12),(k,l,m)}$ (red) and $Y^{(12),(k)}$ (black) using *Gamma* and *NSGamma* probability density functions.

In addition, we investigated whether the observed changes in the raw SPI records for station S-3357 are associated to potential deviations between the distributions of accumulated precipitation in the training and the training, validation, and test sets. In Figure 8, we estimated the density of the 12-month accumulated precipitation for each month. It is clear from the results that the density estimation using the entire data is more skewed to the left compared to the density fitted on the training data, and this is an indication that the variance of accumulated precipitation in the train, validation, and test set is higher than for the entire data.

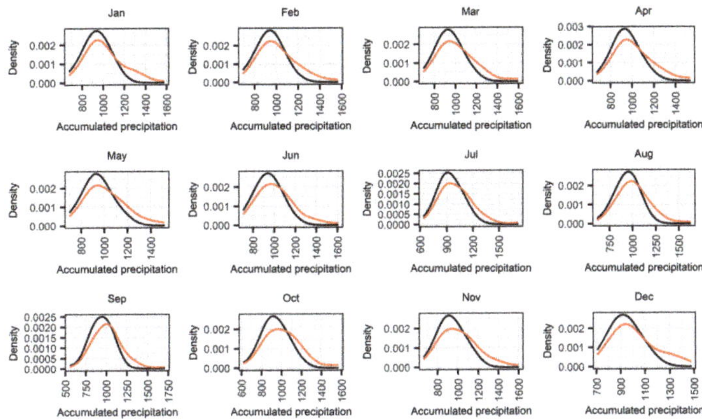

Figure 8. Comparison of the density estimates of the accumulated rainfall between $Y^{(12)(k)}$ (black) and $Y^{(12)(k,l,m)}$ (red).

4.4. Sensitivity Analysis of the Bias at Different SPI Scales

We next address the question *"Is the bias introduced to the training set sensitive to the SPI scale?"* For this, we analyze the bias introduced to the training set at different SPI and NSPI scales (i.e., 3, 6, 9, 12, 24). Results show a positive correlation between the scale of the index and the *information leakage* in the training set (see Figure 9). In particular, results show that between 6.7% (for SPI(3)) and 22.1% (for SPI(24)) of the training records change drought class, corresponding to deviations (in terms of MAD) between 0.07 and 0.27. The same behavior is observed when different scales of NSPI are calculated, affecting between 5.7% (for NSPI(3)) and 19.3% (for NSPI(24)) of the training records corresponding to deviations (in terms of MAD) between 0.06 and 0.22. This observed dependency of the error to the (N)SPI scale is mainly attributed to the fact that in large scales, the index has long memory and accesses long sequences of the data in the validation and test sets. This finding is in agreement with the results of Wu et al. [30], where larger deviations between the probability density functions were observed in larger accumulated periods.

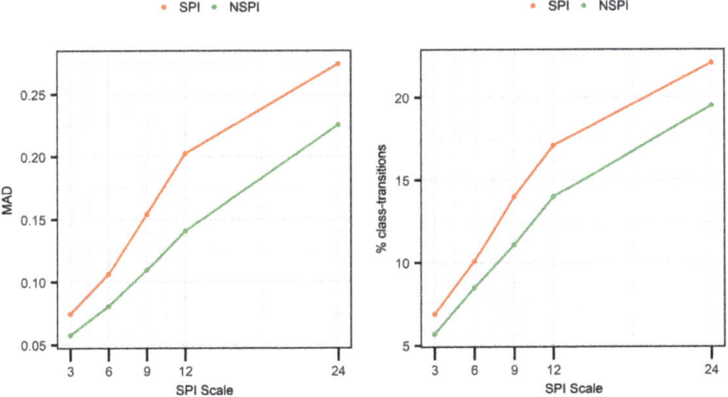

Figure 9. Mean absolute deviation (**left**), percentage of records with drought class transitions (**right**); for different scales of SPI and NSPI across Sweden.

4.5. Bias along a Spatial Gradient

Here, we explore the potential relationship between the bias introduced to the training data and the regional climatic conditions. In this section, we computed the percentage of drought events that change class following the methodology described in Section 2.3.2, and generated the spatial distribution of drought class transitions across the country for different time scales of NSPI (see Figure 10). The percentage of drought class transitions increases with increased NSPI scale and affects up to almost 60% of drought events for certain stations in the southern (snow climate) and northwest part of the country. This result is strongly emphasised at scales NSPI(12) and NSPI(24).

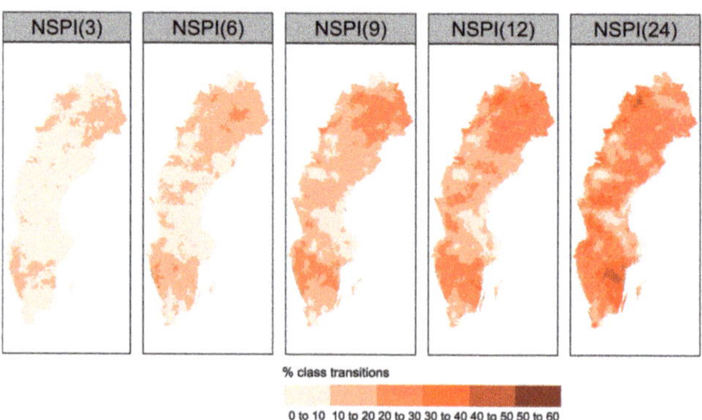

Figure 10. Percentage of drought class transitions for different scales of SPI, using the non-stationary Gamma distribution, across 36,662 basins.

Additionally, the distribution of bias (in terms of MAD) for all 36,662 stations was computed for different time scales and under different climatic conditions. Results in Figure 11 show that the bias increases with increased NSPI scale, while larger deviations are observed in scales NSPI(12) and NSPI(24) and for the snow climatic conditions. This could be attributed to the increase in the mean monthly precipitation during the last decades, however, there might be additional factors, not explicitly addressed in this study, that influence the information leakage issue, such as the physiographic characteristics of the region that could possibly affect the drought spatial variability and its corresponding impacts [44,45].

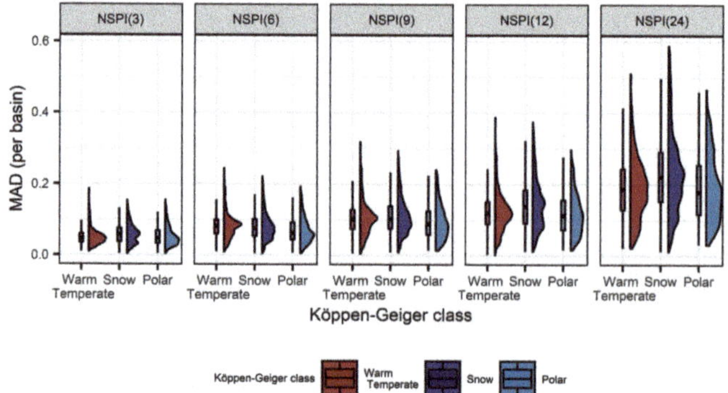

Figure 11. Distribution of the mean absolute deviation (MAD) between the NSPI computed on the training set only and the NSPI computed on the training, validation and test sets for different Köppen-Geiger climatic classes and 36,662 basins in Sweden during 1961–2018.

5. Discussion

5.1. Generalization over a Stronger Spatial Gradient

The results here indicate that change in the climate can be a significant source of bias affecting the training data and, consequently, the learning algorithms that generate drought forecasts. The systematic increase in the mean monthly precipitation in Sweden during the last decades, leads the stationary SPI to underestimate wet events and overestimate dry events in the training set and this is mainly due to the difference in the parameter of the Gamma distribution during estimation (see Figure 5). Similarly, the change in the distribution of accumulated precipitation during the last decades, leads the non-stationary SPI to both overestimate and underestimate wet events. In Europe, the meteorological droughts are associated to decrease in precipitation, especially during the summer period. This precipitation reduction tends to be more severe in the Mediterranean countries that present a different drought regime compared to the rest of Europe [46,47]. Over the next decades, it is projected that temperature will increase more in Europe compared to the global average [48]. A large fraction of Europe is expected to face an increase in the mean temperature of more than 1 °C both during winter and summer. With increased warming, winter precipitation is projected to increase with more frequent precipitation in North Eastern Europe, while in South Eastern Europe, precipitation during summer is expected to decrease.

Under these scenarios, there is a clear indication that the precipitation distribution will change over the next decades in Europe. The learning models that will be used to forecast future droughts will be influenced by the potential misuse of the drought indices and will induce bias in the prediction of future drought events. Based on the insights drawn here, it is expected that the bias in the North Eastern part of Europe will lead to overestimation of dry events in the training set, while in the South Eastern Europe it will lead to an underestimation of dry events, when the stationary SPI is employed. Equivalently, changes in the distribution of accumulated precipitation will lead to biases in the training set during NSPI calculation. To prevent this behavior from future drought forecasting applications, we highlight the need to introduce a drought forecasting framework that deals with these limitations and ensures model generalization capability, especially in areas with extreme climatic conditions, i.e., the Mediterranean.

5.2. Applicability Using Different Drought Indices

Although this study focuses on the identification of bias during model-validation using SPI and NSPI, the methodology is valid to other indices whose calculation is performed similarly. For instance, the standardized precipitation-evapotranspiration index (SPEI) [49] and the effective drought index (EDI) [50] have been used thoroughly to also describe the meteorological droughts [51,52]. Additionally, the Palmer severity drought index (PDSI) [53] and the combined drought indicator (CDI) developed by the European Drought Observatory of the Copernicus Emergency Management Service [54] have been used to characterize agricultural droughts [55,56]. Additionally, characterization of hydrological droughts using streamflow information plays a very important role in drought early warning systems, with the most common indices being the variable threshold (VT), the fixed threshold (FT) and the standardized streamflow index (SSI) [57]. The computation of those indices requires attention, particularly since such indices are widely used in climate services and their misuse could lead to incorrect characterization of drought events and incorrect identification of mitigation and adaptation measures.

6. Conclusions

Herein, we highlighted the importance of correct computation of SPI and NSPI in a drought forecasting setting, and demonstrated the theoretical and numerical implications when the index is computed on the entire dataset, which methodologically neglects model-validation. We quantified the bias introduced to the training set by conducting various experiments for different (N)SPI scales from 3 up to 24 months across 36,662 basins in

Sweden. Two different computational approaches were compared. First, the SPI and NSPI were computed using the training data only (baseline) and second the SPI and NSPI were computed using the training validation and test data (entire dataset). The latter approach is the one that introduces bias in the training set, as it violates the fundamental principles of OOS model-validation. The main conclusions from this study are as follows:

- Climate change coupled with the computation of SPI prior to model-validation can be a significant source of bias in drought forecasting applications. In the case study presented, the increased precipitation during the last decades leads to changes in the distribution parameters of accumulated precipitation for different time scales of the stationary SPI. This phenomenon affects the estimation of drought in the training set and violates the fundamental principles of OOS model-validation;
- NSPI calculation using GAMLSS, involves the estimation of time-varying *location* and *scale* parameters of a Gamma distribution as a function of the increasing trend of accumulated precipitation over time. Although this property results to a trend-free index, still the misuse of the data, introduces biases to the training set;
- The bias introduced to the training data is larger when the stationary SPI is computed. This is mainly because SPI requires fitting the accumulated precipitation records to a time invariant probability density function that incorporates the increasing rainfall trend during SPI calculation. This property leads to a systematic underestimation of wet events in the training data consequently affecting future use of this data in forecasting applications;
- With increased SPI scale, the number of drought class transitions increases and affects up to 22.1% for SPI(24) and 19.3% for NSPI(24) of the available records. This finding is further supported by the MAD metric that indicates increased information leakage with larger SPI and NSPI scales. This is mainly due to the "memory" of the index to access longer sequences of future records during OOS model-validation, thus, leading to increased information leakage issue in the training data;
- The bias introduced due to the incorrect computation of NSPI has spatial dependence, especially in the large scales of the index. The regions affected most are located in the southern (snow climate) and northwest part of the Sweden that exhibit changes in the distribution of accumulated precipitation in the validation and test sets.

Taking into account the findings presented in this paper, we propose that many existing drought forecasting studies that focus on the prediction of SPI should not be applied to real world forecasting applications if the fundamental principles of OOS model-validation are violated. It is expected that the bias introduced to the training set can have a significant impact on the learning algorithms under the drought forecasting setting, especially in larger scales of the index and varying climatic conditions. Even though, the results presented are related to the climatic conditions of Sweden, they could be directly applied to other climatic regions that stronger changes in precipitation have been recorded, i.e., the Mediterranean. It is expected that the bias identified here would be substantial in such climates, and consequently significantly affect the drought predictions and corresponding decision-making.

Author Contributions: Conceptualization, K.M. and D.F.L.; methodology, K.M.; software, K.M.; validation, K.M.; formal analysis, K.M.; investigation, K.M. and D.F.L.; resources, K.M.; data curation, K.M.; writing—original draft preparation, K.M.; writing—review and editing, K.M. and D.F.L.; visualization, K.M.; supervision, D.F.L. All authors have read and agreed to the published version of the manuscript.

Funding: Not applicable.

Institutional Review Board Statement: Not applicable.

Informed Consent Statement: Not applicable.

Data Availability Statement: The meteorological data used for driving the investigation are available from the luftweb portal (https://luftweb.smhi.se, accessed on January 2020), and the vattenweb portal (https://www.smhi.se/data/hydrologi/vattenwebb, accessed on 1 January 2020). The code base of the experiments can be found in a Github repository: https://github.com/mammask/droughtBias, accessed on 9 September 2021.

Acknowledgments: The authors would like to thank Niclas Hjerdt and Ilias G. Pechlivanidis for giving permission to use the data in this investigation.

Conflicts of Interest: The authors declare no conflict of interest.

Appendix A. Stationary Standardized Precipitation Index

The standardized precipitation index (SPI), proposed by [4], defines and monitors drought events. Positive SPI values indicate wet conditions with greater than the median precipitation, while negative SPI values indicate dry conditions with lower than the median precipitation. Table A1 provides the classification of different SPI values.

Table A1. SPI ranges for different meteorological conditions.

SPI Values	Classification
$[2, \inf)$	Extremely Wet
$[1.5, 1.99]$	Very Wet
$[1.0, 1.49]$	Moderately Wet
$[0.99, -0.99]$	Near Normal
$[-1.0, -1.49]$	Moderately Dry
$[-1.5, -1.99]$	Very Dry
$(-\inf, -2]$	Extremely Dry

The computation of SPI requires fitting a probability distribution on monthly aggregated precipitation series at different time scales (e.g., 3, 6, 9, 12 months). Usually, the Gamma distribution fits best precipitation data and is given by the following expression:

$$g(x) = \frac{1}{\beta^{\alpha}\Gamma(\alpha)} x^{\alpha-1} e^{-\frac{x}{\beta}} \tag{A1}$$

for $x > 0$ where $\alpha > 0$ is the shape parameter, $\beta > 0$ is the scale parameter and x is the precipitation amount. The Gamma function is defined by the integral:

$$\Gamma(\alpha) = \int_0^{\infty} y^{\alpha-1} e^{-y} dy \tag{A2}$$

Fitting the Gamma distribution to the monthly precipitation records requires the estimation of the α and β parameters using maximum likelihood estimation [58].

$$\hat{\alpha} = \frac{1}{4A}\left(1 + \sqrt{1 + \frac{4A}{3}}\right) \tag{A3}$$

$$\hat{\beta} = \frac{\overline{x}}{\hat{\alpha}} \tag{A4}$$

where:

$$A = \ln(\overline{x}) - \frac{\sum_{i=1}^{n}}{n} \tag{A5}$$

and n is the number of precipitation records. The resulting parameters are used to compute the cumulative probability of the observed precipitation records for a given period and time scale [58].

$$G(x) = \int_0^x g(x)dx = \frac{1}{\hat{\beta}^{\hat{\alpha}}\Gamma(\hat{\alpha})} \int_0^x x^{\hat{\alpha}-1} e^{-\frac{x}{\hat{\beta}}} dx \tag{A6}$$

The Gamma distribution is undefined for $x = 0$ since there may be no precipitation and by letting $t = \frac{x}{\beta}$ [58] the incomplete Gamma distribution is given by:

$$G(x) = \frac{1}{\beta^{\hat{\alpha}} \Gamma(\hat{\alpha})} \int_0^x t^{\hat{\alpha}-1} e^{-t} dt \tag{A7}$$

Appendix B. Non-Stationary Standardized Precipitation Index (NSPI)

In this study, we computed the non-stationary version of SPI (NSPI) using the GAMLSS framework introduced by Rigby and Stasinopoulos [59]. GAMLSS has been thoroughly used in the past to model non-stationary versions of drought indices [60,61]. It is a semi-parametric regression model, in which a parametric distribution assumption is required for the response variable, and the selected distribution's parameters can vary as a function of explanatory variables or random effects. Within the GAMLSS framework, observations y_t, for $t = 1, 2, \ldots, n$, where n is the length of the observations, are assumed to be independent and fitted to a probability density function $f(y_t|\theta^t)$, conditional on $\theta^t = (\theta_{1t}, \theta_{2t}, \ldots, \theta_{pt})$, where p is the number of distribution parameters at time t. Various distributions are supported by GAMLSS, including, highly skew or kurtotic continuous and discrete distributions. The distribution parameters θ, characterized as location, shape, and scale parameters are related to explanatory variables by monotonic link functions $g_k(\cdot)$, $k = 1, 2, \ldots, p$, given by:

$$g_k(\theta_k) = \eta_k = X_k \beta_k + \sum_{j=1}^{J_k} Z_{jk} \gamma_{jk} \tag{A8}$$

where β_k and η_k are vectors of length n, $\theta_k = (\theta_{1k}, \theta_{2k}, \ldots, \theta_{nk})^T$, $\beta_k = (\beta_{1k}, 1\beta_{2k}, \ldots, \beta_{Jkk})$ is a parameter vector of length J_k, X_k is a fixed known design matrix of order $n \times J_k$, Z_{jk} is a fixed known $n \times q_{jk}$ design matrix, and γ_{jk} is a q_{jk} dimensional random variable. In Equation (A8), for $k = 1, \ldots, p$, η_k, are comprised of a parametric component $X_k \beta_k$ (functions of explanatory variables) and additive components $Z_{jk} \gamma_{jk}$ (random effects). If $J_k = 0$, the model is reduced to a fully parametric GAMLSS model.

Here, we computed NSPI by fitting a GAMLSS on the accumulated precipitation series using different time scales. The accumulated precipitation series were assumed to follow a two-parameter Gamma distribution with its *location* and *scale* parameters, linked to a linear trend that evolves over time, t. The following additive formulation was used in this study:

$$g_a(\mu) = X_k \alpha_k + \sum_{j=1}^{j_k} h_{jk}(x_{jk}) \tag{A9}$$

$$g_b(\sigma) = X_k \beta_k + \sum_{j=1}^{j_k} h_{jk}(x_{jk}) \tag{A10}$$

where μ and σ are the location and scale parameters of the Gamma distribution with the link functions g_a and g_b, respectively. X_k is a matrix of covariates (in our case a linear trend that evolves over time) of order $n \times j_k = n \times 1$, β_k, is a parameter vector of length j_k, and $h_{jk}(\cdot)$ represents the dependence function of the distribution parameters on the linear trend x_{jk}. Mathematically, NSPI is similar to SPI, because they have similar calculation steps, however, NSPI is based on a non-stationary Gamma with time-varying *location* and *scale* parameters.

Appendix C. Comparison of Distribution Parameters

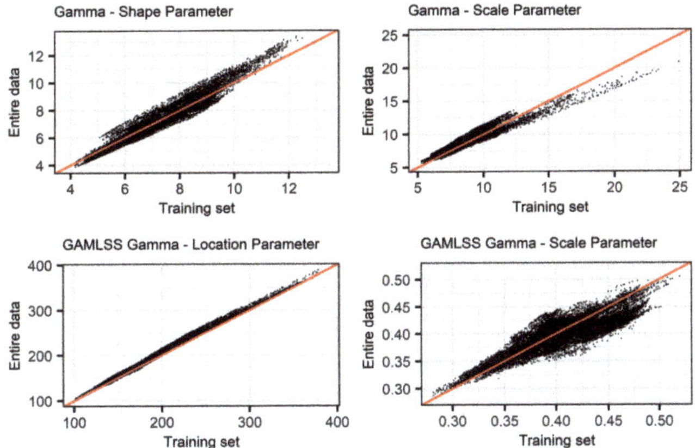

Figure A1. (top) Comparison between the stationary Gamma distribution parameter estimates (SPI); **(bottom)** Aggregated GAMLSS location and scale parameters of non-stationary Gamma (NSPI) of accumulated precipitation between $Y^{(3),(k,l,m)}$ (*y*-axis) and $Y^{(3),(k)}$ (*x*-axis) for 36,662 basins in Sweden during 1961–2018.

References

1. FAO. *The Impact of Disasters on Agriculture and Food Security*; Technical Report; Food and Agriculture Organization of the United Nations: Rome, Italy, 2015.
2. Cammalleri, C.; Naumann, G.; Mentaschi, L.; Formetta, G.; Forzieri, G.; Gosling, S.; Bisselink, B.; De Roo, A.; Feyen, L. *Global Warming and Drought Impacts in the EU*; Joint Research Centre (JRC): Brussels, Belgium, 2020.
3. Vogt, J.; Sepulcre, G.; Magni, D.; Valentini, L.; Singleton, A.; Micale, F.; Barbosa, P. *The European Drought Observatory (EDO): Current State and Future Directions*; EGU General Assembly: Vienna, Austria, 2013; p. EGU2013-7374.
4. McKee, T.B.; Doesken, N.J.; Kleist, J. The relationship of drought frequency and duration to time scales. In Proceedings of the 8th Conference on Applied Climatology, Anaheim, CA, USA, 17–22 January 1993; Volume 17, pp. 179–183.
5. Hao, Z.; AghaKouchak, A.; Nakhjiri, N.; Farahmand, A. Global integrated drought monitoring and prediction system. *Sci. Data* **2014**, *1*, 1–10. [CrossRef]
6. Sutanto, S.J.; Van Lanen, H.A.; Wetterhall, F.; Llort, X. Potential of pan-european seasonal hydrometeorological drought forecasts obtained from a multihazard early warning system. *Bull. Am. Meteorol. Soc.* **2020**, *101*, E368–E393. [CrossRef]
7. Tsakiris, G.; Vangelis, H. Towards a drought watch system based on spatial SPI. *Water Resour. Manag.* **2004**, *18*, 1–12. [CrossRef]
8. Livada, I.; Assimakopoulos, V. Spatial and temporal analysis of drought in Greece using the Standardized Precipitation Index (SPI). *Theor. Appl. Climatol.* **2007**, *89*, 143–153. [CrossRef]
9. Patel, N.R.; Chopra, P.; Dadhwal, V.K. Analyzing spatial patterns of meteorological drought using standardized precipitation index. *Meteorol. Appl.* **2007**, *14*, 329–336. [CrossRef]
10. Karavitis, C.; Alexandris, S.; Tsesmelis, D.; Athanasopoulos, G. Application of the standardized precipitation index (SPI) in Greece. *Water* **2011**, *3*, 787–805. [CrossRef]
11. Van Loon, A.F.; Tijdeman, E.; Wanders, N.; Van Lanen, H.J.; Teuling, A.; Uijlenhoet, R. How climate seasonality modifies drought duration and deficit. *J. Geophys. Res. Atmos.* **2014**, *119*, 4640–4656. [CrossRef]
12. Guenang, G.M.; Kamga, F.M. Computation of the standardized precipitation index (SPI) and its use to assess drought occurrences in Cameroon over recent decades. *J. Appl. Meteorol. Climatol.* **2014**, *53*, 2310–2324. [CrossRef]
13. Wang, H.; Chen, Y.; Pan, Y.; Chen, Z.; Ren, Z. Assessment of candidate distributions for SPI/SPEI and sensitivity of drought to climatic variables in China. *Int. J. Climatol.* **2019**, *39*, 4392–4412. [CrossRef]
14. Stagge, J.H.; Tallaksen, L.M.; Gudmundsson, L.; Van Loon, A.F.; Stahl, K. Candidate distributions for climatological drought indices (SPI and SPEI). *Int. J. Climatol.* **2015**, *35*, 4027–4040. [CrossRef]
15. Angelidis, P.; Maris, F.; Kotsovinos, N.; Hrissanthou, V. Computation of drought index SPI with alternative distribution functions. *Water Resour. Manag.* **2012**, *26*, 2453–2473. [CrossRef]
16. Sienz, F.; Bothe, O.; Fraedrich, K. Monitoring and quantifying future climate projections of dryness and wetness extremes: SPI bias. *Hydrol. Earth Syst. Sci.* **2012**, *16*, 2143–2157. [CrossRef]
17. Alam, N.; Raizada, A.; Jana, C.; Meshram, R.K.; Sharma, N. Statistical modeling of extreme drought occurrence in Bellary District of Eastern Karnataka. *Proc. Natl. Acad. Sci. India Sect. B Biol. Sci.* **2015**, *85*, 423–430. [CrossRef]

18. Shiau, J.T. Effects of gamma-distribution variations on SPI-based stationary and nonstationary drought analyses. *Water Resour. Manag.* **2020**, *34*, 2081–2095. [CrossRef]
19. Russo, S.; Dosio, A.; Sterl, A.; Barbosa, P.; Vogt, J. Projection of occurrence of extreme dry-wet years and seasons in Europe with stationary and nonstationary Standardized Precipitation Indices. *J. Geophys. Res. Atmos.* **2013**, *118*, 7628–7639. [CrossRef]
20. Wang, Y.; Li, J.; Feng, P.; Hu, R. A time-dependent drought index for non-stationary precipitation series. *Water Resour. Manag.* **2015**, *29*, 5631–5647. [CrossRef]
21. Box, G.; Jenkins, E.M. *Time Series Analysis: Forecasting and Control*; Holden-Day, Inc.: San Francisco, CA, USA, 1970.
22. Mishra, A.K.; Desai, V.R. Drought forecasting using stochastic models. *Stoch. Environ. Res. Risk Assess* **2005**, *19*, 326–339. [CrossRef]
23. Han, P.; Wang, P.; Tian, M.; Zhang, S.; Liu, J.; Zhu, D. Application of the ARIMA models in drought forecasting using the standardized precipitation index. *IFIP Adv. Inf. Commun. Technol.* **2013**, *392*, 352–358. [CrossRef]
24. Yeh, H.F.; Hsu, H.L. Stochastic model for drought forecasting in the southern Taiwan basin. *Water* **2019**, *11*, 2041. [CrossRef]
25. Sutanto, S.J.; Wetterhall, F.; Van Lanen, H.A. Hydrological drought forecasts outperform meteorological drought forecasts. *Environ. Res. Lett.* **2020**, *15*, 084010. [CrossRef]
26. Vapnik, V.; Golowich, S.E.; Smola, A.J. Support Vector Method for Function Approximation, Regression Estimation and Signal Processing. In Proceedings of the NIPS'96 9th International Conference on Neural Information Processing Systems, Denver, CO, USA, 3–5 December 1996; MIT Press: Cambridge, MA, USA, 1996; pp. 281–287.
27. McCulloch, W.; Pitts, W. A Logical Calculus of Ideas Immanent in Nervous Activity. *Bull. Math. Biophys.* **1943**, *5*, 115–133. [CrossRef]
28. Belayneh, A.; Adamowski, J. Drought forecasting using new machine learning methods. *J. Water Land Dev.* **2013**, *18*, 3–12. [CrossRef]
29. Mokhtarzad, M.; Eskandari, F.; Vanjani, N.J.; Arabasadi, A. Drought forecasting by ANN, ANFIS, and SVM and comparison of the models. *Environ. Earth Sci.* **2017**, *76*, 729. [CrossRef]
30. Wu, H.; Hayes, M.J.; Wilhite, D.A.; Svoboda, M.D. The effect of the length of record on the standardized precipitation index calculation. *Int. J. Climatol.* **2005**, *25*, 505–520. [CrossRef]
31. Rezaeian-Zadeh, M.; Tabari, H. MLP-based drought forecasting in different climatic regions. *Theor. Appl. Climatol.* **2012**, *109*, 407–414. [CrossRef]
32. Belayneh, A.; Adamowski, J.; Khalil, B.; Ozga-Zielinski, B. Long-term SPI drought forecasting in the Awash River Basin in Ethiopia using wavelet neural network and wavelet support vector regression models. *J. Hydrol.* **2014**, *508*, 418–429. [CrossRef]
33. Deo, R.C.; Kisi, O.; Singh, V.P. Drought forecasting in eastern Australia using multivariate adaptive regression spline, least square support vector machine and M5Tree model. *Atmos. Res.* **2017**, *184*, 149–175. [CrossRef]
34. Yaseen, Z.M.; Ali, M.; Sharafati, A.; Al-Ansari, N.; Shahid, S. Forecasting standardized precipitation index using data intelligence models: Regional investigation of Bangladesh. *Sci. Rep.* **2021**, *11*, 1–25. [CrossRef]
35. Rhee, J.; Im, J. Meteorological drought forecasting for ungauged areas based on machine learning: Using long-range climate forecast and remote sensing data. *Agric. For. Meteorol.* **2017**, *237*, 105–122. [CrossRef]
36. Tashman, L.J. Out-of-sample tests of forecasting accuracy: An analysis and review. *Int. J. Forecast.* **2004**, *16*, 437–450. [CrossRef]
37. Bergmeir, C.; Benitez, J.M. On the Use of Cross-validation for Time Series Predictor Evaluation. *Inf. Sci.* **2012**, *191*, 192–213. [CrossRef]
38. Hastie, T.; Tibshirani, R.; Friedman, J. *The Elements of Statistical Learning: Prediction, Inference and Data Mining*, 2nd ed.; Springer: New York, NY, USA, 2009; pp. 241–260.
39. Mishra, A.K.; Desai, V.R.; Singh, V.P. Drought forecasting using a hybrid stochastic and neural network model. *Hydrol. Eng. ASCE* **2007**, *12*, 626–638. [CrossRef]
40. Kottek, M.; Grieser, J.; Beck, C.; Rudolf, B.; Rubel, F. World map of the Köppen-Geiger climate classification updated. *Hydrol. Earth Syst. Sci* **2006**. [CrossRef]
41. Girons Lopez, M.; Crochemore, L.; Pechlivanidis, I.G. Benchmarking an operational hydrological model for providing seasonal forecasts in Sweden. *Hydrol. Earth Syst. Sci.* **2021**, *25*, 1189–1209. [CrossRef]
42. *Guide, Svoboda, Hayes, and Wood*; World Meteorological Organization: Geneva, Switzerland, 2012.
43. Arheimer, B.; Lindström, G. Climate impact on floods: Changes in high flows in Sweden in the past and the future (1911–2100). *Hydrol. Earth Syst. Sci.* **2015**, *19*, 771–784. [CrossRef]
44. Van Loon, A.; Van Huijgevoort, M.; Van Lanen, H. Evaluation of drought propagation in an ensemble mean of large-scale hydrological models. *Hydrol. Earth Syst. Sci.* **2012**, *16*, 4057–4078. [CrossRef]
45. Van Loon, A.; Laaha, G. Hydrological drought severity explained by climate and catchment characteristics. *J. Hydrol.* **2015**, *526*, 3–14. [CrossRef]
46. Blauhut, V.; Stahl, K.; Stagge, J.; Tallaksen, L.; De Stefano, L.; Vogt, J. Estimating drought risk across Europe from reported drought impacts, drought indices, and vulnerability factor. *Hydrol. Earth Syst. Sci.* **2016**, *20*, 2779–2800. [CrossRef]
47. Hanel, M.; Rakovec, O.; Markonis, Y.; Maca, P.; Samaniego, L.; Kysely, J.; Kumar, R. Revisiting the recent European droughts from a long-term perspective. *Sci. Rep.* **2018**, *8*, 1–11. [CrossRef] [PubMed]
48. Dosio, A. *Mean and Extreme Climate in Europe under 1.5, 2, and 3 °C Global Warming*; JRC PESETA IV Project—Task 1; Publications Office of the European Union: Luxembourg, 2020. [CrossRef]

49. Vicente-Serrano, S.M.; Beguería, S.; Lopez-Moreno, J.I. A multiscalar drought index sensitive to global warming: The standardized precipitation evapotranspiration index. *Water Resour. Manag.* **2010**, *23*, 1696–1718. [CrossRef]
50. Byun, H.R.; Wilhite, D.A. Objective quantification of drought severity and duration. *J. Clim.* **1999**, *12*, 2747–2756. [CrossRef]
51. Beguería, S.; Vicente-Serrano, S.M.; Reig, F.; Latorre, B. Standardized precipitation evapotranspiration index (SPEI) revisited: parameter fitting, evapotranspiration models, tools, datasets and drought monitoring. *Int. J. Climatol.* **2014**, *34*, 3001–3023. [CrossRef]
52. Deo, R.C.; Byun, H.R.; Adamowski, J.F.; Begum, K. Application of effective drought index for quantification of meteorological drought events: A case study in Australia. *Theor. Appl. Climatol.* **2017**, *128*, 359–379. [CrossRef]
53. Palmer, W.C. *Meteorological Drought*; US Department of Commerce, Weather Bureau: Silver Spring, MD, USA, 1965; Volume 30.
54. Sepulcre-Canto, G.; Horion, S.; Singleton, A.; Carrao, H.; Vogt, J. Development of a Combined Drought Indicator to detect agricultural drought in Europe. *Nat. Hazards Earth Syst. Sci.* **2012**, *12*, 3519–3531. [CrossRef]
55. Vicente-Serrano, S.M.; Beguería, S.; López-Moreno, J.I. Comment on "Characteristics and trends in various forms of the Palmer Drought Severity Index (PDSI) during 1900–2008" by Aiguo Dai. *J. Geophys. Res. Atmos.* **2011**, *116*. [CrossRef]
56. Balint, Z.; Mutua, F.; Muchiri, P.; Omuto, C.T. Monitoring drought with the combined drought index in Kenya. In *Developments in Earth Surface Processes*; Elsevier: Amsterdam, The Netherlands, 2013; Volume 16, pp. 341–356.
57. Sutanto, S.J.; Van Lanen, H.A. Streamflow drought: Implication of drought definitions and its application for drought forecasting. *Hydrol. Earth Syst. Sci. Discuss.* **2021**, *25*, 3991–4023. [CrossRef]
58. Thom, H.C.S. A note on gamma distribution. *Mon. Weather Rev.* **1958**, *86*, 117–122. [CrossRef]
59. Rigby, R.A.; Stasinopoulos, D.M. Generalized additive models for location, scale and shape. *J. R. Stat. Soc. Ser. C (Appl. Stat.)* **2005**, *54*, 507–554. [CrossRef]
60. Villarini, G.; Smith, J.A.; Napolitano, F. Nonstationary modeling of a long record of rainfall and temperature over Rome. *Adv. Water Resour.* **2010**, *33*, 1256–1267. [CrossRef]
61. Bazrafshan, J.; Hejabi, S. A non-stationary reconnaissance drought index (NRDI) for drought monitoring in a changing climate. *Water Resour. Manag.* **2018**, *32*, 2611–2624. [CrossRef]

Article

Hydrological Drought Assessment Based on the Standardized Streamflow Index: A Case Study of the Three Cape Provinces of South Africa

Christina M. Botai [1,*], Joel O. Botai [1,2,3,4,*], Jaco P. de Wit [1], Katlego P. Ncongwane [1,5], Miriam Murambadoro [1], Paul M. Barasa [4] and Abiodun M. Adeola [1]

[1] South African Weather Service, Private Bag X097, Pretoria 0001, South Africa; Jaco.deWit@weathersa.co.za (J.P.d.W.); Katlego.Ncongwane@weathersa.co.za (K.P.N.); Miriam.Murambadoro@weathersa.co.za (M.M.); Abiodun.Adeola@weathersa.co.za (A.M.A.)

[2] Department of Geography, Geoinformatics and Meteorology, University of Pretoria, Private Bag X020, Hatfield, Pretoria 0028, South Africa

[3] Department of Information Technology, Central University of Technology, Free State, Private Bag X200539, Bloemfontein 9300, South Africa

[4] Centre for Transformative Agricultural and Food Systems, School of Agricultural, Earth and Environmental Sciences, University of KwaZulu-Natal, Pietermaritzburg 3209, South Africa; pmmtoni@gmail.com

[5] School of Geography and Environmental Science, University of KwaZulu-Natal, Durban 4041, South Africa

[*] Correspondence: Christina.Botai@weathersa.co.za (C.M.B.); Joel.Botai@weathersa.co.za (J.O.B.); Tel.: +27-12-367-6269 (C.M.B.); +27-367-6070 (J.O.B.)

Citation: Botai, C.M.; Botai, J.O.; de Wit, J.P.; Ncongwane, K.P.; Murambadoro, M.; Barasa, P.M.; Adeola, A.M. Hydrological Drought Assessment Based on the Standardized Streamflow Index: A Case Study of the Three Cape Provinces of South Africa. *Water* **2021**, *13*, 3498. https://doi.org/10.3390/w13243498

Academic Editor: Alina Barbulescu

Received: 9 November 2021
Accepted: 4 December 2021
Published: 8 December 2021

Publisher's Note: MDPI stays neutral with regard to jurisdictional claims in published maps and institutional affiliations.

Copyright: © 2021 by the authors. Licensee MDPI, Basel, Switzerland. This article is an open access article distributed under the terms and conditions of the Creative Commons Attribution (CC BY) license (https://creativecommons.org/licenses/by/4.0/).

Abstract: Global impacts of drought conditions pose a major challenge towards the achievement of the 2030 Sustainable Development Goals. As a result, a clarion call for nations to take actions aimed at mitigating the adverse negative effects, managing key natural resources and strengthening socioeconomic development can never be overemphasized. The present study evaluated hydrological drought conditions in three Cape provinces (Eastern, Western and Northern Cape) of South Africa, based on the Standardized Streamflow Index (SSI) calculated at 3- and 6-month accumulation periods from streamflow data spanning over the 3.5 decades. The SSI features were quantified by assessing the corresponding annual trends computed by using the Modified Mann–Kendall test. Drought conditions were also characterized in terms of the duration and severity across the three Cape provinces. The return levels of drought duration (DD) and drought severity (DS) associated with 2-, 5-, 10-, 20- and 50-year periods were estimated based on the generalized extreme value (GEV) distribution. The results indicate that hydrological drought conditions have become more frequent and yet exhibit spatial contrasts throughout the study region during the analyzed period. To this end, there is compelling evidence that DD and DS have increased over time in the three Cape provinces. Return levels analysis across the studied periods also indicate that DD and DS are expected to be predominant across the three Cape provinces, becoming more prolonged and severe during the extended periods (e.g., 20- and 50-year). The results of the present study (a) contribute to the scientific discourse of drought monitoring, forecasting and prediction and (b) provide practical insights on the nature of drought occurrences in the region. Consequently, the study provides the basis for policy- and decision-making in support of preparedness for and adaptation to the drought risks in the water-linked sectors and robust water resource management. Based on the results reported in this study, it is recommended that water agencies and the government should be more proactive in searching for better strategies to improve water resources management and drought mitigation in the region.

Keywords: hydrological drought; water resources; mitigation; streamflow; return levels; trends; GEV distribution

1. Introduction

Drought is a naturally recurring hazard associated with a decrease in water availability over time within a region. Such conditions are often attributed to anomalous weather conditions associated with the decreasing intensity or a deficiency of precipitation; the changes in the onset and cessation of precipitation; and in climatological parameters, such as temperature, relative humidity and evapotranspiration [1–3]. Drought is also influenced by natural global circulation changes; the long-term abnormal Sea Surface Temperature (SST), particularly in the tropics; and El Nino Southern Oscillation (ENSO) associated with below-normal rainfall [4]. In addition, drought over Africa (especially in the Sahel region) could be associated with the southward shift of the warm SST in the Atlantic and the warming in the Indian Ocean [5].

South Africa, similarly to many semi-arid and arid countries globally, suffers from frequent occurrence of drought conditions. In recent years, some parts of the country have been declared disaster areas; however, some regions have since recovered or are slowly recovering. Some regions have experienced what has been termed the "worst drought in over 35 years" [6], including the 2015 drought that was partly attributed to a rainfall deficit that reached the lowest annual average since 1904 [7]. The southwestern and southeastern regions of the country, particularly the Western, Eastern and Northern Cape provinces, have been experiencing diverse impacts of drought, resulting from the common classified categories (e.g., meteorological, agricultural, hydrological and socioeconomic) conditions [8,9]. The drought conditions had profound and negative implications on the economies of these adjacent provinces. In particular, a persistent physical phenomenon attributed to the first three classes of drought (meteorological, agricultural and hydrological) has caused substantial and irreversible socioeconomic challenges in the Eastern Cape, resulting in significant low water levels, bringing uncertainty to the most vulnerable communities. The Western Cape province has also experienced a water crisis, particularly in 2015/2016 and 2018/19, which was attributed to past drought conditions in the province [10]. In addition, severe drought in the Northern Cape province has affected farming activities and cost the provincial government millions of Rands in drought-related funding relief to mitigate the inherent effects.

Persistent drought conditions in the three Cape provinces have led the National Disaster Management Centre to reclassify drought as a national drought disaster in a bid to mobilize the necessary resources to support the affected communities, including farmers. While this declaration is essential and provides a short-term solution, there is a need to explore long-term solutions to alleviate the negative impacts of drought in these regions. Effective drought monitoring and early warning systems form the basis for reducing vulnerability, as well as developing proactive policies to enhance adaptive capacity and drought mitigation measures. Drought indices are some of the tools that can be used to evaluate and monitor drought. Examples of existing drought indices that have been applied in drought-related studies include the Standardized Precipitation Index (SPI) [11], Standardized Precipitation and Evapotranspiration Index (SPEI) [12], Palmer Drought Severity Index (PDSI) [13], Effective Drought Index (EDI) [14], Surface Water Supply Index (SWSI) [15], Streamflow Drought Index (SDI) [16] and Standardized Streamflow Index (SSI) [17], among others. The selection and application of these drought indices are dependent on various factors, such as the nature of the index, local conditions and data requirements and availability. Some of these indices have been extensively applied in South Africa for drought-related studies. These include drought characteristics over the Western Cape province [10], in the Free State and North West provinces [18], in Mpumalanga, Kwa-Zulu Natal, Free State, and North West provinces [19] and the Southern Africa [20]. Moreover, drought propagation using SPI and SSI was reported by Reference [21], as well as drought-risk assessment in the Eastern Cape province [22], hydroclimatic extremes in the Limpopo River Basin [23] and drought characteristics based on the SPI and EDI in the Free State province [24], among others.

Most of these studies have primarily focused on a meteorological drought, where they have analyzed precipitation data, presumably because rainfall is the leading driver to drought. Despite the fact that, in most parts of South Africa, the impacts of drought are manifested in water resources, e.g., increased water demand for domestic, irrigation and industrial use, among others, studies on hydrological drought are rather limited. The purpose of this study was to characterize and understand the extent of hydrological drought conditions that have adversely affected the water resources in the three Cape provinces of South Africa, using the SSI. The specific study objectives were to (1) assess characteristics of the past hydrological drought in terms of historical trends, duration and severity and (2) determine the return levels of drought duration and severity across selected periods. The results can contribute towards the implementation of drought monitoring and early warning systems that could support planning, preparedness and innovation to improve the regions' adaptive capacity for water resources' supply and demand.

2. Study Area, Materials and Methods

2.1. Study Area

The three Cape provinces of South Africa, (the Northern Cape (NC), Eastern Cape (EC), and Western Cape (WC)), as shown in Figure 1, are located in the southwest of the country and account for more than half (55%) of the country's total geographical area [25,26]. The WC province is situated on the southwestern tip of South Africa between the Indian and Atlantic oceans. It is characterized by a warm temperate Mediterranean climate, with cool, wet winters and relatively dry, warm summers [27]. The province experiences diverse climate conditions, including dominant rainfall in austral winter and early spring. In general, the WC province exhibits two dominant rainfall zones, e.g., the winter (west coast) and the bi-modal (spring and autumn in the southern regions) rainfall features [27]. All-season rainfall is received in the south-coast regions of the province [27]. The received rainfall is highly variable, ranging between the lowest of 60 mm to the highest of 3345 mm per year [28]. Most areas in the province receive annual rainfall between 350 and 1000 mm per year [28]. Coastal and inland temperatures range between 15 and 27 °C and between 3 and 5 °C, respectively [28]. The WC province is responsible for a large share of national output in agriculture, forestry and fisheries, with a large proportion of the agricultural activities mostly concentrated in the Cape Winelands [29].

The EC province is a coastal province that lies on the country's southeastern coast bordering the subtropical KwaZulu-Natal and Mediterranean WC province. The province is characterized by a bi-modal type of rainfall, meaning that it receives both summer (most southern) and winter (southwestern) rainfall [30]. Some pocket areas in the western coast of the province are classified as all-season rainfall regions. In general, the EC province has an arid to semi-arid climate, with annual rainfall ranging between 350 and 550 mm/year, mean annual temperature of 17.6 °C and daily maximum temperatures of up to 40 °C in the summer [31].

The NC province is South Africa's largest province, covering approximately 31% of the country's land area, and has the country's lowest population density of all provinces [32]. In general, the NC province is mostly a desert-to-semi-desert area. The region is characterized by fluctuating temperatures and varying landscapes [33]. The western areas of the province receive winter rainfall, whereas summer rainfall with thunderstorms is experienced in the eastern regions. The mean annual rainfall is sparse, ranging between 50 and 400 mm per annum, while the temperature in summer often reaches maximum of 40 °C [34].

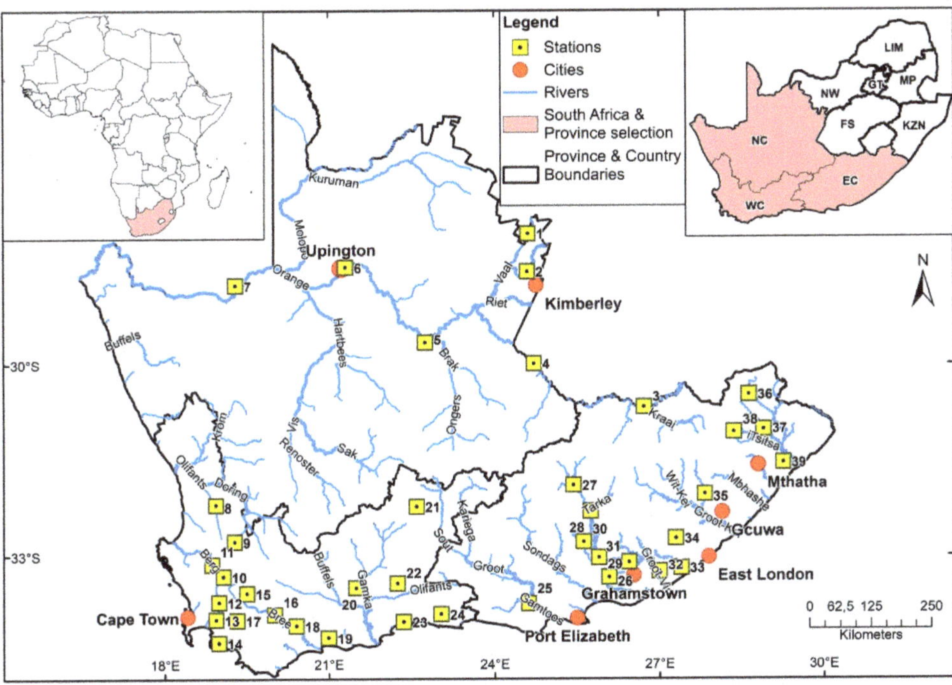

Figure 1. Central area: study area map and distribution of streamflow gauge stations. Top left panel: map of Africa with South Africa highlighted in brown. Top right panel: map of South Africa with its nine provinces. The 3 Cape provinces are highlighted in brown, where NC, WC and EC are abbreviations for the Northern Cape, Western Cape and Eastern Cape provinces, respectively.

2.2. Materials

The daily observed streamflow datasets were acquired from the Department of Water and Sanitation, South Africa, on https://www.dws.gov.za/Hydrology/Verified/hymain. aspx (accessed on 6 September 2021). The present study considered datasets from 39 stations distributed across the three Cape provinces, with provincial distribution as follows: 6 in the NC, 16 in the EC and 17 in the WC (see Figure 1). Table 1 gives a summary of the selected flow gauge stations, including the location, catchment area and average flow. The stations were selected based on the availability of continuous datasets for the period spanning from 1985 to 2020 and less than 5% gaps. As shown in Figure 1, the WC has a dense network of streamflow stations, followed by the EC province. Notwithstanding harboring few streamflow stations, the northern parts of the NC province have Kalahari Desert conditions. As a result, a better characterization of the drought conditions in the province was deemed necessary.

Table 1. Selected river stations, with basic information including annual mean flow. The letters in station names (e.g., C, D, G, . . . , K) represent the drainage region, while the middle H letter stands for hydrological data.

Station No.	Station Name	Location	Latitude	Longitude	Catchment Area (km^2)	Mean Flow (m^3/s)
1	C3H007	NC	-27.9032	24.61514	23.900	5.52
2	C9H009	NC	-28.5162	24.60069	121.220	23.85
3	D1H003	EC	-30.6783	26.71638	36.975	116.63
4	D3H012	NC	-29.9911	24.72027	89.755	126.80
5	D7H002	NC	-29.6517	22.74591	337.690	130.17

Table 1. *Cont.*

Station No.	Station Name	Location	Latitude	Longitude	Catchment Area (km^2)	Mean Flow (m^3/s)
6	D7H005	NC	−28.4579	21.23923	361.530	156.61
7	D8H004	NC	−28.7359	19.30553	856.400	117.90
8	E1H006	WC	−32.2117	18.93666	160	1.31
9	E2H007	WC	−32.7803	19.28333	265	1.35
10	G1H008	WC	−33.3139	19.07472	393	2.05
11	G1H013	WC	−33.1308	18.86277	2936	17.13
12	G1H020	WC	−33.7078	18.99111	628	10.07
13	G2H005	WC	−33.9736	18.93805	31	0.52
14	G4H007	WC	−34.3294	18.98833	464	6.65
15	H2H006	WC	−33.5708	19.50611	707	2.67
16	H5H004	WC	−33.8978	20.01166	6713	23.39
17	H6H010	WC	−33.9836	19.32916	15	0.05
18	H7H006	WC	−34.0675	20.40555	9842	32.77
19	H8H001	WC	−34.2517	20.99194	790	2.88
20	J2H005	WC	−33.4897	21.48944	253	0.19
21	J2H018	WC	−32.2403	22.58583	98	0.03
22	J3H014	WC	−33.4211	22.24083	151	0.44
23	K3H003	WC	−34.0067	22.35027	145	0.70
24	K5H002	WC	−33.8911	23.02944	133	0.76
25	L7H006	EC	−33.731	24.61794	29.560	1.85
26	P1H003	EC	−33.3294	26.07775	1479	0.18
27	Q2H002	EC	−31.9042	25.43061	1713	0.35
28	Q4H013	EC	−32.3142	25.74111	4742	0.54
29	Q7H005	EC	−33.0276	25.8933	19.130	7.04
30	Q8H008	EC	−32.785	25.61483	1512	0.62
31	Q9H012	EC	−33.0983	26.44475	23.054	7.12
32	Q9H018	EC	−33.2378	26.99486	29.743	11.17
33	R1H015	EC	−33.1854	27.39075	2530	3.27
34	R2H001	EC	−32.7319	27.29361	29	0.31
35	S5H002	EC	−32.0443	27.82238	2359	3.91
36	T3H002	EC	−30.4828	28.62083	2101	9.61
37	T3H005	EC	−31.0318	28.8845	2597	14.59
38	T3H009	EC	−31.0717	28.35361	307	3.60
39	T7H001	EC	−31.5512	29.2438	315	1.19

2.3. Methods

2.3.1. Standardized Streamflow Index

The SSI developed by Vicente-Serrano [17] is considered a useful index for the assessment and characterization of hydrological drought. Hydrological drought is associated with a reduction in the groundwater and/or surface-water resources, including river flows, reservoir storage and acquires [35], consequently affecting water supply for various purposes, including domestic, agriculture, power generation and recreation, among others. The SSI has been applied in various studies globally for the assessment of hydroclimatic extremes [23], characterization of anomalies in streamflow data [17,36], drought assessment [37], etc. While there has been no consensus on the exact methodology to use in the computation of the SSI, two approaches have been proposed in the literature by Modarres [36] and Vecente-Serrano et al. [17]. In studies by Vecente-Serrano et al. [17], the SSI considers a probability distribution fitting that best represents the streamflow variations over the analyzed domain. This concept is statistically similar to the calculation of the Standardized Precipitation Index (SPI) [11].

In this study, the SPI concept was used to compute SSI to characterize the hydrological drought conditions across the three Cape provinces of South Africa. An important aspect when computing a drought index is the ability to select the most suitable probability distribution function that best fits the datasets of hydroclimatic variables. For instance, the gamma probability distribution has been identified as the best and hence is widely used in computing SPI [11]. In the case of SSI, no single probability distribution has been identified as being suitable for its computation [17]. However, according to References [38,39], the gamma distribution is also appropriate and can be used to fit long-term streamflow data and calculate SSI. In this regard, the gamma probability distribution was used in the present study to fit the aggregated monthly streamflow records.

A gamma distribution in which a random variable, x, is continuous can be expressed as follows [11,40]:

$$g(x, \alpha, \beta) = \frac{1}{\beta^\alpha \Gamma(\alpha)} x^{\alpha-1} e^{-x/\beta} \tag{1}$$

where $\alpha > 0$ and $\beta > 0$ are the estimated shape and scale parameters, respectively; $x > 0$ is the streamflow (m^3/s); and $\Gamma(\alpha)$ is the gamma function defined by the following:

$$\Gamma(\alpha) = \int_0^\infty x^{\alpha-1} e^{-x} dx \tag{2}$$

The gamma distribution is used to compute the cumulative probability function given as follows:

$$G(x) = \int_0^x g(x) dx = \int_0^x \frac{1}{\beta^\alpha \Gamma(\alpha)} x^{\alpha-1} e^{-x/\beta} dx = \frac{1}{\Gamma(a)} \int_0^x t^{a-1} e^{-1} dt \tag{3}$$

If $x = 0$ and $q = P(x = 0) > 0$, where $P(x = 0)$ is the probability of zero streamflow, then the gamma distribution is undefined. In this regard, the cumulative probability density can be described by the following:

$$H(x) = q + (1-q)G(x) \tag{4}$$

The cumulative probability distribution function is transformed into a normal distribution, with an average and standard deviation of zero (0) and one (1), respectively, resulting in SSI time series. The resulting SSI consists of both negative and positive values which represent drought/dry (i.e., a period having negative/below zero values) and non-drought/wet (i.e., a period with positive/above zero values) events, respectively.

In this study, the SSI was computed for three (e.g., SSI-3) and six (SSI-6) accumulation periods and categorized by using the classification criteria of SPI, recommended by the World Meteorological Organization standards [41] as given in Table 2. The selected accumulated time-steps cover both seasonal (or agricultural) and supra-seasonal, where water resources are likely to be affected due to insufficient and/or delayed precipitation. Annual features of the SSI were calculated based on the hydrological calendar year (e.g., October–September).

Table 2. Classification of drought based on the Standardized Streamflow Index estimated values.

Non-Drought	Mild Drought	Moderate Drought	Severe Drought	Extreme Drought
SSI $\geq$ 0.0	$-1.0 \leq$ SSI < 0.0	$-1.5 \leq$ SSI < -1.0	$-2.0 \leq$ SSI < -1.5	SSI ≤ -2.0

2.3.2. Drought Duration and Severity

The SSIs calculated for the 3- and 6-month time scales were used to characterize hydrological drought in terms of drought duration (DD) and drought severity (DS) across the three Cape provinces. Following various studies (e.g., References [42,43]), the following procedure was used to calculate the DD and DS:

(a) A drought event (epoch) was determined when 2 or more consecutive months exhibited negative SSI values.

(b) For each drought event, DD represents the number of months of the drought event.

(c)   The DS was computed as the absolute sum of the SSI, and is given in Equation (5) [43,44]:

$$DS_e = \left| \sum_{j=1}^{DD} SSI_j \right| \tag{5}$$

where j represents a drought month, and DD corresponds to the duration of a drought event e.

2.3.3. The Modified Mann–Kendall Trend Analysis

Trends analysis for streamflow, as well as the selected drought features (SSI, DD and DS), across the study area, was carried out by using the Modified Mann–Kendall (MMK) test [45,46]. The MMK test, as the name implies, is the modified version of the original form of the Mann–Kendall (MK) test [47–49]. The MK non-parametric test is known to be flexible with all distributions [50] and has been extensively used in research fields, such as meteorology, hydrology and climatology, among others, to detect shifts and changes in climatic [51] and hydrologic variables [52,53]. Despite this, the MK test has drawbacks in handling issues relating to autocorrelation in the datasets, as is commonly the case of streamflow data. In this regard, the MMK trend test was used in the present study to account for the presence of autocorrelation in the streamflow data.

The MK test statistic (S) is defined by Equation (6), as described in Malik and Kumar [46]:

$$S = \sum_{i=1}^{n-1} \sum_{j=1}^{n} \text{sgn}\left(X_j - X_i\right) \tag{6}$$

where n represents the number of datasets, X_i is the rank for the ith datasets ($i = 1, 2, 3, \ldots,$ $n - 1$) and X_j is the rank for the jth datasets ($j = i + 1, 2, \ldots, n$). The sign function, sgn, is computed as per Equation (7) [46]:

$$\text{sgn}\left(X_j - X_i\right) = \begin{cases} 1; & \text{if } \left(X_j - X_i\right) > 0 \\ 0; & \text{if } \left(X_j - X_i\right) = 0 \\ -1; & \text{if } \left(X_j - X_i\right) < 0 \end{cases} \tag{7}$$

The MMK test is normally applied to a significant autocorrelation coefficient of time-series data, with the modified variable computed by using Equation (8):

$$Var(S)^* = Var(S)\frac{n}{n_s} \tag{8}$$

where $Var(S)^*$ is the modified variance of the MMK test, $Var(S)$ represents the variance of the MK test series and n/n_s is a correction attributed to autocorrelation within the data, as is given by Equation (9):

$$\frac{n}{n_s} = 1 + \frac{2}{n(n-1)(n-2)} \sum_{k=1}^{n-1} (n-k)(n-k-1)(n-k-2)r_k \tag{9}$$

In Equation (9), n corresponds to an actual number of datasets, and r_k is a significant autocorrelation coefficient at lag-k, with its value ranging between negative and positive one. The standardized MMK test statistics are estimated as follows:

$$Z_{MMK} = \begin{cases} \frac{S-1}{\sqrt{Var(S)^*}}; & \text{if } S > 0 \\ 0; & \text{if } S = 0 \\ \frac{S+1}{\sqrt{Var(S)^*}}; & \text{if } S < 0 \end{cases} \tag{10}$$

The Z_{MMK} satisfies a standard normal distribution, where the mean and variance are zero (0) and one (1), respectively. In this study, the null hypothesis (e.g., no trend) and the opposite hypothesis (trend) were tested at a 95% confidence level, e.g., trends were considered statistically significant whenever the *p*-value was less or equal to 0.05.

2.3.4. Drought Return Levels Analysis

The DD and DS values were analyzed by using the Generalized Extreme Value (GEV) distribution. The GEV is a family of three distributions (e.g., Gumbel, Frechet and Weibull) used for analyzing extreme events, including modeling the block maxima [54,55]. Using the GEV, the distribution of the magnitudes of DD and DS can be approximated as per Equation (11):

$$F(x) = exp - \left\{ -\left[1 - \xi\left(\frac{z-\mu}{\sigma}\right)\right]^{\frac{1}{\xi}}\right\} \tag{11}$$

where μ, σ and ξ are the location, scale and shape GEV parameters (all dimensionless), respectively. These parameters are estimated based on the maximum likelihood estimation method [55], where the likelihood function is given by the following:

$$L = \prod_{i=1}^{N}\left\{ \frac{1}{\sigma}\left[1 - \xi\left(\frac{x_i - \mu}{\sigma}\right)\right]^{\frac{1}{\xi}-1} e^{-[1-\xi\frac{(x_i-\mu)}{\sigma}]^{\frac{1}{\xi}}}\right\} \tag{12}$$

where N is the number of observations. In this study, the predicted DD and DS return levels associated with 2, 5, 10 and 20 years were calculated by using Equation (13):

$$Z_p = \hat{\mu} - \frac{\hat{\sigma}}{\xi}\left[1 - \left\{ -log\left(1 - \frac{1}{T}\right)\right\}^{-\xi}\right] \tag{13}$$

where Z_p is the estimated return level corresponding to *T*-years return periods (e.g., *T* = 2, 5, 10, 20 and 50).

3. Results

3.1. Historical Trends of Streamflow

Figure 2a–c depicts annual streamflow (in m^3/s) for selected gauge stations in the WC, NC and EC provinces, respectively. Streamflow is highly variable across the provinces, ranging between ~2 and ~50 m^3/s in the WC, ~1 to ~20 m^3/s in EC and ~30 to ~420 m^3/s in NC province. The highest annual streamflow was recorded in 1992, 1993, 1997, 2008 and 2013 in the WC region. In the NC region, the highest annual streamflow was recorded in 1988, 1997, 2000, 2002, 2006 and 2011. Similarly, the EC province recorded the highest annual streamflow in 1988, 2006 and 2011. In general, the EC has received the lowest streamflow, whereas a greater streamflow was received in the NC province over the period 1985–2020. It is also noted that streamflow has reduced in the WC and EC from 2015 to 2020. On the other hand, the NC province has experienced reduced annual streamflow from 2012 to 2020. The distribution of averaged mean streamflow (m^3/s) across the stations and study site is depicted in Table 1. Most of the stations in both the WC and EC provinces show significantly small averaged mean streamflow values.

Historical trends of streamflow from 1985 to 2020 across the three Cape provinces were assessed based on the MMK test. The results are presented in Figure 3, where the upward and downward triangles correspond to positive and negative trends, and the blue and black dots inside the triangles represent significant and non-significant trends, tested at 95% significant level, respectively. In general, the study area demonstrates notable drying trends. In particular, 75%, 88% and 67% of the stations in the EC, WC and NC provinces, respectively, depict negative annual trends in streamflow. Overall, 80% of the stations exhibit a negative trend in streamflow across the three provinces. In addition, 38% of the detected negative trends are statistically significant at a 95% significant level. While most areas in the WC province are known to be "all-year rainfall" [10,28], there has been either a shift (where there was no rainfall in some seasons) or delayed/insufficient rainfall to benefit water resources. These results are consistent with the continued drought conditions that persist to burden the supply of water resources, particularly in the EC and WC provinces.

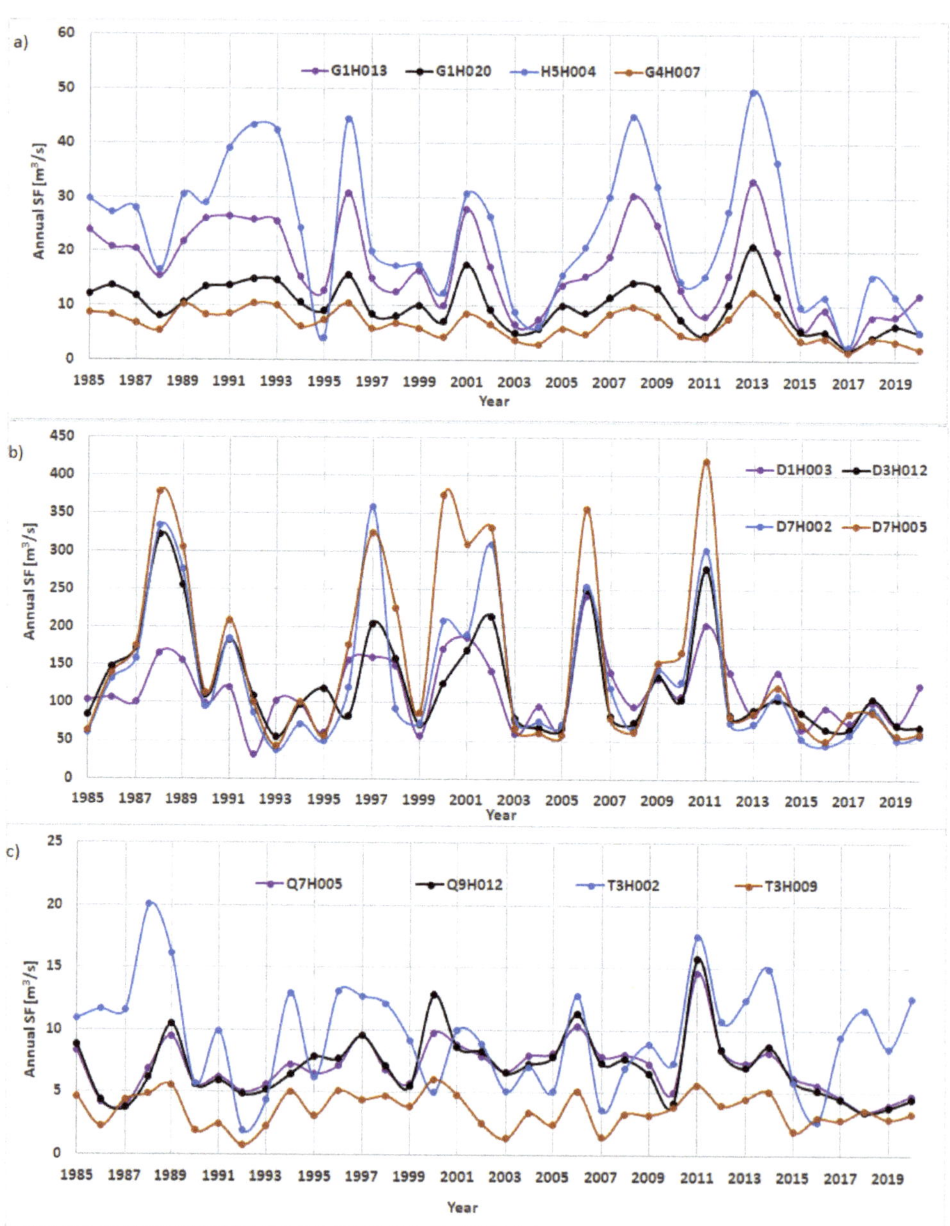

Figure 2. Annual streamflow time series for selected stream-gauge stations in (**a**) the Western Cape, (**b**) Northern Cape and (**c**) Eastern Cape study sites over the period 1985–2020. The abbreviations appearing on the station names represent drainage regions where the stations are located, and H in the middle gives the classification of data type—in this case, hydrological.

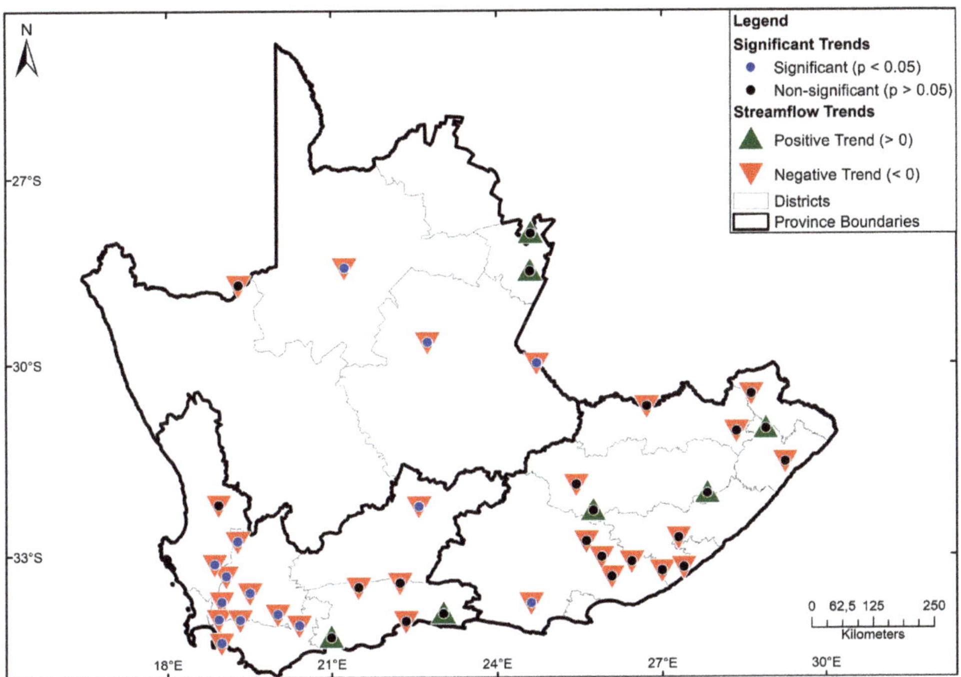

Figure 3. Trends in streamflow across the study site.

3.2. Characteristics of Standardized Streamflow Index

Figure 4a–c depicts SSI-3 and -6 annual mean time series for selected stations in the WC, EC and NC provinces. Based on the results, most of the stations have experienced mild drought over the study period. The longest drought in the EC and NC provinces was observed in the years 1991–1996, 2003–2005 and 2015–2020, mostly lying between mild and moderate categories. The WC province experienced the most prolonged drought in the years 1995–2000, 2003–2006, 2010–2012 and 2015–2020, mostly fluctuating between mild and moderate drought categories. The WC province also experienced severe drought in 2017 across all the selected stream-gauge stations.

The MMK trend test results of the SSI computed from streamflow data at 3- and 6-month accumulation time-steps across the study sites are shown in Figure 5. In this figure, the top and bottom panels correspond to trends based on 3- and 6-month cumulative periods, respectively. Positive (negative) and significant (non-significant) trends are shown by green (red) triangles and blue (black) dots, respectively. The test results depict similar annual trend patterns across the accumulation periods and stations within the study area. Dry conditions are evident across the study area, with all the stations in the WC exhibiting negative trends for both SSI-3 and SSI-6 accumulation epochs. Similarly, most of the NC regions, where gauge stations are located, depict negative trends, reflecting a long-term hydrological drought pattern. On the contrary, most of the stations (approximately 68%) in the EC province depict notable increasing trends in SSI values across the time-steps for the period 1985–2020. These trends are, however, not statistically significant at the 95% confidence level.

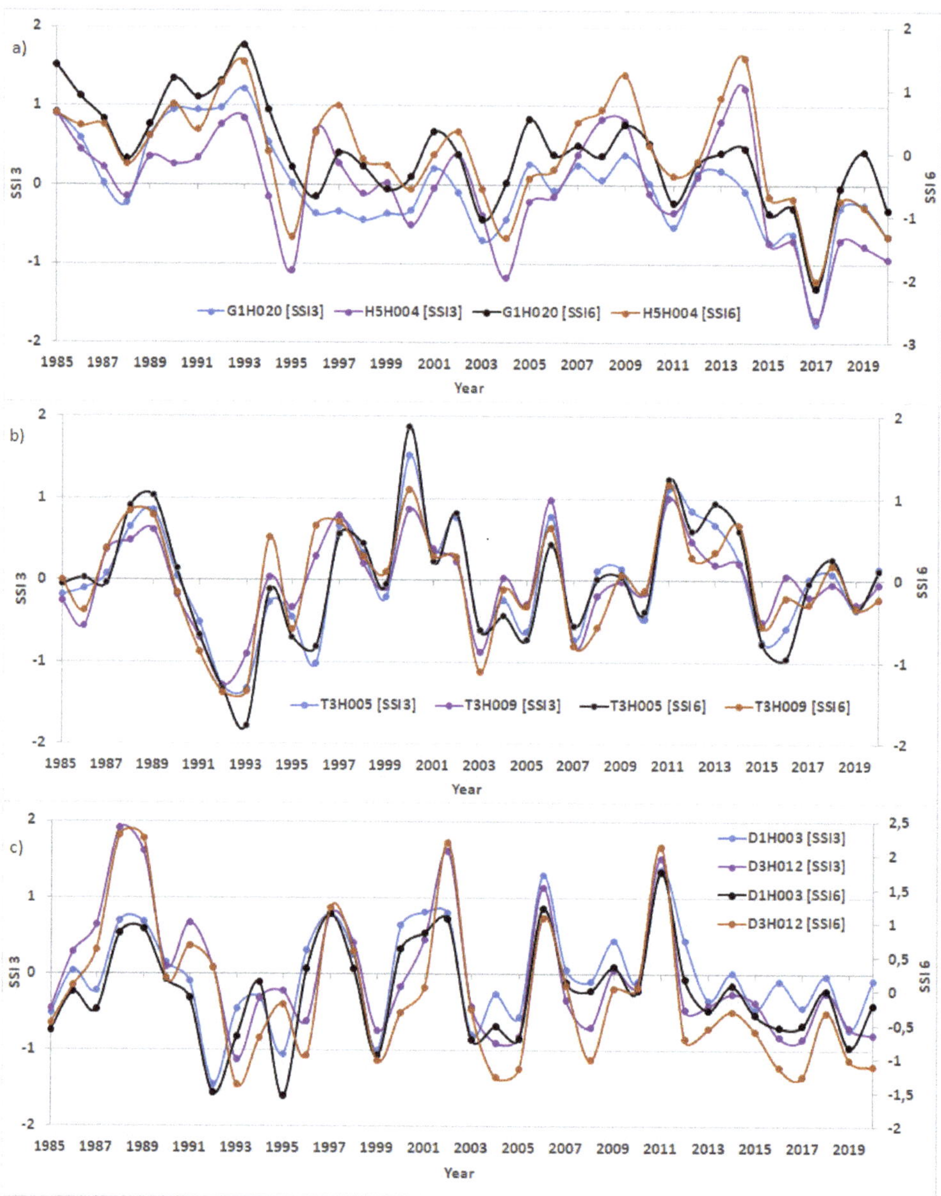

Figure 4. Annual SSI time series at 3- and 6-month accumulation time-steps for selected stations in (**a**) the Western Cape, (**b**) Eastern Cape and (**c**) Northern Cape provinces over the period 1985–2020.

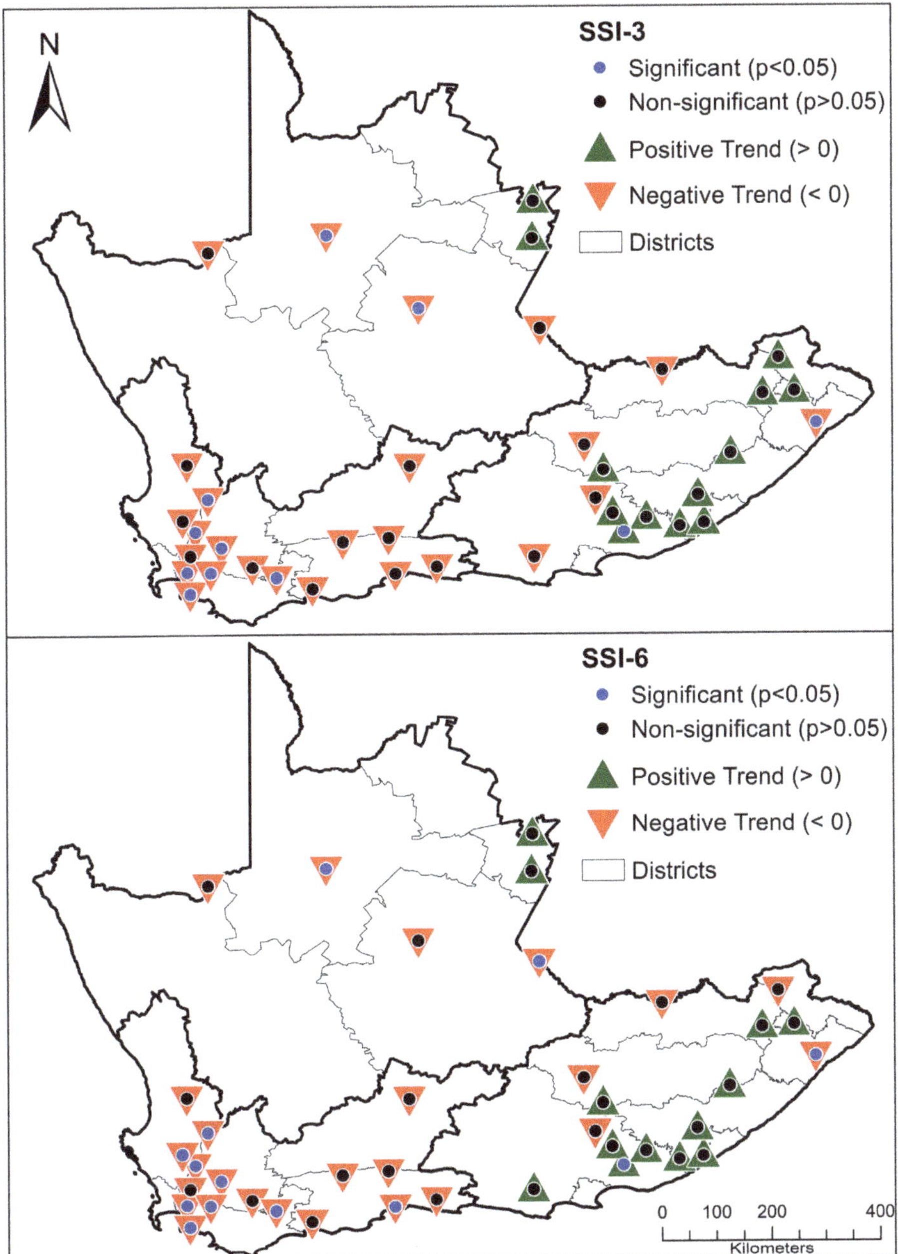

Figure 5. Trends in the Standardized Streamflow Index at 3- (**top**) and 6-month (**bottom**) accumulation periods.

3.3. Characteristics of Drought Duration and Severity

Figure 6 illustrates the spatial distribution of the average duration and severity of drought across the three Cape provinces during the study period (1985–2020). The average DD ranges between 5 and ~9 months across the provinces. While most parts of the WC experienced average DD ranging between ~5 and ~7 months, the parts of the EC and

NC provinces experience the longest DD lasting for ~9 months. In particular, the WC exhibits the lowest DD ranging from 5 to 7 across the stations and accumulation periods. Similarly, greater DD values are observed in the NC and EC provinces, ranging from 6 to 8 months across the accumulation time-steps. The EC and NC provinces DD range from approximately 6 to 9 months across the accumulation time-steps. Severe drought is observed across the provinces, with estimated maximum average DS of 5.0 noticed across most of the stations, and in both the analysis of 3- and 6-month accumulation periods.

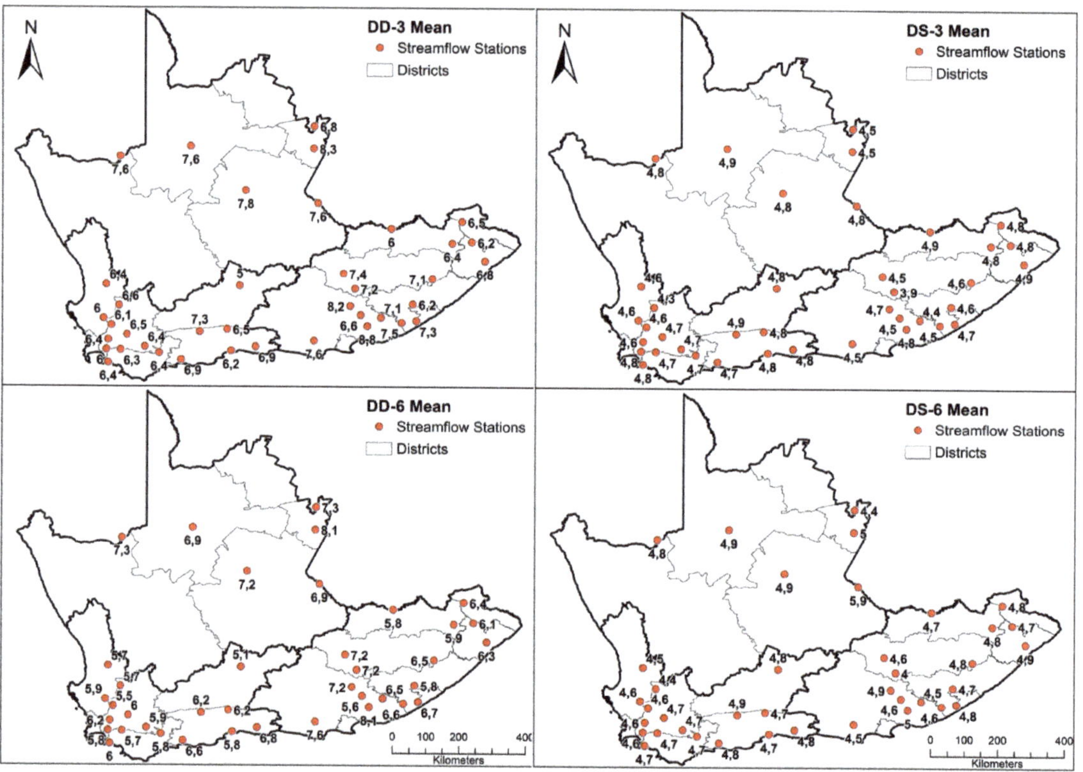

Figure 6. Annual mean of drought duration (in months) and severity based on the SSI-3 and SSI-6 analysis shown in left and right panels, respectively.

The trend analysis results indicate that DD has significantly increased across the study area, with exceptions to parts of the EC province (see the first column of Figure 7). Approximately 47% of the observed trends in DD are statistically significant at the 95% statistically significance level. The severity of drought (second column of Figure 7) has increased during the study period (1985–2020), particularly in the WC and NC provinces. Overall, about 78% of the stations exhibit positive trends in DS, with 48% of the observed trends being at the statistically significant 95% significance level. In the EC province, only 18% of the stations indicated an increase in the DS, while the majority of the stations (82%) displayed a decrease in the severity of drought over the 1985–2020 study period.

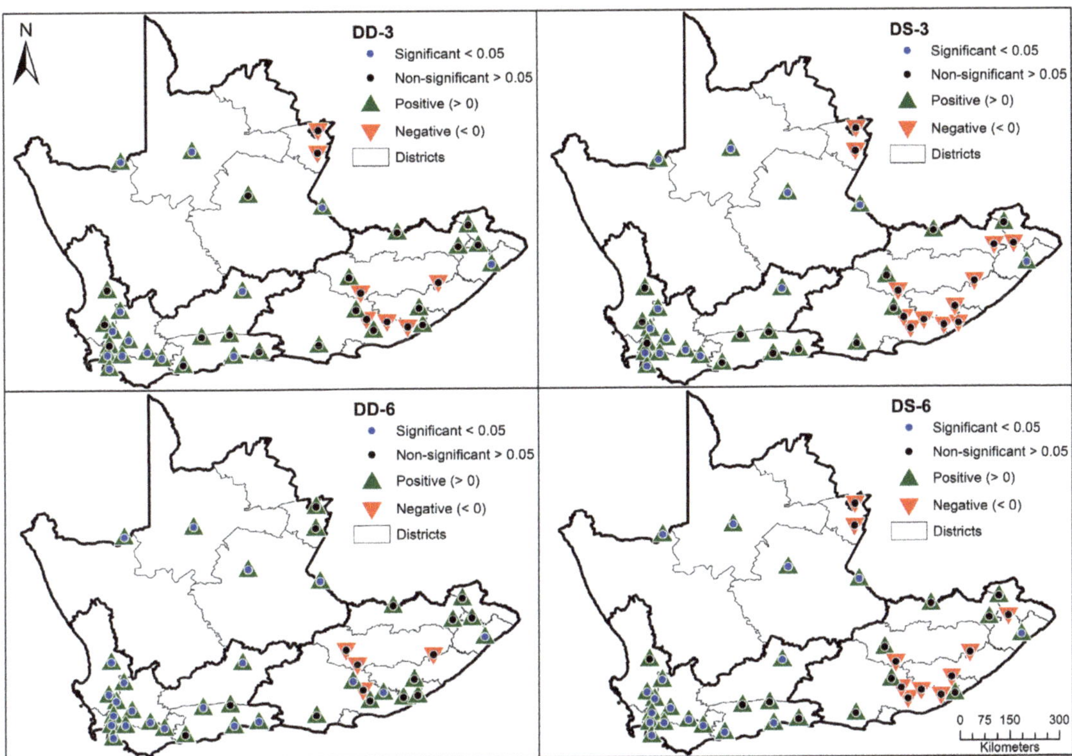

Figure 7. Trends in drought duration (**left**) and severity (**right**) at 3- and 6-month accumulation periods.

3.4. Return Levels of Drought Duration and Severity

Figure 8 depicts the spatial distribution of drought events estimated from the SSI at 3- and 6-month accumulation time-steps across the three Cape provinces. These events were estimated based on the condition that the SSI exhibited continued negative values for 2 months or more. The results of SSI at 3- and 6-month accumulation time-steps at six stations in the NC region indicate that there were 21–30 and 14–25 drought events during the period 1985–2020, respectively. The EC province experienced 24–36 and 14–30 drought episodes based on the analysis of SSI at 3- and 6-month accumulation periods, respectively. In the WC province, drought episodes ranged from the seven, which was the lowest, to 39, the highest, for the SSI-3 analysis. Drought episodes were reduced when considering the SSI-6 analysis, with the events ranging from five at the lowest to 27 at the highest. The J2H018 in the WC region is the solely stream-gauge station that has recorded the lowest drought events during the 3.5 decades. Such a huge deviation requires further inspection of streamflow data for this station. Overall, the results indicate that SSI-3 exhibited greater drought events as compared to the SSI-6 accumulation period. The drought events in Figure 8 were used to estimate the return levels of DD and DS at both 3- and 6-month accumulation time-steps.

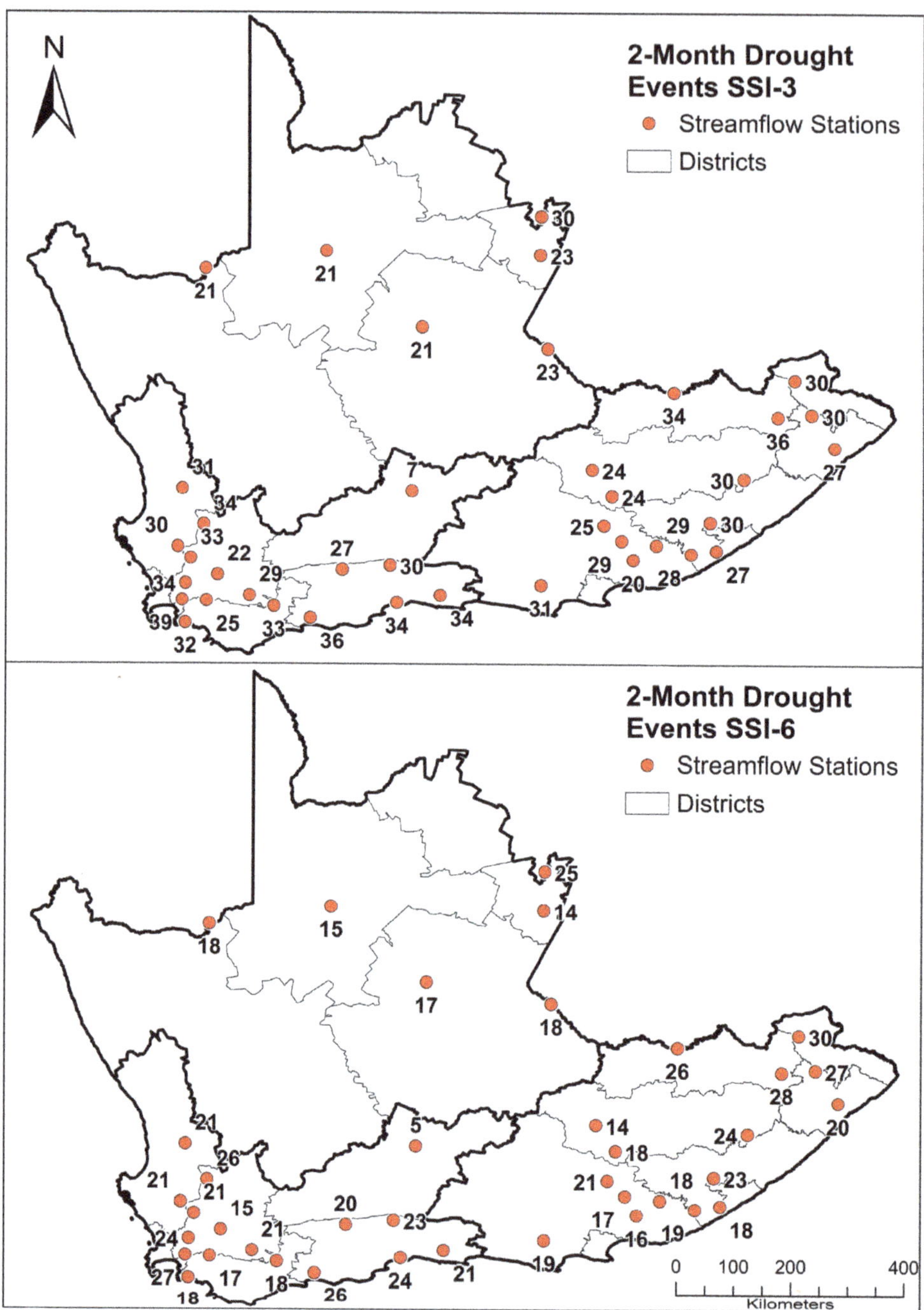

Figure 8. Drought episodes across the stations and study site. These events are defined when the standardized streamflow index at 3- (**top**) and 6-month (**bottom**) accumulation periods exhibited continuous negative values for 2 and more months.

Figure 9 depicts the spatial distribution of estimated return levels of DD associated with 2-, 5-, 10-, 20- and 50-year periods computed based on 3- (first column) and 6-month (second column) accumulation time-steps. In Figure 9, DD return levels for the NC province study site are shown in the first row, whereas the second and third rows represent the output for the WC and EC provinces, respectively. As noted in the results, the estimated return levels exhibit localized and spatial variability across the studied periods and accumulation time-steps. In both the DD at 3- and -6-month accumulation periods' analyses, the return levels increase in subsequent periods. In the NC province, hydrological drought is expected to occur with duration ranging between 5 and 8 and between 10 and 19 months across the province for the 2- and 5-year periods, respectively. Prolonged hydrological drought is expected to range from 28 to 66 and from 53 to 148 months in the next 20- and 50-year periods, respectively. Hydrological drought is likely to be short (2–5 months) in the WC during the next 2 years, followed by a rapid increase in subsequent periods, with duration in the range of 15–72 and 22–115 months for the 20- and 50-year periods, respectively. The J2H018 is the only stream-gauge station depicting extreme return levels across the periods, resulting in a slight deviation pattern from the rest of the stations. Similarly, hydrological drought is expected to last between 4 and 9 months in a shorter period and reach a maximum of 138 in 50 years' time. Results for the 6-month accumulation period depict a similar increasing pattern across the stations and the provinces. Moreover, the results indicate that the NC is likely to experience prolonged DD, followed by the EC province. The WC province is expected to experience short DD, as compared to the other Cape provinces. Overall, the results of the return levels suggest that the three Cape provinces are likely to experience persistent drought with localized duration. Moreover, the duration of drought is expected to increase in subsequent periods.

The results for the return levels of DS associated with the 2-, 5-, 10-, 20- and 50-year periods are shown in Figure 10 for 3- and 6-month accumulation periods. Similar to DD, the estimated return levels for DS increase with increasing periods. The 2-year return levels for DS range from the lowest of 1.4 to maximum severity of 5 in the WC to 1.2–4.6 in the EC and to 2.1–3.5 in the NC provinces. The severity doubles for the 5-year period across the stations and provinces. A significant increase in drought severity is recorded for the 20- and 50-year periods, reaching a maximum severity of 55.6, 55.0 and 48.8 for the 20-year period and in the NC, WC and EC provinces, respectively. The severity of hydrological drought is expected to double in magnitude in 50 years, as compared to the 20-year analysis period. A similar increasing pattern of DS return levels is observed for the 6-month accumulation time-step. While drought is expected to become severe in the future, the WC is likely to experience less severity. On the contrary, drought is likely to be more severe in the NC region, followed by the EC province.

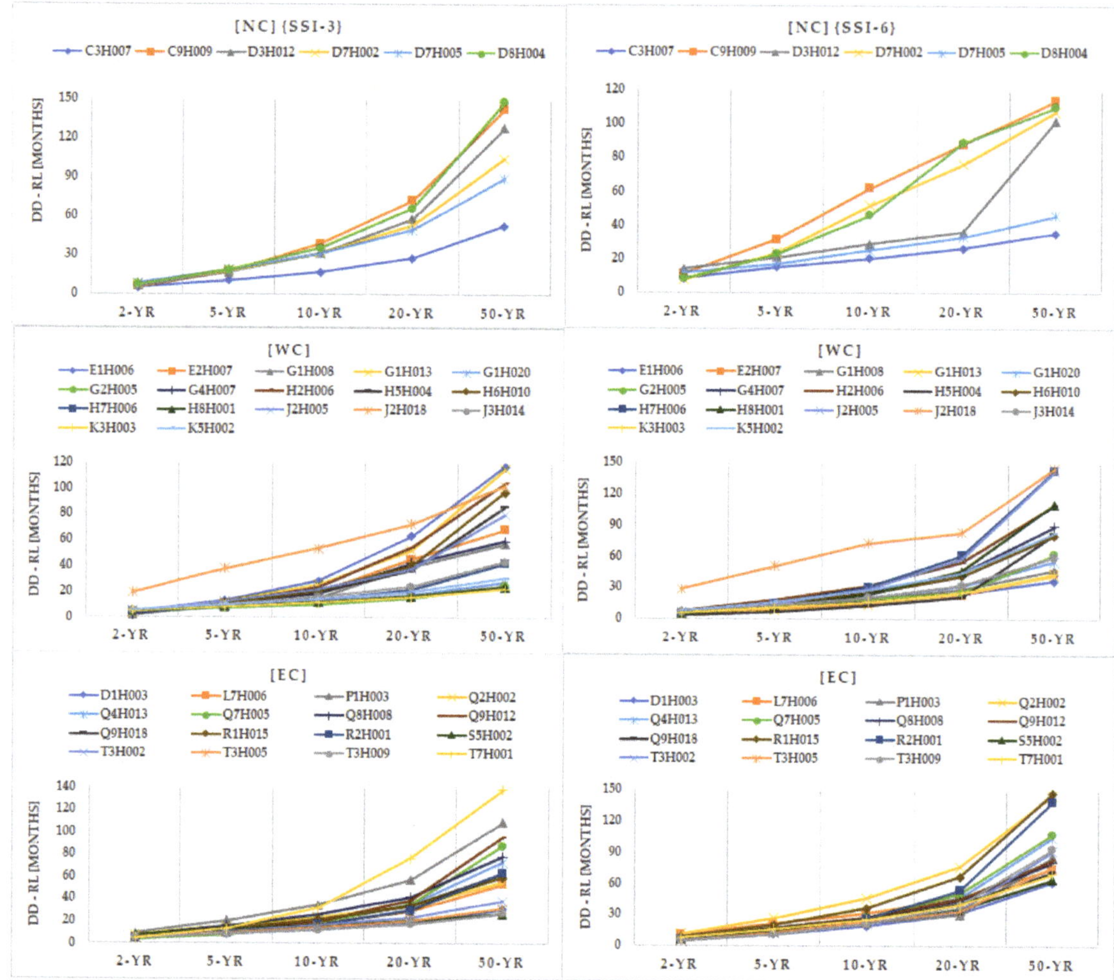

Figure 9. Return levels for drought duration for studied periods based on the SSI-3 and SSI-6 analysis, as depicted in the first and second columns, respectively.

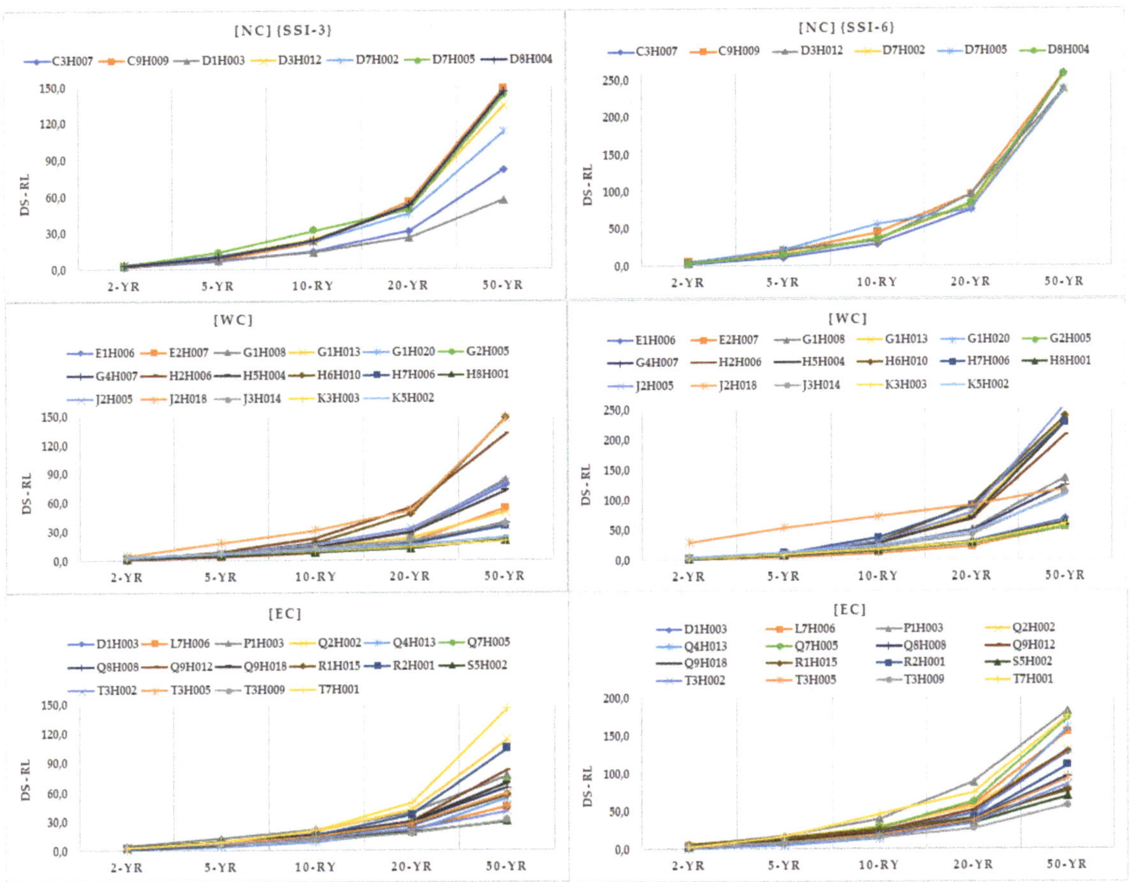

Figure 10. Return levels for drought severity for studied periods based on the SSI-3 and SSI-6 analysis, as depicted in the first and second columns, respectively.

4. Discussion

The present study investigated past hydrological drought conditions in the three Cape provinces of South Africa. Streamflow observations spanning 3.5 decades were used to compute SSI and its features at 3- and 6-month accumulation epochs. Based on the results, streamflow across most stations (~80%) exhibited a decreasing trend pattern. These findings collaborate with various research conducted in the WC and EC provinces. For instance, studies by Lakhraj-Govender and Grab [56] reported a decreasing trend in annual and winter rainfall and river flow in the WC province for the period spanning from 1987 to 2017. Similar studies focusing on rainfall characteristics have been reported by Botai et al. [10] in the WC, Botai et al. [22] in the EC and Mahlalela et al. [57] in the Southwestern Cape regions. The observed downward trend can be attributed to seasonality effects, manifested as reduced precipitation, coupled with increased temperature, as well as other influential factors, such as relative humidity and wind speed. These climatic variables, particularly the temperature, atmospheric humidity and wind speed, have moderating effects on other hydroclimatic variables, for instance, influencing potential evaporation [58,59]. Precipitation fundamentally affects water availability and supply within a region [59]. Consequently, it is clear that the change in these climatic variables has modified the surface water balance, contributing to the drying conditions in the three

Cape provinces in the past three and a half decades (1985–2020). Moreover, conditions in the EC province are further complicated by its climatic conditions, e.g., the western parts lie near the transition zone between the summer and winter rainfall regions towards the southeastern part of the country [60]. The EC regions are also influenced by mid-latitude and tropical systems, which lead to complex meteorology that involves interlinkages with the neighboring Agulhas current or landscape [61], contributing to a decrease in mean summer rainfall along the EC coast [62]. The streamflow in the WC province is also influenced by regional climate, including fluctuation in rainfall and evaporation rates (due temperature changes), as well as enhanced water withdrawal (and demand) for consumption [59].

Prolonged drought episodes are notable across the three provinces. These drought episodes include the 1991–1996 (EC), 1995–2000 (WC) and 2015–2020 across the three Cape provinces. These results agree with previously reported drought events that led to severe water crisis in the region [10,22,57,63]. Negative trends in SSI-3 and SSI-6 were detected in approximately 67% of the stream-gauge stations across the provinces. While the SSI values in the EC province depict positive trends in approximately 68% of the stations, water resources have remained excessively under pressure during the study period. This condition is still largely experienced currently, with provincial average dam levels in 2020 reaching the lowest of less than 50% full storage capacity. The ongoing drought in the EC province has left the most vulnerable communities with great uncertainty of facing the famously so-called "day zero" state [64], where plans, such as increasing water restrictions and the possibility of shutting off taps, are being considered. Most of the major dams that supply water in the region are currently below 50% [65]. Therefore, these suggest that, while some areas in the EC province might have received increased precipitation during the study period, the received amount could not have been significant to enhance ground-water recharge and compensate the growing demand for water resources.

The WC province has also experienced severe drought conditions that resulted in a decline in dam storage capacity. Significant reduction in water levels can be attributed to the 2015–2018 drought conditions, which were reportedly a manifestation of past droughts in the province [10]. During this period, dam levels reached the lowest of ~30% full storage capacity, particularly in 2017 and ~40% in 2018, posing threats of an impending "day zero" state. The current results suggest that the WC province is still experiencing acute drought conditions. Such conditions can be attributed to factors such as climate change, deficit/decrease in rainfall and lack of monitoring systems [66]. Persistent drought in the WC regions is likely to exacerbate water shortages, particularly in vulnerable urban and rural areas, affecting agricultural activities and other activities that highly depend on water resources. The NC province is a generally arid area, and, therefore, it is inherently prone to drought conditions which increase the vulnerability and coping capacity of economic sectors, such as agriculture and mining.

The duration and severity of drought conditions over the three provinces have increased during the 1985–2020 study period, as evidenced from the observed upward trends of DD and DS across the 3- and 6-month accumulation epochs. While the study area is prone to drought conditions, the duration and severity features vary spatially across the provinces. For instance, the southeastern regions covering the WC and the EC provinces experienced DD lasting for 5–7 months on average, whereas the NC, which exhibits a unique geographical location (e.g., mostly aridity) and climate conditions, experienced prolonged drought, reaching a maximum of ~9 months. Furthermore, drought has become more severe over time, reaching a maximum severity of 5.0 in most parts of the three Cape provinces. The observed DD and DS for the period 1985–2020 are expected to continue, according to results from drought return levels, associated with 2-, 5-, 10-, 20- and 50-year periods. Results indicate that DD and DS will slightly increase for 2-year return period, followed by a rapid increase in subsequent periods, particularly in the next 20 and 50 years. These increases will be localized and vary according to geographical location, as well as the climate conditions, as observed in the historical features.

Increased return levels of extreme events are likely to have significant impacts on key water-linked sectors, such as agriculture and tourism, as well as manufacturing and production industries, which sustain socioeconomic development in the three provinces. Such extreme events are expected to add more pressure to the already burdened water resources, thereby increasing water stress and impact on agricultural activities and their value chain (e.g., livestock production, wheat, fruit and wine production). According to Kalaba [67], the country's economy only grew by a mere 1.1% per annum between 2015 and 2017, a period when the country was gripped with one of the worst droughts. Such a growth rate is not sufficient to help tackle the triple challenges of poverty, unemployment and inequality and have implications on efforts to achieve national and global development goals (e.g., National Development Plan and SDG). Persistent drought conditions and the corresponding extreme events which are likely to reverberate into the future will probably exacerbate the triple challenges in the three Cape provinces. Additionally, the drought conditions would impact employment and especially the remuneration for low skilled and seasonal workers in sectors, such as agriculture and agro-processing, whereby productivity is likely to be affected by the availability of water for crop irrigation. In general, as evaluated in the current study, hydrological drought is a reality that requires concerted proactive actions by water agencies within the provinces, as well as the surrounding provinces that share catchments. Information derived from this study can support disaster management efforts at provincial and national levels to reduce spending on disaster relief, which often diverts money away from other fiscal activities. Such drought-risk information can also support the management of water resources in catchments at risk management and reduction, food price inflation, the food-import bill and economic growth in the future.

5. Conclusions

In this contribution, the SSI computed from 3.5 decades of streamflow data was used to evaluate hydrological drought conditions in the southwestern and southeastern parts of South Africa, covering the WC, NC and EC provinces. A decreasing streamflow trend was detected in most of the stream-gauge stations across the three Cape provinces. Hydrological drought analysis based on the SSI for 3- and 6-months accumulation periods indicate that most of the study regions experienced pronounced, yet localized drought conditions during the studied period. More specifically, negative trends in SSI-3 and SSI-6 were detected in approximately 67% of the stations across the provinces. Such a downward trend can be attributed to reduced streamflow, influenced by reduced precipitation or a shift in seasonal precipitation, coupled with increased temperature, among other factors. Based on the estimated return levels, increased DD is expected to occur within the study regions, with shortened and prolonged duration in the southeastern and southwestern areas, respectively. Severe drought events are also expected in the future, following a similar pattern as the estimated DD. In general, the historical trends analyses presented in this study form the basis for streamflow and drought projections in the three provinces. The present study and its findings are crucial for the implementation of appropriate policies and strategies for effective water-resource planning and management in the WC, EC and NP provinces. Based on the key findings of this study, it is recommended that water agencies and the government should be more proactive in searching for strategies to improve water resource management and drought mitigation in these regions.

Author Contributions: Conceptualization, C.M.B. and J.O.B.; methodology, C.M.B. and J.O.B.; software, J.O.B.; formal analysis, C.M.B. and J.O.B.; investigation, C.M.B. and J.O.B.; data curation, C.M.B. and J.P.d.W.; writing—original draft preparation, C.M.B., K.P.N., M.M., P.M.B. and A.M.A.; writing—review and editing, All Authors.; visualization, C.M.B., J.O.B. and J.P.d.W.; supervision, J.O.B. All authors have read and agreed to the published version of the manuscript.

Funding: This research was funded by the Water Research Commission of South Africa, grant number C2019/2020-00017.

Institutional Review Board Statement: Not applicable.

Water **2021**, 13, 3498

Informed Consent Statement: Not applicable.

Data Availability Statement: Not applicable.

Acknowledgments: The authors wish to thank the anonymous reviewers for providing constructive and detailed comments that assisted to improve the quality of the manuscript.

Conflicts of Interest: The authors declare no conflict of interest.

References

1. Bordi, I.; Frigio, S.; Parenti, P.; Speranza, A.; Sutera, A. The analysis of the Standardized Precipitation Index in the Mediterranean area: Large-scale patterns. *Ann. Geophys.* **2001**, *44*, 965–978. [CrossRef]
2. Vicente-Serrano, S. Spatial and temporal analysis of droughts in the Iberian Peninsula (1910–2000). *Hydrol. Sci. J.* **2006**, *51*, 83–97. [CrossRef]
3. Adhyani, N.L.; June, T.; Sopaheluwakan, A. Exposure to drought: Duration, severity and intensity (Java Bali and Nusa Tenggara). *IOP Conf. Ser. Earth Environ. Sci.* **2017**, *58*, 012040. [CrossRef]
4. Niang, I.; Ruppel, O.C.; Abdrabo, M.A.; Essel, A.; Lennard, C.; Padgham, J.; Urquhart, P. Africa. In *Climate Change 2014: Impacts, Adaptation, and Vulnerability. Part B: Regional Aspects. Contribution of Working Group II to the Fifth Assessment Report of the Intergovernmental Panel on Climate Change*; Barros, V.R., Field, C.B., Dokken, D.J., Mastrandrea, M.D., Mach, K.J., Bilir, T.E., Chatterjee, M., Ebi, K.L., Estrada, Y.O., Genova, R.C., et al., Eds.; Cambridge University Press: Cambridge, UK; New York, NY, USA, 2014; pp. 1199–1265.
5. Dai, A. Characteristics and trends in various forms of the Palmer Drought Severity Index during 1900–2008. *J. Geophys. Res. Atmos.* **2011**, *116*, D12115. [CrossRef]
6. Hornby, D.; Vanderhaeghen, Y.; Versfeld, D.; Ngubane, M. A harvest of dysfunction: Rethinking the approach to drought, its causes and impacts in South Africa. In *Report for Oxfam South Africa*; Oxfam South Africa: Johannesburg, South Africa, 2016.
7. Bureau for Food and Agriculture Policy. *Policy Brief on 2015/2016 Drought*; BFAP: Cape Town, South Africa, 2016.
8. Wilhite, D.A.; Glantz, M.H. Understanding the drought phenomenon: The role of definitions. *Water Int.* **1985**, *10*, 111–120. [CrossRef]
9. Yang, W. Drought Analysis under Climate Change by Application of Drought Indices and Copulas. Master's Thesis, Portland State University, Portland, OR, USA, 2010.
10. Botai, C.M.; Botai, J.O.; de Wit, J.P.; Ncongwane, K.P.; Adeola, A.M. Drought characteristics over the Western Cape Province, South Africa. *Water* **2017**, *9*, 876. [CrossRef]
11. McKee, T.B.; Doesken, N.J.; Kleist, J. The relationship of drought frequency and duration to time scales. In Proceedings of the 8th Conference on Applied Climatology, Anaheim, CA, USA, 17–22 January 1993; American Meteorological Society: Boston, MA, USA, 1993; pp. 179–183.
12. Vicente-Serrano, S.M.; Beguería, S.; Lopez-Moreno, I. A Multiscalar Drought Index Sensitive to GlobalWarming: The Standardized Precipitation Evapotranspiration Index. *J. Clim.* **2010**, *23*, 1696–1718. [CrossRef]
13. Palmer, W.C. *Meteorological Drought*; Weather Bureau Research Paper, No. 45; Department of Commerce: Washington, DC, USA, 1965; p. 58.
14. Byun, H.; Wilhite, D.A. Objective quantification of drought severity and duration. *J. Clim.* **1999**, *12*, 2747–2756. [CrossRef]
15. Shafer, B.; Dezman, L. Development of a SurfaceWater Supply Index (SWSI) to assess the severity of drought conditions in snowpack runoff areas. In Proceedings of the 50th Western Snow Conference, Reno, NV, USA, 19–23 April 1982; pp. 164–175.
16. Nalbantis, I.; Tsakiris, G. Assessment of hydrological drought revisited. *Water Resour. Manag.* **2009**, *23*, 881–897. [CrossRef]
17. Vicente-Serrano, S.M.; Lopez-Moreno, J.I.; Beguería, S.; Lorenzo-Lacruz, J.; Azorin-Molina, C.; Moran-Tejeda, E. Accurate computation of a streamflow drought index. *J. Hydrol. Eng.* **2012**, *17*, 318–332. [CrossRef]
18. Botai, C.M.; Botai, J.O.; Dlamini, L.C.; Zwane, N.S.; Phaduli, E. Characteristics of droughts in South Africa: A case study of the Free State and North West provinces. *Water* **2016**, *8*, 439. [CrossRef]
19. Adisa, O.M.; Botai, J.O.; Adeola, A.M.; Botai, C.M.; Hassen, A.; Darkey, D.; Tesfamariam, E.; Adisa, A.T.; Adisa, A.F. Analysis of drought conditions over major maize producing provinces of South Africa. *J. Agric. Meteorol.* **2019**, *75*, 173–182. [CrossRef]
20. Rouault, M.; Richard, Y. Intensity, and spatial extent of drought in southern Africa. *Geophys. Res. Lett.* **2005**, *32*, 1–4. [CrossRef]
21. Botai, J.O.; Botai, M.C.; de Wit, J.P.; Muthoni, M.; Adeola, A.M. Analysis of drought progression physiognomies in South Africa. *Water* **2019**, *11*, 299. [CrossRef]
22. Botai, C.M.; Botai, J.O.; Adeola, A.M.; de Wit, J.P.; Ncongwane, K.P.; Zwane, N.N. Drought risk analysis in the Eastern Cape Province of South Africa: The copula lens. *Water* **2020**, *12*, 1938. [CrossRef]
23. Botai, C.M.; Botai, J.O.; Zwane, N.N.; Hayombe, P.; Wamiti, E.K.; Makgoale, T.; Murambadoro, M.D.; Adeola, A.M.; Ncongwane, K.P.; de Wit, J.P.; et al. Hydroclimatic extremes in the Limpopo River Basin, South Africa, under changing climate. *Water* **2020**, *12*, 3299. [CrossRef]
24. Adeola, O.M.; Masinde, M.; Botai, J.O.; Adeola, A.M.; Botai, C.M. Ana analysis of precipitation extreme events based on the SPI and EDI values in the Free State Province, South Africa. *Water* **2021**, *13*, 3058. [CrossRef]
25. StatsSA. *Census 2011 Provincial Profile: Eastern Cape*; Report 03-01-71; Statistics South Africa: Pretoria, South Africa, 2011. Available online: http://www.statssa.gov.za/publications/Report-03-01-71/Report-03-01-712011.pdf (accessed on 29 October 2021).

26. Pocket Guide to South Africa. Provinces 2016–2017. Available online: https://www.gcis.gov.za/sites/default/files/pictures/provinces1617.pdf (accessed on 1 August 2021).

27. Zwane, E.M. Impact of climate change on primary agriculture, water sources and food security in Western Cape, South Africa. *Jamba J. Disaster Risk Stud.* **2019**, *11*, a562. [CrossRef]

28. Gasson, B. *The Biophysical Environment of the Western Cape Province in Relation to its Economy and Settlements*; School of Architecture and Planning, University of Cape Town for the Department of Local Government and Housing (Directorate of Development Promotion) of the Province of the Western Cape: Cape Town, South Africa, 1998.

29. Partridge, A.; Morokong, T. Western Cape Agricultural Sector Profile: 2018 Western Cape Department of Agriculture Muldersvlei Road Elsenburg 7607 South Africa. *Tech. Rep.* **2018**, 1–48. [CrossRef]

30. Department of Environmental Affairs: Republic of South Africa. *Draft–South Africa's 2nd National Communication under the United Nations Framework Convention on Climate Change*; Department of Environmental Affairs (DEA): Pretoria, South Africa, 2010.

31. Zengeni, R.; Kakembo, V.; Nkongolo, N. Historical rainfall variability in selected rainfall stations in Eastern Cape, South Africa. *S. Afr. Geogr. J.* **2016**, *98*, 118–137. [CrossRef]

32. StatsSA 2019. P0302-Mid-Year Population Estimates, 2020. 9 July 2020; Statistical Release. Available online: http://www.statssa.gov.za/?page_id=1854&PPN=P0302&SCH=72634 (accessed on 1 August 2021).

33. Jordaan, A.J.; Sakulski, D.; Jordaan, A.D. *Drought Risk Assessment for Northern Cape Province*; Unpublished Report; Department of Agriculture and Rural Development: Kimberley, South Africa, 2011.

34. Jordaan, A.J.; Sakulski, D.; Jordaan, A.D. Interdisciplinary drought risk assessment for agriculture: The case of communal farmers in the Northern Cape Province, South Africa. *S. Afr. J. Agric. Ext.* **2013**, *41*, 44–58.

35. Tallaksen, L.M.; van Lanen, H.A.J. *Hydrological Drought-Processes and Estimation Methods for Streamflow and Groundwater*; Elsevier: Amsterdam, The Netherlands, 2004.

36. Modarres, R. Streamflow drought time series forecasting. *Stoch. Environ. Res. Risk Assess.* **2007**, *21*, 223–233. [CrossRef]

37. Tijdeman, E.; Stahl, K.; Tallaksen, L.M. Drought characteristics derived based on the Standardized Streamflow Index: A large sample comparison for parametric and nonparametric methods. *Water Resour. Res.* **2020**, *56*, e2019WR026315. [CrossRef]

38. Shamshirband, S.; Hashemi, S.; Salimi, H.; Samadianfard, S.; Asadi, E.; Shadkani, S.; Kargar, K.; Mosavi, A.; Nabipour, N.; Chau, K.-W. Predicting Standardized Streamflow index for hydrological drought using machine learning models. *Eng. Appl. Comput. Fluid Mech.* **2020**, *14*, 339–350. [CrossRef]

39. Salimi, H.; Asadi, E.; Darbandi, S. Meteorological and hydrological drought monitoring using several drought indices. *Appl. Water Sci.* **2021**, *11*, 11. [CrossRef]

40. Karavitis, C.A.; Alexandris, S.; Tsesmelis, D.E.; Athanasopoulos, G. Applications of the Standardized Precipitation Index (SPI) in Greece. *Water* **2011**, *3*, 787–805. [CrossRef]

41. WMO. *Standardized Precipitation Index User Guide*; Svoboda, M., Hayes, M., Wood, D., Eds.; WMO-No. 1090; WMO: Geneva, Switzerland, 2012.

42. Lee, T.; Modarres, R.; Quarda, T.B.M.J. Data-based analysis of bivariate copula tail dependence for drought duration and severity. *Hydrol. Process.* **2013**, *27*, 1454–1463. [CrossRef]

43. Shiau, J.T.; Feng, S.; Nadaraiah, S. Assessment of hydrological droughts for the Yellow River, China, using copulas. *Hydrol. Process.* **2007**, *21*, 2157–2163. [CrossRef]

44. Tan, C.; Yang, J.; Li, M. Temporal-spatial variation of drought indicated by SPI and SPEI in Ningxia Hui Autonomous region, China. *Atmosphere* **2015**, *6*, 1399–1421. [CrossRef]

45. Hamed, K.H.; Rao, R. A modified Mann-Kendall trend test for autocorrelated data. *J. Hydrol.* **1998**, *204*, 182–196. [CrossRef]

46. Malik, A.; Kumar, A. Spatio-temporal trend analysis of rainfall using parametric and non-parametric tests: Case study in Uttarakhand, India. *Theor. Appl. Climatol.* **2020**, *140*, 183–207. [CrossRef]

47. Mann, H.B. Nonparametric tests against trend. *Econometrica* **1945**, *13*, 245–259. [CrossRef]

48. Kendall, M.G. *Rank Correlation Methods*; Charles Griffin: London, UK, 1975.

49. Gilbert, R.O. *Statistical Methods for Environmental Pollution Monitoring*; Wiley: Hoboken, NY, USA, 1987.

50. Wang, J.; Lin, H.; Huang, J.; Jiang, C.; Xie, Y.; Zhou, M. Variations of drought tendency, frequency, and characteristics and their responses to climate change under CMIP5 RCP scenarios in Huai River Basin, China. *Water* **2019**, *11*, 2174. [CrossRef]

51. Liang, L.; Li, L.; Liu, Q. Temporal variation of reference evapotranspiration during 1961–2005 in the Taoer River basin of Northeast China. *Agric. For. Meteorol.* **2010**, *150*, 298–306. [CrossRef]

52. Ahn, K.H.; Merwade, V. Quantifying the relative impact of climate and human activities on streamflow. *J. Hydrol.* **2014**, *515*, 257–266. [CrossRef]

53. Wang, X.; He, K.; Dong, Z. Effects of climate change and human activities on runoff in the Beichuan River Basin in the northeastern Tibetan Plateau, China. *Catena* **2019**, *176*, 81–93. [CrossRef]

54. Coles, S.G. *An Introduction to Statistical Modeling of Extreme Values*; Springer: London, UK, 2001; Volume 208.

55. Beirlant, J.; Goegebeur, Y.; Segers, J.; Teugels, J. *Statistics of Extremes: Theory and Applications*; Wiley Series in Probability and Statistics; Wiley: Chichester, UK, 2004.

56. Lakhraj-Governder, R.; Grab, S.W. Rainfall and river flow trends for the Western Cape Province, South Africa. *S. Afr. J. Sci.* **2019**, *115*, 1–6.

57. Mahlalela, P.; Blamey, R.C.; Reason, C.J.C. Mechanisms behind early winter rainfall variability in the southwestern Cape, South Africa. *Clim. Dyn.* **2019**, *53*, 21–39. [CrossRef]

58. Allen, R.G.; Pereira, L.S.; Raes, D.; Smith, M. *Crop Evapotranspiration: Guidelines for Computing Crop Water Requirements*; Food and Agriculture Organization of the United Nations: Rome, Italy, 1998.

59. Liu, M.; Xu, X.; Sun, A.Y.; Wang, K. Decreasing spatial variability of drought in southwest China during 1959–2013. *Int. J. Climatol.* **2017**, *37*, 4610–4619. [CrossRef]

60. Reason, C.J.C.; Rouault, M.; Melice, J.-L.; Jagadheesha, D. Interannual winter rainfall variability in SW Africa and large scale ocean-atmosphere interactions. *Meteorol. Atmos. Phys.* **2002**, *80*, 19–29. [CrossRef]

61. Singleton, A.T.; Reason, C.J.C. Numerical simulations of a severe rainfall event over the Eastern Cape coast of South Africa: Sensitivity to sea surface temperature and topography. *Tellus A* **2006**, *58*, 355–367. [CrossRef]

62. Jury, M.R.; Valentine, H.R.; Lutjeharms, J.R.E. Influence of the agulhas current on summer along the southeast coast of South Africa. *J. Appl. Meteorol.* **1993**, *32*, 1282–1287. [CrossRef]

63. Burls, N.J.; Blamey, R.C.; Cash, B.A.; Swenson, E.T.; al Fahad, A.; Bopape, M.-J.M.; Straus, D.M.; Reason, C.J.C. The Cape Town "Day Zero" drought and Hadley cell expansion. *NPJ Clim. Atmos. Sci.* **2019**, *2*, 27. [CrossRef]

64. Kouga Municipality. Defeat Day Zero in Kouga. 2021. Available online: https://www.kouga.gov.za/article/dam-levels-just-over-16 (accessed on 25 August 2021).

65. Department of Water and Sanitation Weekly Report. Available online: https://www.dws.gov.za/Hydrology/Weekly/Weekly.pdf (accessed on 8 November 2021).

66. Orimoloye, I.R.; Olalade, O.O.; Mazinyo, S.P.; Kalumba, A.M.; Ekundayo, O.Y.; Busayo, E.T.; Akinsanola, A.A.; Nel, W. Spatial assessment of drought severity in Cape Town area, South Africa. *Heliyon* **2019**, *5*, e02148. [CrossRef]

67. Kalaba, M. How will Droughts Affect South Africa's Broader Economy. 2019. Available online: https://theconversation.com/how-droughts-will-affect-south-africas-broader-economy-111378. (accessed on 25 August 2021).

 *water*

Article

Development of an Objective Low Flow Identification Method Using Breakpoint Analysis

Krzysztof Raczyński [1,*] and Jamie Dyer [1,2]

[1] Northern Gulf Institute, Mississippi State University, 2 Research Blvd, Starkville, MS 39759, USA; jamie.dyer@msstate.edu

[2] Department of Geosciences, Mississippi State University, 200A Hilbun Hall, Starkville, MS 39762, USA

* Correspondence: chrisr@ngi.msstate.edu

Abstract: Low flow events (a.k.a. streamflow drought) are described as episodes where stream flows are lower or equal to a specified minimum threshold level. This threshold is usually predefined at the methodological stage of a study and is generally applied as a chosen flow percentile, determined from a flow duration curve (FDC). Unfortunately, many available methods for choosing both the percentile and FDCs result in a large range of potential thresholds, which reduces the ability to statistically compare the results from the different methods while also losing the natural character of the phenomenon. The aim of this work is to introduce a new approach for low flow threshold calculation through the application of an objective approach using breakpoint analysis. This method allows for the identification of an environmental moment of river transition, from atmospheric feed flows to base flow, which characterizes the moment at the beginning of the hydrological drought. The method allows for not only the capture of the genesis of a low flow event but, above all, unifies the approach toward threshold levels and completely excludes the impact of the subjective researcher's decisions, which occur at the methodological stage when selecting the threshold criteria or when choosing a respective percentile. In addition, the method can be successfully used in datasets characterized by a high level of discretization, such as numerical model data, where the subsurface runoff component is not described in sufficient detail. Results of this work show that the objective identification method is better able to capture the occurrence of a low flow event, improving the ability to identify hydrologic drought conditions. The proposed method is published together with the Python module *objective_thresholds* for broad use in other studies.

Keywords: low flow; national water model; objective; threshold; breakpoint; low flow identification; streamflow drought

check for **updates**

Citation: Raczyński, K.; Dyer, J. Development of an Objective Low Flow Identification Method Using Breakpoint Analysis. *Water* **2022**, *14*, 2212. https://doi.org/10.3390/w14142212

Academic Editor: Alina Barbulescu

Received: 22 May 2022
Accepted: 11 July 2022
Published: 13 July 2022

Publisher's Note: MDPI stays neutral with regard to jurisdictional claims in published maps and institutional affiliations.

Copyright: © 2022 by the authors. Licensee MDPI, Basel, Switzerland. This article is an open access article distributed under the terms and conditions of the Creative Commons Attribution (CC BY) license (https://creativecommons.org/licenses/by/4.0/).

1. Introduction

Drought, as a long-term hydrologic phenomenon, is difficult to numerically define and parameterize. Especially in the case of surface waters, where it is difficult to define the moment of drought initiation as the conditions are generally a result of numerous surface and atmospheric factors [1]. The lowering of river water levels is usually due to prolonged dry conditions; therefore, river low flows are generally considered an indicator of hydrological drought progression [2]. This indicator is somewhat easier to parameterize than other types of droughts due to a direct relation with river levels; however, one still needs some kind of criterion to define when flows are considered "low". In response to this, Yevjevich [3] introduced the threshold level method (TLM), which was based on a threshold approach to the phenomenon. The TLM method has been widely accepted by the scientific community, and now, even in the current literature, low flow events (a.k.a., streamflow droughts) are defined as periods when river discharge is not higher than the defined threshold level [4,5]. Such a definition, however, contributed to large discrepancies in later studies [5–11].

Since the definition of low flows based on a predefined flow threshold was established, the methods for determining it have been extended to include statistical, hydrological, economic, and other criteria [12]. For the hydrological criteria, the value of the threshold has included: (1) lowest average annual flow, (2) highest annual minimum flow, (3) average annual minimum flow, (4) a division using one of the prior conditions using seasonal values or n-day annual minima, and others [4,12–15]. When applying statistical criteria, the most common approach is to use a specific flow percentile provided by the flow duration curve (FDC); however, the chosen value may differ depending on the study, with exceedance probabilities of $p = 70, 80, 90, 95\%$ having been used previously [2,10,12,16–20]. Such a range of criteria for selecting the threshold, which is a critical value in the context of the subsequent analysis of hydrological drought, results in a substantial heterogeneity of the resulting analyses and the inability to compile and compare the results [21]. This issue is compounded by the application of different low flow criteria for varying purposes (e.g., reservoir operations, water resource management, water quality, etc.). Seeing as defining and quantifying hydrologic drought is a common application of low flow analysis, developing an objective method for that purpose provides a useful tool that adds value to many associated scientific and management approaches.

Various levels of probability or graphical and other methods used to identify low flows have further contributed to the development of intermediate criteria. As strictly numerical values do not necessarily carry an environmental context, especially in complex hydrologic environments, additional criteria have begun to emerge that allow for connecting and separating periods of low flow that could have the same origin but were separated as a result of an external event (e.g., storm, reservoir, wastewater drop, etc.). Zelenhasić and Salvai [22] introduced the inter-event time criterion (IT) in which they introduced another parameter to the definition of a low flow: the maximum duration of flows that would not separate events [23]. In terms of IT, it is up to the researcher to determine the critical time (e.g., five days), which indicates the same genesis of successive low flows. Madsen and Rosbjerg [24] modified the criterion to a so-called inter-event time and volume-based criterion (IC) by supplementing the definition with the maximum volume of water supplied during the threshold exceedance time. This affected the method of determining the basic parameters of the low flow event, as accounting for the excess threshold time and associated volume to include within low flow episode parameters became a requirement. Vogel and Stedinger [25] approached this issue differently, considering the low flow in terms of hydrological drought and the related runoff deficiencies. They concluded that the appropriate criterion for the distribution of independent events should be based on the amount of water deficit created in the environment during the low flow and not on the basis of the duration of the threshold exceedance. Vogel and Stedinger [25] adapted the reservoir algorithm concept to the needs of low flow analysis, leading to the sequent peak algorithm (SPA) method.

The existence of various criteria for the division and parameterization of low flows in the literature still contributes to large methodological discrepancies. For the IT and IC criteria, depending on the study, authors use different criteria, e.g., 3-day periods [21], 5- or 6-day periods [9], or even one-month periods [26]. Furthermore, Yahiaoui et al. [27] recommended that the period should be selected each time depending on the needs of the analysis. In addition to the above criteria, the SPA algorithm has also seen wide application in the analysis of low flows [28,29]. Yet another criterion used by researchers with the same frequency is the method based on the value of the "smoothed" n-day moving average [9,30] or the minimum annual mean n-day flow [31].

The evolution of the low flow definition methods described above served to improve the definition of a low flow event to reflect the actual low flow conditions in a river, which is overall useful and encouraged. At the same time, however, the multitude of definition criteria, parameterizations, and assumptions introduces numerous combinations, and it is up to the researcher to choose a specific method appropriate to the research application. These choices have direct consequences on the impact and applicability of the results, as using a common criterion ensures the comparability of results while adjusting the criterion to the problem under study (e.g., temporal, spatial, or environmental dependence) makes the results potentially more locally representative but not directly comparable [32]. Although statistical criteria are currently the most popular, especially the Q_{10} flow (corresponding to the 10th percentile of the flow), also referred to as Q_{90} if the cumulative distribution function is used [33] or the $7Q_{10}$ criterion (as the 10th percentile of the 7-day average flows) [34], the development of modern methods of numerical modeling introduces another issue: discretization and uniqueness of data.

In the case of observed streamflow, the primary disadvantage is data stationarity [35] or the completeness of the dataset. More often, however, observations suffer from insufficient spatial density, mainly due to the costs of their acquisition and spatial coverage. For this reason, other sources such as statistical or physical model simulations become necessary. In the case of physical models, parameterizations related to complex and/or unknown variables, such as subsurface runoff, lead to differences in accuracy and the representativeness of related processes, which during droughts, are crucial for the correct calculation of the baseline runoff and low flows in the river [6,36,37]. This sometimes leads to a large discretization of the flow values, i.e., the lowest flows can have repeated values in a dataset. If the data uniqueness is greater than 90%, then the use of the Q_{10} as the threshold will represent statistical information; however, in some cases, for example with the National Water Model (NWM) retrospective data, minimum streamflow values are often repeated for extended periods [38]. In this case, the FDC flattens out on the lower flows, with the 10th percentile being equivalent to higher percentile values (i.e., 15th or 20th percentile). In this case, the use of percentiles as thresholds leads to the separation of values from the respective environmental information or even false statistical results if the threshold is equal to all of the other lowest flows in the data series. Such episodes will have zero volume based on the TLM definition, which further translates into issues in parameterizing low flow episodes.

All the above-mentioned problems lead to the same conclusion: there is a need to develop a criterion for the identification of low flows by means of TLM, which will allow for (1) the minimization (and ideally exclusion) of the subjective assessment of the researcher in the process of selecting threshold criteria, (2) the ability to account for the state of the environment in the process of identifying the onset of hydrological drought, and (3) the application of the methodology in a data series with a significant degree of value discretization. The aim of this work is to develop such a method without also introducing complex or computationally expensive criteria.

The remainder of the article is organized into the following Section 2: overview of the data used, Section 3: description of the method, Section 4: comparison of the newly introduced method to low flows defined using the Q_{10} method, and Section 5: conclusions. Additionally, Supplementary Materials are available, which include information about how to obtain, install, and use the developed algorithm (a Python module termed *objective_thresholds*), which will allow researchers to directly apply the method without the need to reconstruct the methodology.

2. Data

The streamflow data used to develop the low flow algorithm came from the National Water Model (NWM) retrospective v.2.1 dataset (NOAA National Water Center, version: Retrospective 2.1, Tuscaloosa, AL, USA), which has a period of record from 1979–2020 [39]. The NWM was developed by the National Oceanic and Atmospheric Administration (NOAA) Office of Water Prediction (OWP) in 2016 to improve the accuracy and spatial coverage of hydrological predictions over the continental United States (CONUS) and is based on the Weather Research and Forecasting–Hydrological Modeling System (WRF-Hydro). While the NWM has undergone numerous upgrades and revisions since its release, the most recent retrospective simulation was used for this project, which utilizes the analysis of record for calibration (AORC) for initialization. Unlike the operational version of the NWM, however, the retrospective data do not incorporate data assimilation.

The study area was limited to the Southeast US, defined by the USGS Region 3 hydrologic unit (South Atlantic-Gulf (SAG)), which constitutes 338,037 NWM retrospective stream nodes (hereafter referred to as nodes) along waterways that ultimately drain into the Atlantic Ocean within and between Virginia and Florida, or the Gulf of Mexico running within and between Florida and Louisiana (Figure 1). After the evaluation of data completeness, a decision was made to include only stations of Strahler order three and higher since 80% of the lower-order nodes showed streamflow values of 0 m^3/s, which, from the perspective of drought analysis, introduced the risk of a misrepresentative calculations for both threshold level and drought event statistics. The study dataset, therefore, consisted of 73,891 nodes, from which daily mean flow values were calculated from the hourly model values. Additional criteria of no more than 5% of zero or null data were introduced to avoid computational bias, which resulted in 61,948 nodes.

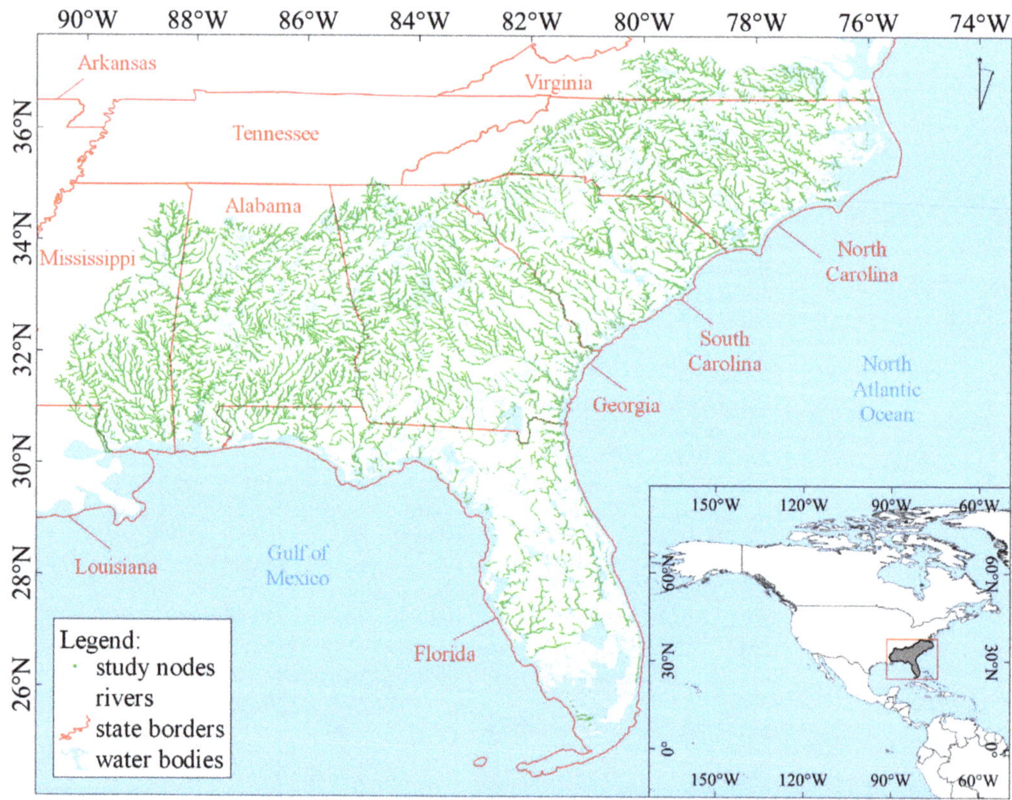

Figure 1. Study area showing NWM nodes used for analysis.

For the NWM nodes retained for analysis, analysis of FDCs revealed that the average percentage of unique flow values above the 90th percentile was 98%, while below the 10th percentile, it was roughly 50%. Additionally, the nodes below the 10th percentile were characterized by a low variability (Table 1, Figure 2). In these nodes, therefore, there is a situation where the lowest 10% (or even 20% in some cases) of the flow values are identical. This is likely because the model is not able to adequately reflect the influence of groundwater inflow on river discharge, especially along smaller river segments, causing predicted annual low flow thresholds to be the same values as annual or monthly minimums and/or overestimating baseflow characteristics [40].

Table 1. Characteristics of the lowest 10% of values from the study data. n—number of data points; n_u—ratio of unique values; Q_m—mean flow [m^3/s]; *std*—standard deviation [m^3/s]; *var*—variance; *Cv*—coefficient of variation; *IQR*—inter-quartile range [m^3/s].

	n	n_u	Q_m	*std*	*Cv*	*IQR*
mean	527.7	0.554	0.079	0.014	0.252	0.022
std	333.8	0.353	0.231	0.039	0.196	0.062
var	111438.8	0.125	0.053	0.002	0.038	0.004
Cv	0.633	0.637	2.917	2.825	0.779	2.831
IQR	665	0.703	0.032	0.007	0.133	0.011

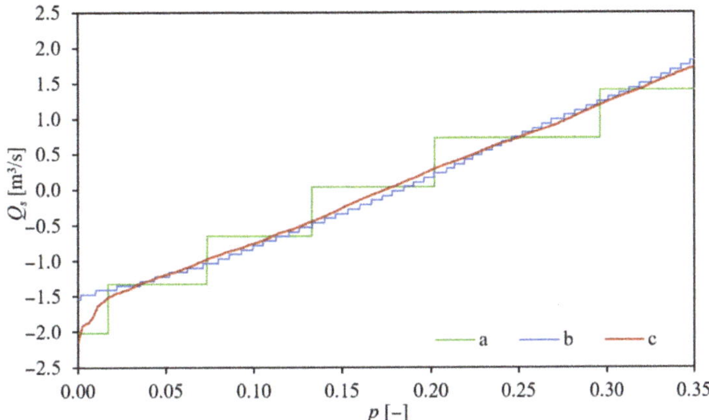

Figure 2. Example of standardized flow values (Q_s) for the lower range (to 35th percentile) of the FDC for a sample series with high discretization (**a**: unique values constitute less than 0.01% of the lower FDC range), moderate discretization (**b**: unique values constitute 1% of the lower FDC range), and close to natural distribution (**c**: unique values constitute around 90% of the lower FDC range).

After consideration, a decision was made to exclude 1198 nodes that were characterized by constant minimum flow values (not shown), leading to a final dataset consisting of daily flows for the period from 1 February 1979 to 31 December 2020 for 60,750 nodes (2,551,500 stream years).

3. Methods

3.1. Breakpoint Approach to Low Flow Identification

According to the generally accepted pattern of drought generation and evolution, hydrological drought is the last stage of environmental drought development [5]. The moment when primary river inflow changes from surface or shallow subsurface runoff (which characterizes mean conditions) to groundwater is an indicator of drought initiation; therefore, this transition can be identified as the beginning of low flow conditions. Based on this, then, an underlying assumption for the methodology presented here is that low flow begins at the moment when primary river inflow changes sources and baseflow conditions are reached [5,41–43]. To define this transition objectively, one can follow the so-called curve breakpoint method—where the breakpoint is identified as a change in the slope of a trend line within the hydrograph—as the most ecologically relevant moment [44,45]. This approach is often used in flood analysis, where the breakpoint is most often interpreted as water levels reaching bankfull conditions [46,47], but can also be used for drought studies if applied correctly. Considering the range of low flows in a series of sorted values, the breakpoint of the curve will be the moment of change in the river supply from surface runoff to groundwater [43,48]. Tomaszewski [48] proposed this approach by determining the minimum annual (or monthly) flows from a series of data. The moment of supply change is then defined as a point on the curve in which there is a significant change in the slope of the regression line, signifying a change in the series (Figure 3). This point can be used as the estimator of the threshold level for hydrological drought.

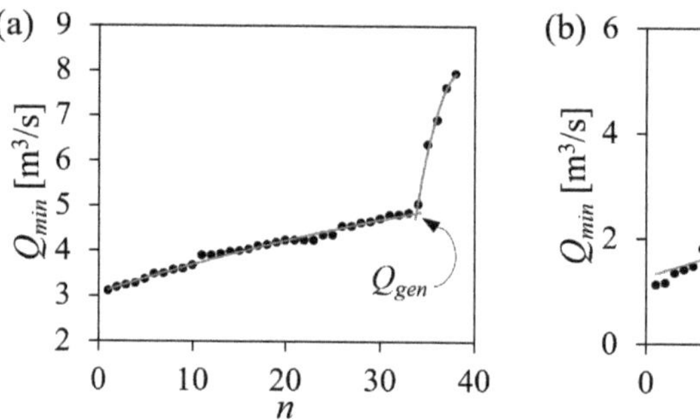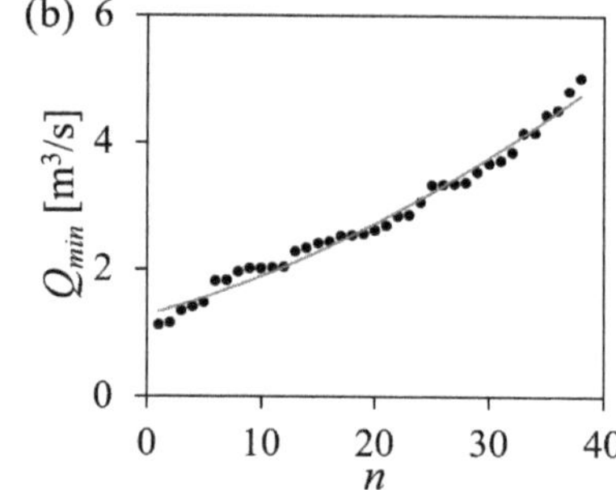

Figure 3. Trend change method for finding the threshold (Q_{gen}) based on Tomaszewski's method [48], on the example series of minimum annual flow values (Q_{min}) sorted increasingly (**a**) and the example series for which no clear trend change is present (**b**).

When examining various cases of the distribution of annual minima, one may encounter cases in which the change of the trend is inconclusive or absent (Figure 3) [49]. In this case, Tomaszewski [43] suggests adopting the highest annual minimum flow as the threshold. According to observations from an area in eastern Poland, the value of the highest annual minimum flow in catchments where the trend changes were not found coincides with the 30th percentile, while other studies indicated the 20th or 10th percentile [49,50]. This discrepancy may be related to the length of the data series being analyzed when using annual values, as the shorter the data series, the fewer the points from which to derive curves and find the trend change. This approach also fails in the case of data series characterized by a high degree of discretization, like in the case of NWM, as explained in Section 2.

While these issues can be addressed by incorporating the full flow time series in the calculation of the low flow breakpoint, which maximizes the number of data points while minimizing discretization, such an approach introduces the possibility of the breakpoint being defined at the upper end of the FDC in relation to other environmental factors such as over-bank flow conditions. As a result, it is important to truncate the full time series to only include the lower end of the FDC to focus the method on only low flow patterns. To that end, one can assume that low-intensity (a.k.a. shallow) low flows are those related to general environmental water shortage where researchers set limits of low flow around the 30th percentile (or highest of the lowest annual flows). In the case of high-intensity (a.k.a. deep) low flows, which are indicative of severe hydrologic drought, the threshold is usually around the 10th percentile (or the mean of the lowest annual flows or similar methods). This suggests that the expected optimal threshold may be somewhere from the 10th–30th flow percentiles, with any values below the 50th percentile reflective of dry conditions [8,51].

3.2. Breakpoint Algorithm Selection

To determine the breakpoint of the data series corresponding to each node in the study data, several of the most recognized breakpoint algorithms were analyzed, including:

- Fisher-Jenks algorithm (FJ) [52];
- Dynamic programming (DP) [53,54];

- Kernel change detection (KCD) [55,56];
- Binary segmentation (BiS) [57];
- Bottom-up segmentation (BUS) [54,58];
- Window sliding segmentation (WS) [54,59,60];
- K-means algorithm (KM) [61,62];
- Ward method (WA) [63,64].

Apart from the K-means algorithm, all methods were able to replicate the same obtained threshold result over 10 repetitions of the calculations. For KM, only 0.56% of the nodes achieved the same value over the 10 repetitions. This is due to the assumption of the method starting point as each time, the method randomly initiates the cluster centers and then progresses until the stabilization of the distance matrix occurs. Changing the method to start from the same space every time to unify the outputs would add an additional source of researcher intervention into defining the initial conditions, which stands in opposition with developing objective research guidelines.

The highest values of the low flow thresholds were obtained by the KM and KCD algorithms, while the lowest were obtained by the WS algorithm (Figure 4). Among the tested methods, the WS and KM algorithms were characterized by the largest deviations from the mean value, while the FJ and WA algorithms were characterized by the values closest to the mean of all the tested algorithms, with the mean difference not exceeding 0.2% (Figure 4). The execution time (per hundred nodes) of the DP, KCD, BiS, and BUS algorithms exceeded 10 s, whereas WS and KM required around eight seconds, and FJ only four seconds (Figure 4). Considering the above, the FJ algorithm was selected as the preferred breakpoint detection algorithm as it resulted in values closest to the mean of all tested algorithms and had short execution times.

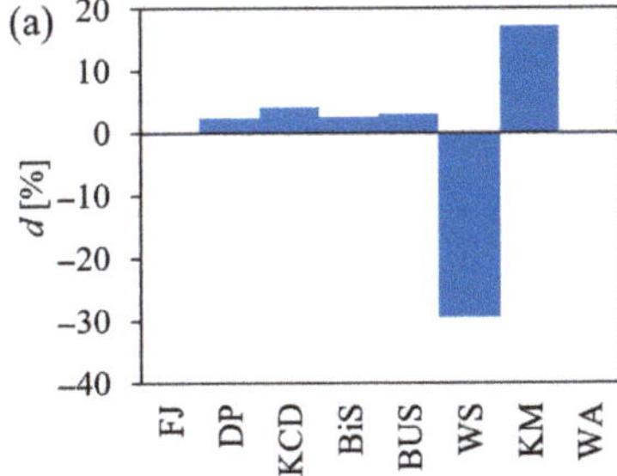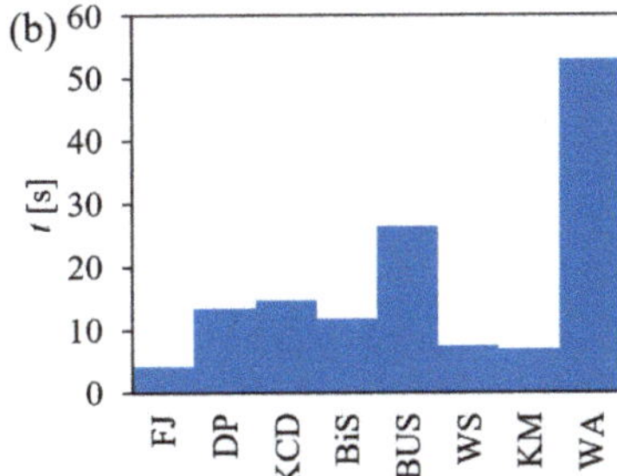

Figure 4. (**a**) Average difference (*d*) between threshold values returned by specific algorithms and the mean from all the methods, and (**b**) the mean time (*t*) of algorithm execution per 100 nodes.

In the absence of a discrete breakpoint in the dataset, represented by the lower range of the FDC, all the methods showed a tendency to select a specific point in the dataset. For example, the KM algorithm tended to choose the highest value as the breakpoint, while WS generally chose the lowest value. Four algorithms (FJ, DP, KCD and BiS) tended to choose a value close to the median of the distribution. To test if the results of the breakpoint algorithms were statistically different from the medians, a T-test was run using series representing the objective threshold based on the FJ algorithm, the median, and the percentile representing the middle probability value of a given range (e.g., for a FDC range of 30% the values of Q_{15} was considered the middle probability). Since all series were considered part of the same data distribution, a dependent T-test was used. The differences between the objective thresholds, median values, and middle-range flow percentiles were shown to be statistically significant, meaning the series did not result in the same values most of the time (Table 2). This shows that although the FJ method showed a tendency to define the breakpoint of the FDC as a value close to the median, the resulting Q_{obj} remained statistically significant in its difference from the median.

Table 2. Results of T-tests for the comparison of series representing objective thresholds defined using the FJ algorithm (Q_{obj}), median value, and middle percentile flow (Q_p) for multiple FDC ranges.

Relation	Q_{obj}–Median	Q_{obj}–Q_p	Median–Q_p
FDC range		20%	
statistics	−54.0508	−59.3941	4.6209
p-value	0.0000	0.0000	0.0000
FDC range		25%	
statistics	−51.2043	−58.6077	4.4995
p-value	0.0000	0.0000	0.0000
FDC range		30%	
statistics	−39.3186	−44.1184	4.0386
p-value	0.0000	0.0000	0.0001
FDC range		35%	
statistics	−22.8904	−25.4409	4.0794
p-value	0.0000	0.0000	0.0000
FDC range		40%	
statistics	8.5917	13.9009	3.4002
p-value	0.0000	0.0000	0.0007
FDC range		45%	
statistics	38.5965	41.6176	3.2249
p-value	0.0000	0.0000	0.0013
FDC range		50%	
statistics	49.8975	50.6962	3.1834
p-value	0.0000	0.0000	0.0015

3.3. Objective Threshold Approach Description

Based on the above considerations, a method was derived that takes into account the range of flows from the lower part of the FDC to estimate the breakpoint corresponding to the moment of change when river flow is based on predominantly atmospheric input (in the range of average flows) to groundwater input (or other non-atmospheric supplies, such as reservoir discharge), characteristic of lower flow ranges. The calculation method consists of the following steps:

1. Determination of the number (n) of points in the daily flow series needed to calculate the breakpoint based on the lower FDC range (by default: below the 35th percentile, as described further in the Results section):

$$Q: \{Q \in R + \mid Q \leq Q_{35}\}$$

2. Implementation of the Fisher–Jenks algorithm to define the breakpoint [52] by minimizing deviation of each class from the class mean, while maximizing the deviation of each class from the means of the other classes:

   a. Order flow data series in increasing order and assigning weights (w):

$$w: i \in \{1, \dots, n\}$$

   b. Compute the diameter matrix Di,j to store the distance between all pairs of n observations, such that:

$$1 \leq i \leq j \leq n$$

c. Populate the error matrix with variance of n observations when classified into two classes (one class for atmospheric driven resources, representing mean flow conditions (FDC part above breakpoint), and second for the drought conditions and baseflow (FDC part below breakpoint)):

$$E[P_{i,L}] = \min\left(D_{1,j-1} + E[P_{j-1,L-1}]\right)$$

d. Locate the optimal partition from the error matrix by maximizing inter-class variance and minimizing intra-class variance:

$$E[P_{n,2}] = E[P_{j-1,1}] + D_{j,n}$$

3. Application of the defined breakpoint (Q_t) as the low flow threshold for further analysis of low flow distribution, streamflow droughts, or for water management systems at the alert point, according to the following relation:

$$Q_{lf} = \begin{cases} 0, & if\ Q > Q_t \\ Q, & if\ Q \le Q_t \end{cases}$$

where: Q—flow in a specific moment,

Q_t—threshold flow determined by Fisher-Jenks algorithm,
Q_{lf}—flow identified as low flow.

The calculated flow breakpoint value can be directly applied as the threshold in the TLM for low flow or hydrological drought analysis.

The above method can be applied directly in other research by using the *lowflow* module from the *objective_thresholds* Python package. More information on the installation and usage of this can be found in Supplementary Materials, available with this article or at the package repository website.

4. Results

Analysis carried out on the study data shows that the 35th percentile of daily flows (indicated in Section 3.3) is sufficient enough to find the curve breakpoint, indicating a change in the river supply, as all of the low flow thresholds identified by the objective breakpoint method did not exceed the 30th percentile of flow, and in most cases (around 83%) the threshold fell within the 15th and 20th percentiles (Figure 5).

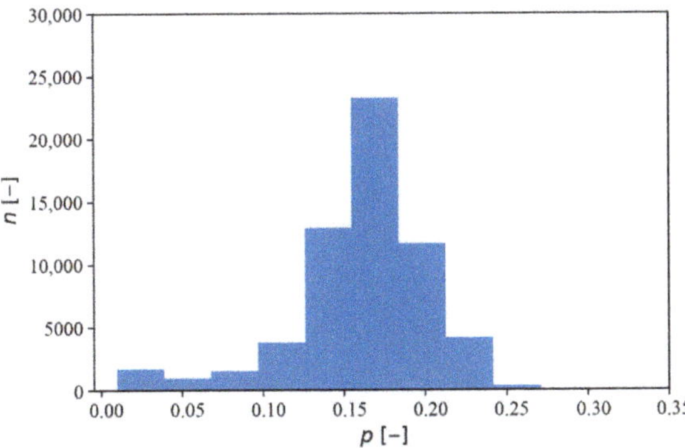

Figure 5. Threshold low flow percentiles determined by the objective breakpoint method.

The relation between Q_{10} and Q_{obj} can be represented by a linear relation (Figure 6), with R^2 values of around 0.998 for the study area rivers. This relationship reveals that, on average, Q_{obj} is 1.17 times higher than the Q_{10} threshold. Higher threshold values relate to an increase in low flow parameters; however, because the values fall in the 10th–30th percentile range, they remain in the range of "shallow" and "deep" streamflow drought as indicated in Section 3. Less than 1% of the cases in the tested data sample had threshold values lower than the associated Q_{10}, although in about 90% of the nodes, the increase did not exceed 100% of the Q_{10} threshold value (Figure 6). In a few cases, the ratio of Q_{obj} to Q_{10} exceeded three; however, these cases corresponded to situations when the threshold value determined by the Q_{10} was low (~0.01–0.03 m^3/s) due to the flattening of the FDC at the lower range (multiple repetitive values), while the threshold determined by the objective breakpoint method was around 0.05–0.10 m^3/s.

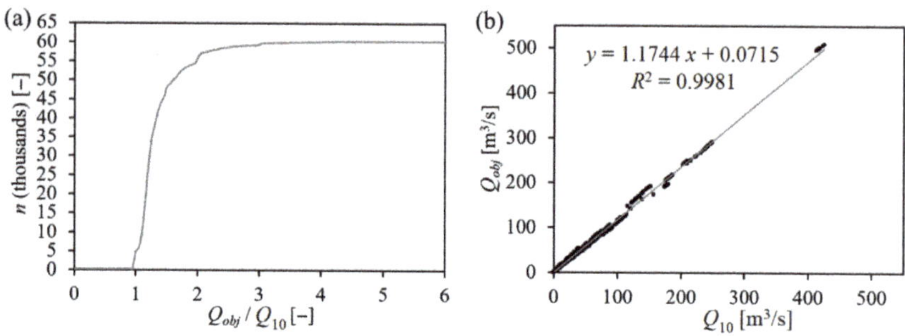

Figure 6. (**a**) Ratio of Q_{obj} to Q_{10} in sample and (**b**) linear relation between thresholds.

When comparing the average annual number of days with low flows, determined by the classical Q_{10} method and the objective breakpoint method, a different distribution of the density function occurred (Figure 7). Concerning the Q_{10} method, in most cases the duration of low flow events averaged around 30–35 days, with a low variance around this value (Table 2). For low flow duration based on Q_{obj}, the distribution has a higher mean and variance. In most cases, the average number of low flow days each year is about 60; however, due to a more normal distribution of values in the series, it is possible to better capture the specific environmental conditions occurring in each catchment area individually. Both distributions are left skewed, indicating that there are nodes with a lower number of days with low flow. For Q_{10}, the kurtosis of the distribution was 4.0, while for Q_{obj}, it was 2.6 (Table 3), implying a much more leptokurtic distribution for Q_{10} (as shown in Figure 7).

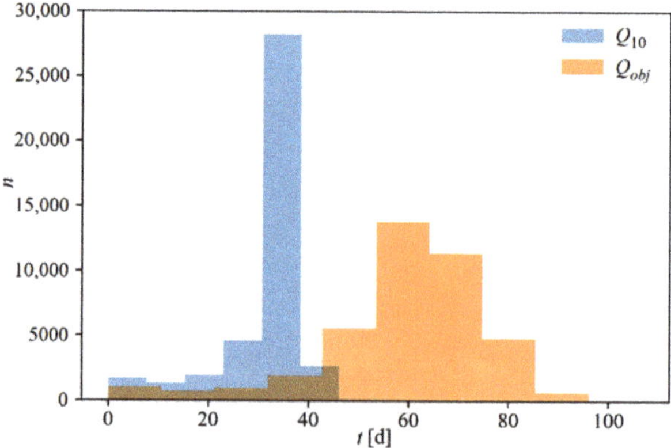

Figure 7. Annual number of days with low flow according to Q_{obj} and Q_{10} thresholds.

Table 3. Descriptive statistics for the distribution of the annual number of days with low flow for Q_{obj} and Q_{10} methods: μ—mean [m^3/s], m—median [m^3/s], σ—standard deviation [m^3/s], $\beta2$—kurtosis, S_{kp}—skewness, $n_{\sigma1,2,3}$—percent of values within one, two and three σ from μ.

	μ	m	σ	$\beta2$	S_{kp}	$n_{\sigma1}$	$n_{\sigma2}$	$n_{\sigma3}$
Q_{obj}	59.64	62.05	16.03	2.635	−1.315	76.76	94.49	97.45
Q_{10}	32.40	36.45	8.816	3.979	−2.012	85.62	92.74	96.49

Due to the general increase in the low flow threshold value, using the Q_{obj} method relative to Q_{10}, the basic parameters of low flow (e.g., number of events, duration, and volume) change accordingly. In the case of the number of low flow events determined using Q_{10}, for each of the analyzed nodes, about 50–100 low flow events were observed during the 42-year study period (upper range was 175). These values increased substantially when using Q_{obj}, where both the mean and the median increased by about 50 (Figure 8). The maximal number of episodes increased from 200 for Q_{10} to 300 for Q_{obj}, which translates to an average of 8.4 days per episode for Q_{obj} and 6.3 days per episode for Q_{10} per year. The low flows identified by the objective method are longer, which allows for the inclusion of periods occurring in streamflow, even when additional criteria, of a minimal time of 7 days, are applied [12,38].

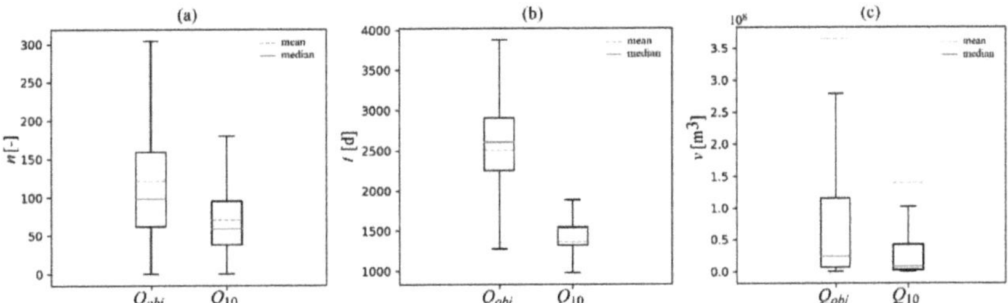

Figure 8. Box and whisker plots of the number of low flow events. (**a**): Total duration of low flows and (**b**) total volume of low flows (**c**) over the study period; graphs do not include outliers.

The duration of the low flows varied slightly more than the number of events. In the case of Q_{10}, in most cases the low flows did not last more than 2000 days in total over the study period. However, the mean and median values are close to the lower and upper IQR limits, respectively (Figure 8), which indicates that, while outliers shift the median towards the upper limits, the considerable number of low values (around 1000 days) shifts the mean to the lower limit. This introduces inconsistency to the spatial distribution of low flows (Figure 8). When using Q_{obj}, the range of values is higher, but this corresponds to the percentile range indicated earlier (convergence in relation to the 15–20th percentile) such that as the percentile value is doubled, the duration of low flows is doubled. The mean and median of the distribution are closer to each other and oscillate within the center of the IQR, as in the cases of normally distributed series (Figure 8).

Although the median low flow volume for both methods oscillate around a similar value, it is relatively low. This is due to the large share of low-order rivers in the studied dataset, in which no considerable outflow deficiencies developed. Nodes of Strahler order three and four constitute around 73% of the total nodes, which affects the shifting of the median volume of low flows to a lower range (Figure 8). However, when analyzing the distribution of the values, the wider distribution of the volume observed with Q_{obj} better captures the diversity of environmental conditions leading to the formation of outflow deficiencies with varying intensity.

Due to the way in which the definition of the TLM method is constructed, where the low flow is a period with a flow equal to or lower than the adopted threshold, the phenomenon of a zero-volume low flow event might occur. This problem is mostly associated with model data, where a small number of unique values in the lower FDC range are present due to high discretization. This leads to the inclusion of days that meet the mathematical criteria for low flow, but due to the occurrence of the same value in the associated outflow hydrograph, the threshold value corresponds to a volume of zero.

In terms of spatial relationships, the analysis was conducted based on Strahler stream order division. Within the study dataset, the highest river order is eight, which included two rivers: the Mobile River and the Apalachicola River. With Q_{obj}, there is a clear distinction between low flow volumes between these two rivers, while Q_{10} shows similar volume ranges for both rivers (Figure 9). A similar pattern exists for the distribution of total low flow duration time, where the Mobile River has shorter durations, and the Apalachicola River has longer durations distributed along the reach. For Q_{10}, the duration of low flows is similar among the two rivers, albeit with some outliers showing no distinct spatial pattern along the reach. The distribution of low flow volumes in rivers of order seven is similar for both methods, with four rivers having higher volumes when using Q_{10} (Figure 9); however, the length of low flows is different with Q_{10}, resulting in no spatial differentiation (with some outliers), while Q_{obj} varies spatially. In general, most rivers have longer total low flow durations in their upper reaches that decrease downstream, which reflects the natural tendency of smaller tributaries to have a faster response in river levels to environmental events that drive streamflow. This pattern becomes more pronounced at lower Strahler stream orders, where the biggest differences are noted in the spatial distribution of low flow. Along these river reaches, the highest low flow volumes and times occur within the eastern part of the study area in North and South Carolina, as well as central parts of Georgia and Alabama. This relation is, however, not reflected in the Q_{10} method, where the spatial distribution of low flow volumes and times is relatively equal throughout the study area.

The above observations are highlighted when considering the relation between the duration of low flows and their volumes (Figure 10). In general, Q_{obj} results in a wider spread of values relative to Q_{10}, representing a greater difference in environmental conditions. This means that either the change in duration times does not affect the volumes, or changes in the volumes are not reflected in the changes in duration. Additionally, Q_{obj} results in a lower number of nodes with volumes close to 0. It is worth mentioning that for nodes with higher Strahler stream orders (e.g., seven and eight), the relationship changes between the two thresholds. For Q_{obj}, the volumes are usually close when there are small changes in duration time, while for Q_{10}, the durations are close when volumes are prone to change. This is a direct result of the statistical character of Q_{10} and the consequences defined earlier. The strength of the relationship between low flow time and volume depends on the stream order; however, when considering the mean correlation values, they are higher for the objective method by approximately 0.22 (Q_{obj} $\bar{r} = 0.57$ and Q_{10} $\bar{r} = 0.35$).

Q_{10} is unable to accurately represent spatial relations and differences, and due to its statistical nature, results in constant, undifferentiated low flow patterns across the study area (with some randomly occurring outliers). At the same time, Q_{obj} is able to distinguish spatially varying river characteristics, such that low flows identified by this threshold vary spatially and along the course of individual rivers. Q_{obj} allows for the accurate capture of the natural character of events like streamflow droughts and introduces the environmental aspects to the analysis, taking into account the specificity of a given river in the studied node. As the objective threshold (Q_{obj}) fell within the 10th–30th percentile range for all nodes used in this study, it is important to investigate the relationship between not only Q_{obj} and Q_{10} but also between Q_{obj} and Q_{30} to better understand the pattern of the objective threshold values relative to the static statistical criteria. As shown in Figure 11, while the correlations between Q_{obj} and both Q_{10} and Q_{30} have a strong linear relationship, the slope of the resulting regression lines shows opposite values relative to the 1:1 trend line. In other words, for Q_{obj} compared to Q_{30}, instead of exceeding the objective threshold value for a given percentile, there is a decrease in value relative to the percentile. This is expected as statistical thresholds inherently maintain a constant frequency of events and always result in the same part of the dataset considered as an event (for Q_{10} this will be 10% of data and for Q_{30}, 30% of data, regardless of the environmental aspects of the river).

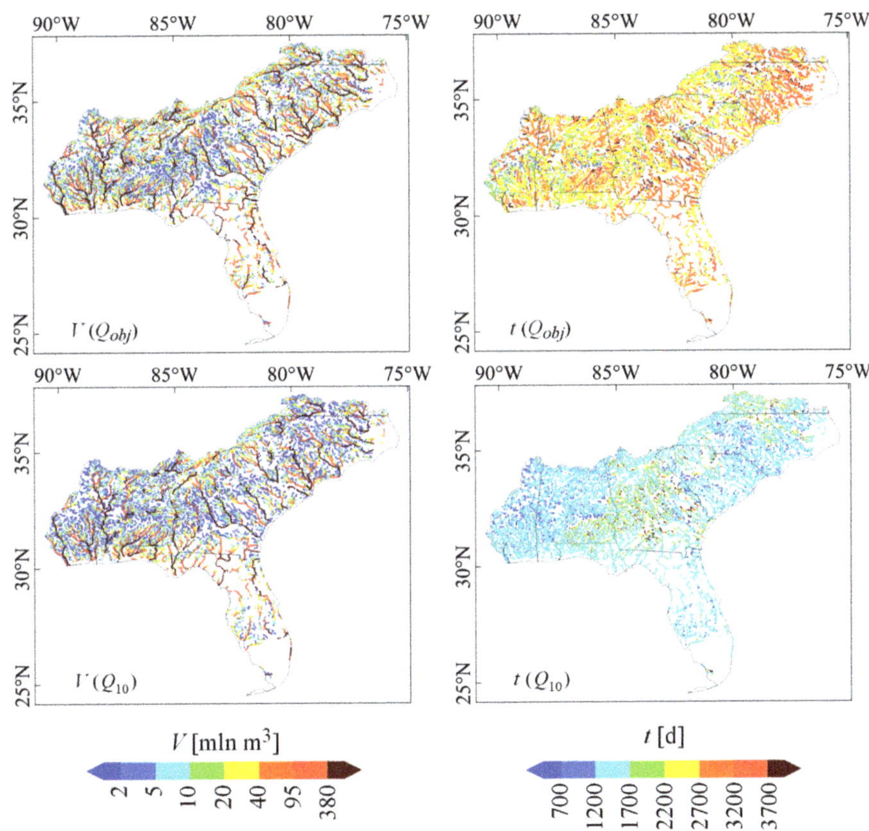

Figure 9. Spatial variability of volumes (*V*) and duration time (*t*) of low flows at each node, according to stream order and the threshold method used.

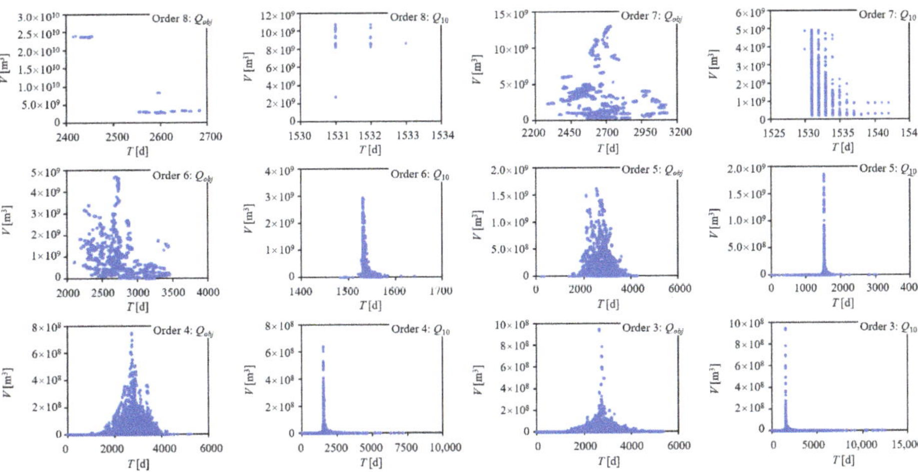

Figure 10. Duration time (*T*) with volume (*V*) for low flow relations in the studied rivers, according to their order and the threshold method used.

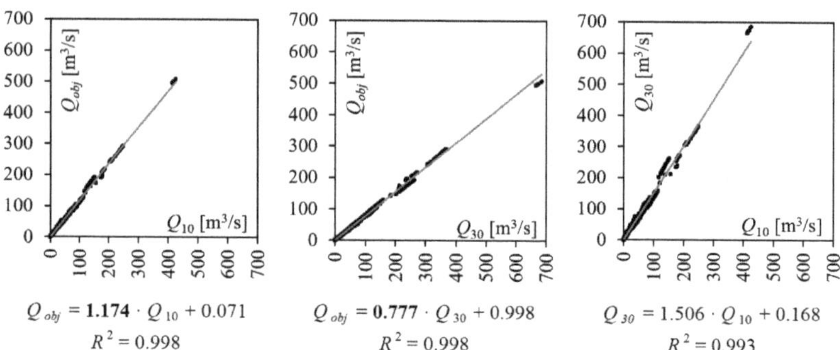

$$Q_{obj} = \mathbf{1.174} \cdot Q_{10} + 0.071 \qquad Q_{obj} = \mathbf{0.777} \cdot Q_{30} + 0.998 \qquad Q_{30} = 1.506 \cdot Q_{10} + 0.168$$
$$R^2 = 0.998 \qquad\qquad R^2 = 0.998 \qquad\qquad R^2 = 0.993$$

Figure 11. Linear relations between the thresholds defined by Q_{10}, Q_{30}, and Q_{obj}.

5. Conclusions

Augmenting available hydrological data with numerical model data provides additional information about the state of local-scale environments, and supplements the spatial deficiencies resulting from the limitations of existing river observation networks. However, the inclusion of model data contributes to issues with data quality and interpretation, which are related to the quality of the models themselves, their complexity, and the level of discretization of the resulting data. The computational capabilities of modern computers allow for the use of more advanced and effective computational methods in research; therefore, it is worthwhile to define those methods that can be used in turn to define and describe hydrological conditions, irrespective of the source of the data, especially for the analysis of extreme events such as floods or droughts. For the latter, definitions have historically been based on methods introduced decades ago, and although effective, there are distinct limitations related to the subjective decisions of researchers about the threshold level of low flows or the use of statistical criteria in defining a low flow (e.g., Q_{10}, Q_{30}, $7Q_{10}$, etc.). This article presents a new way of defining the low flow threshold based on an objective approach, utilizing a breakpoint method derived from a given streamflow time series, which is more representative of environmental criteria.

The introduced method is based on the use of the Fisher–Jenks algorithm to find the breakpoint of a curve constructed from 35% of the lowest flow values, which corresponds to the lower FDC range. The resulting breakpoint corresponds to the moment of change of the way a river input is derived from primarily atmospheric (representative of normal conditions) to groundwater sources (representative of drought conditions). The use of the objective breakpoint approach allows for the inclusion of these inherent environmental conditions into the TLM method, which then excludes subjective researcher decisions regarding the low flow threshold value or percentile. This allows for a more robust, data-driven approach to defining low flow thresholds that can be applied to both observed and simulated hydrologic time series.

The comparison between Q_{obj} and the widely used Q_{10} threshold reveals that Q_{10} is unable to differentiate spatial patterns, resulting in a similar range of defined low flow events, with skewed, widely spread distributions of low flow parameters. Based on the same data, Q_{obj} is able to better capture the natural characteristics of rivers, allowing for spatial recognition of the drivers responsible for streamflow drought occurrence. The objective threshold approach outperforms set statistical criteria (e.g., 10th percentile) in terms of spatial pattern recognition by introducing environmental factors into the threshold definition. Additionally, low flow parameters such as duration and volume are closer to a normal distribution when defined using Q_{obj}, with fewer outliers and volumes close to zero. The correlation between low-flow duration and volume depends on the stream order. On average, stream order to T and V correlation is higher by 0.22 for Q_{obj}, compared to Q_{10}.

The computational methodology presented in this article can be applied directly to other research by importing a Python module called *objective_thresholds*. Details on how to install and use the module are available in the Supplementary Materials available with this article and in the module's documentation in the Python library and its repository.

Supplementary Materials: The following supporting information can be downloaded at: https://www.mdpi.com/article/10.3390/w14142212/s1, documentation, installation, and usage notes of *objective_thresholds* Python package.

Author Contributions: Conceptualization, K.R. and J.D.; methodology, K.R.; software, K.R.; validation, K.R. and J.D.; formal analysis, K.R.; investigation, K.R.; resources, J.D.; data curation, K.R. and J.D.; writing—original draft preparation, K.R. and J.D.; writing—review and editing, K.R. and J.D.; visualization, K.R.; supervision, J.D.; project administration, J.D.; funding acquisition, J.D. All authors have read and agreed to the published version of the manuscript.

Funding: This research was funded by the National Oceanic and Atmospheric Administration (NOAA), grant number NA19OAR4590411.

Institutional Review Board Statement: Not applicable.

Informed Consent Statement: Not applicable.

Data Availability Statement: [Python package] Krzysztof Raczyński, Jamie Dyer. 2022. objective_thresholds; https://github.com/chrisrac/objective_thresholds (accessed on 15 May 2022) and/or: https://pypi.org/project/objective-thresholds/ Version 0.0.1 (accessed on 15 May 2022).

Conflicts of Interest: The authors declare no conflict of interest. The funders had no role in the design of the study; in the collection, analyses, or interpretation of data; in the writing of the manuscript, or in the decision to publish the results.

References

1. Van Loon, A.F.; Gleeson, T.; Clark, J.; Van Dijk, A.I.J.M.; Stahl, K.; Hannaford, J.; Di Baldassarre, G.; Teuling, A.J.; Tallaksen, L.M.; Uijlenhoet, R.; et al. Drought in the Anthropocene. *Nat. Geosci.* **2016**, *9*, 89–91. [CrossRef]
2. Tokarczyk, T. Classification of low flow and hydrological drought for a river basin. *Acta Geophys.* **2013**, *61*, 404–421. [CrossRef]
3. Yevjevich, V. An Objective Approach to Definitions and Investigations of Continental Hydrologic Droughts. Colorado State University. 1967. Available online: https://mountainscholar.org/bitstream/handle/10217/61303/HydrologyPapers_n23.pdf (accessed on 15 May 2022).
4. Stahl, K.; Vidal, J.-P.; Hannaford, J.; Tijdeman, E.; Laaha, G.; Gauster, T.; Tallaksen, L.M. The challenges of hydrological drought definition, quantification and communication: An interdisciplinary perspective. *Proc. Int. Assoc. Hydrol. Sci.* **2020**, *383*, 291–295. [CrossRef]
5. Van Loon, A.F. Hydrological drought explained. *WIREs Water* **2015**, *2*, 359–392. [CrossRef]
6. Van Lanen, H.A.J.; Wanders, N.; Tallaksen, L.M.; Van Loon, A.F. Hydrological drought across the world: Impact of climate and physical catchment structure. *Hydrol. Earth Syst. Sci.* **2013**, *17*, 1715–1732. [CrossRef]
7. Van Loon, A.F.; Van Lanen, H.A.J. A process-based typology of hydrological drought. *Hydrol. Earth Syst. Sci.* **2012**, *16*, 1915–1946. [CrossRef]
8. Wong, W.K.; Beldring, S.; Engen-Skaugen, T.; Haddeland, I.; Hisdal, H. Climate Change Effects on Spatiotemporal Patterns of Hydroclimatological Summer Droughts in Norway. *J. Hydrometeorol.* **2011**, *12*, 1205–1220. [CrossRef]
9. Fleig, A.K.; Tallaksen, L.M.; Hisdal, H.; Demuth, S. A global evaluation of streamflow drought characteristics. *Hydrol. Earth Syst. Sci.* **2006**, *18*, 535–552. [CrossRef]
10. Hisdal, H.; Stahl, K.; Tallaksen, L.M.; Demuth, S. Have streamflow droughts in Europe become more severe or frequent? *Int. J. Climatol.* **2001**, *21*, 317–333. [CrossRef]
11. Hisdal, H. Hydrological drought characteristics. *Dev. Water Sci.* **2004**, *48*, 139–198.
12. Smakhtin, V.U. Low flow hydrology: A review. *J. Hydrol.* **2001**, *240*, 147–186. [CrossRef]
13. Grosser, P.F.; Schmalz, B. Low Flow and Drought in a German Low Mountain Range Basin. *Water* **2021**, *13*, 316. [CrossRef]
14. Cammalleri, C.; Vogt, J.; Salamon, P. Development of an operational low-flow index for hydrological drought monitoring over Europe. *Hydrol. Sci. J.* **2016**, *62*, 346–358. [CrossRef]
15. Holmes, S. *Overwiev of Drought and Hydrologic Conditions in the United States and Southern Canada: Water Years 1986–1990*; US Department of The Interior, US Geological Survey: Washington, DC, USA, 1992.
16. Sutanto, S.J.; Van Lanen, H.A.J. Streamflow drought: Implication of drought definitions and its application for drought forecasting. *Hydrol. Earth Syst. Sci.* **2021**, *25*, 3991–4023. [CrossRef]
17. Floriancic, M.G.; van Meerveld, I.; Smoorenburg, M.; Margreth, M.; Naef, F.; Kirchner, J.W.; Molnar, P. Spatio-temporal variability in contributions to low flows in the high Alpine Poschiavino catchment. *Hydrol. Process.* **2018**, *32*, 3938–3953. [CrossRef]

18. Van Loon, A.F.; Van Huijgevoort, M.H.J.; Van Lanen, H.A.J. Evaluation of drought propagation in an ensemble mean of large-scale hydrological models. *Hydrol. Earth Syst. Sci.* **2012**, *16*, 4057–4078. [CrossRef]
19. Gustard, A.; Demuth, S. Manual of Low-flow. Estimation and Prediction. *Oper. Hydrol. Rap.* **2008**, *50*, 138.
20. Hisdal, H.; Tallaksen, L.M. Drought Event Definition. *Tech. Rep.* **2000**, *6*, 45.
21. Sung, J.H.; Chung, E.-S. Development of streamflow drought severity–duration–frequency curves using the threshold level method. *Hydrol. Earth Syst. Sci.* **2014**, *18*, 3341–3351. [CrossRef]
22. Zelenhasić, E.; Salvai, A. A method of streamflow drought analysis. *Water Resour. Res.* **1987**, *23*, 156–168. [CrossRef]
23. Svensson, C.; Kundzewicz, W.Z.; Maurer, T. Trend detection in river flow series: 2. Flood and low-flow index series/Détection de tendance dans des séries de débit fluvial: 2. Séries d'indices de crue et d'étiage. *Hydrol. Sci. J.* **2005**, *50*, 6. [CrossRef]
24. Madsen, H.; Rosbjerg, D. On the modelling of extreme droughts. *AHS Publ. Ser. Proc. Rep. Int. Assoc. Hydrol. Sci.* **1995**, *231*, 377–386.
25. Vogel, R.M.; Stedinger, J.R. Generalized storage-reliability-yield relationships. *J. Hydrol.* **1987**, *89*, 303–327. [CrossRef]
26. Yang, T.; Zhou, X.; Yu, Z.; Krysanova, V.; Wang, B. Drought projection based on a hybrid drought index using Artificial Neural Networks: A New drought index and drought projection in tarim river basin. *Hydrol. Process.* **2015**, *29*, 2635–2648. [CrossRef]
27. Yahiaoui, A.; Touaïbia, B.; Bouvier, C. Frequency analysis of the hydrological drought regime. Case of oued Mina catchment in western of Algeria. *Rev. Nat. Technol.* **2009**, *1*, 3–15.
28. Van Lanen, H.A.J.; Kundzewicz, W.Z.; Tallaksen, L.M.; Hisdal, H.; Fendekova, M.; Prudhomme, C. Indices for Different Types of Droughts and Floods at Different Scales; Water and Global Change, Technical Report no. 11. 2008. Available online: https://www.academia.edu/15920058/INDICES_FOR_DIFFERENT_TYPES_OF_DROUGHTS_AND_FLOODS_AT_DIFFERENT_SCALES (accessed on 15 May 2022).
29. Kjeldsen, T.R.; Lundorf, A.; Rosbjerg, D. Use of a two-component exponential distribution in partial duration modelling of hydrological droughts in Zimbabwean rivers. *Hydrol. Sci. J.* **2000**, *45*, 285–298. [CrossRef]
30. Hisdal, H.; Tallaksen, L.M.; Frigessi, A. Handling non-extreme events in extreme value modelling of streamflow droughts. In *FRIEND 2002: Regional Hydrology: Bridging the Gap between Research and Practice*; IAHS Publication: Wallingford, UK, 2002; Volume 274, pp. 281–288.
31. Pyrce, R. *Hydrological Low Flow Indices and Their Uses*; Watershed Science Centre: Peterborough, ON, USA, 2004.
32. Raczyński, K.; Dyer, J. Simulating low flows over a heterogeneous landscape in southeastern Poland. *Hydrol. Process.* **2021**, *35*, e14322. [CrossRef]
33. Wanders, N.; Wada, Y.; Van Lanen, H.A.J. Global hydrological droughts in the 21st century under a changing hydrological regime. *Earth Syst. Dyn.* **2015**, *6*, 1–15. [CrossRef]
34. Ryu, J.H.; Lee, J.H.; Jeong, S.; Park, S.K.; Han, K. The impacts of climate change on local hydrology and low flow frequency in the Geum River Basin, Korea. *Hydrol. Process.* **2011**, *25*, 3437–3447. [CrossRef]
35. Milly, P.C.D.; Betancourt, J.; Falkenmark, M.; Hirsch, R.M.; Kundzewicz, Z.W.; Lettenmaier, D.P.; Stouffer, R.J. Stationarity Is Dead: Whither Water Management? *Science* **2008**, *319*, 573–574. [CrossRef]
36. WMO. *Manual on Low-Flow Estimation and Prediction: Operational Hydrology Report No. 50*; WMO: Geneva, Switzerland, 2008.
37. Jones, R.N.; Chiew, F.H.S.; Boughton, W.C.; Zhang, L. Estimating the sensitivity of mean annual runoff to climate change using selected hydrological models. *Adv. Water Resour.* **2006**, *29*, 1419–1429. [CrossRef]
38. Dyer, J.; Mercer, A.; Raczyński, K. Identifying spatial patterns of hydrologic drought over the southeast US using retrospective National Water Model simulations. *Water* **2022**, *14*, 1525. [CrossRef]
39. NOAA. The National Water Model. 2022. Available online: https://water.noaa.gov/about/nwm (accessed on 15 May 2022).
40. Lahmers, T.M.; Hazenberg, P.; Gupta, H.; Castro, C.; Gochis, D.; Dugger, A.; Yates, D.; Read, L.; Karsten, L.; Wang, Y.-H. Evaluation of NOAA National Water Model Parameter Calibration in Semi-Arid Environments Prone to Channel Infiltration. *J. Hydrometeorol.* **2021**, *22*, 2939–2969. [CrossRef]
41. Raczyński, K.; Dyer, J. Multi-annual and seasonal variability of low-flow river conditions in southeastern Poland. *Hydrol. Sci. J.* **2020**, *65*, 2561–2576. [CrossRef]
42. Tsakiris, G.; Nalbantis, I.; Vangelis, H.; Verbeiren, B.; Huysmans, M.; Tychon, B.; Jacquemin, I.; Canters, F.; Vanderhaegen, S.; Engelen, G.; et al. A System-based Paradigm of Drought Analysis for Operational Management. *Water Resour. Manag.* **2013**, *27*, 5281–5297. [CrossRef]
43. Tomaszewski, E. *Multiannual and Seasonal Dynamics of Low Flows in Rivers of Central Poland*; Wydawnictwo Uniwersytetu Łódzkiego: Łódź, Poland, 2012.
44. Higginbottom, T.P.; Symeonakis, E. Identifying Ecosystem Function Shifts in Africa Using Breakpoint Analysis of Long-Term NDVI and RUE Data. *Remote Sens.* **2020**, *12*, 1894. [CrossRef]
45. Horion, S.; Prishchepov, A.V.; Verbesselt, J.; de Beurs, K.; Tagesson, T.; Fensholt, R. Revealing turning points in ecosystem functioning over the Northern Eurasian agricultural frontier. *Glob. Chang. Biol.* **2016**, *22*, 2801–2817. [CrossRef]
46. Scott, D.T.; Gomez-Velez, J.D.; Jones, N.C.; Harvey, J.W. Floodplain inundation spectrum across the United States. *Nat. Commun.* **2019**, *10*, 5194. [CrossRef]
47. Dodds, W.K.; Clements, W.H.; Gido, K.; Hilderbrand, R.H.; King, R.S. Thresholds, breakpoints, and nonlinearity in freshwaters as related to management. *J. N. Am. Benthol. Soc.* **2010**, *29*, 988–997. [CrossRef]
48. Tomaszewski, E. Defining the threshold level of hydrological drought in lake catchments. *Limnol. Rev.* **2011**, *11*, 81–88. [CrossRef]

49. Raczyński, K. Threshold levels of streamflow droughts in rivers of the Lublin region. *Ann. Univ. Mariae Curie-Sklodowska Sect. B Geogr. Geol. Mineral. Petrogr.* **2015**, *30*, 117–129. [CrossRef]
50. Tokarczyk, T. *Low Flow as Indicator of Hydrological Drought*; Instytut Meteorologii i Gospodarki Wodnej: Warszawa, Poland, 2010.
51. Sarailidis, G.; Vasiliades, L.; Loukas, A. Analysis of streamflow droughts using fixed and variable thresholds. *Hydrol. Process* **2019**, *33*, 414–431. [CrossRef]
52. Rey, S.J.; Stephens, P.; Laura, J. An evaluation of sampling and full enumeration strategies for Fisher Jenks classification in big data settings. *Trans. GIS* **2017**, *21*, 796–810. [CrossRef]
53. Guédon, Y. Exploring the latent segmentation space for the assessment of multiple change-point models. *Comput. Stat.* **2013**, *28*, 2641–2678. [CrossRef]
54. Truong, C.; Oudre, L.; Vayatis, N. Selective review of offline change point detection methods. *Signal Process.* **2020**, *167*, 107299. [CrossRef]
55. Celisse, A.; Marot, G.; Pierre-Jean, M.; Rigaill, G.J. New efficient algorithms for multiple change-point detection with reproducing kernels. *Comput. Stat. Data Anal.* **2018**, *128*, 200–220. [CrossRef]
56. Arlot, S.; Celisse, A.; Harchaoui, Z. A Kernel Multiple Change-point Algorithm via Model Selection. *J. Mach. Learn. Res.* **2019**, *20*, 1–56.
57. Korkas, K.; Fryzlewicz, P. Multiple change-point detection for non-stationary time series using Wild Binary Segmentation. *Stat. Sin.* **2017**, *27*, 287–311. [CrossRef]
58. Alodah, A.; Seidou, O. Assessment of Climate Change Impacts on Extreme High and Low Flows: An Improved Bottom-Up Approach. *Water* **2019**, *11*, 1236. [CrossRef]
59. Keogh, E.; Chu, S.; Hart, D.; Pazzani, M. An online algorithm for segmenting time series. In Proceedings of the 2001 IEEE International Conference on Data Mining, San Jose, CA, USA, 29 November–2 December 2001; pp. 289–296. [CrossRef]
60. BenYahmed, Y.; Bakar, A.A.; RazakHamdan, A.; Ahmed, A.; Abdullah, S.M.S. Adaptive sliding window algorithm for weather data segmentation. *J. Theor. Appl. Inf. Technol.* **2015**, *80*, 322–333.
61. Shobha, N.; Asha, T. Monitoring Weather based Meteorological Data: Clustering approach for Analysis. In Proceedings of the International Conference on Innovative Mechanisms for Industry Applications, Bengaluru, India, 21–23 February 2017; pp. 75–81. [CrossRef]
62. Sinaga, K.P.; Yang, M.-S. Unsupervised K-Means Clustering Algorithm. *IEEE Access* **2020**, *8*, 80716–80727. [CrossRef]
63. Murtagh, F.; Contreras, P. Algorithms for hierarchical clustering: An overview. *WIREs Data Min. Knowl. Discov.* **2012**, *2*, 86–97. [CrossRef]
64. Murtagh, F.; Legendre, P. Ward's Hierarchical Agglomerative Clustering Method: Which Algorithms Implement Ward's Criterion? *J. Classif.* **2014**, *31*, 274–295. [CrossRef]

 *water*

Article

Variability of Annual and Monthly Streamflow Droughts over the Southeastern United States

Krzysztof Raczynski [1,*] and Jamie Dyer [2]

[1] Northern Gulf Institute, Mississippi State University, 2 Research Blvd, Starkville, MS 39759, USA
[2] Department of Geosciences, Mississippi State University, 200A Hilbun Hall, Mississippi State, MS 39762, USA
* Correspondence: chrisr@ngi.msstate.edu

Abstract: Understanding the patterns of streamflow drought frequency and intensity is critical in defining potential environmental and societal impacts on processes associated with surface water resources; however, analysis of these processes is often limited to the availability of data. The objective of this study is to quantify the annual and monthly variability of low flow river conditions over the Southeastern United States (US) using National Water Model (NWM) retrospective simulations (v2.1), which provide streamflow estimates at a high spatial density. The data were used to calculate sums of outflow deficit volumes at annual and monthly scales, from which the autocorrelation functions (ACF), partial autocorrelation functions (PACF) and the Hurst exponent (H) were calculated to quantify low flow patterns. The ACF/PACF approach is used for examining the seasonal and multiannual variation of extreme events, while the Hurst exponent in turn allows for classification of "process memory", distinguishing multi-seasonal processes from white noise processes. The results showed diverse spatial and temporal patterns of low flow occurrence across the Southeast US study area, with some locations indicating a strong seasonal dependence. These locations are characterized by a longer temporal cycle, whereby low flows were arranged in series of several to dozens of years, after which they did not occur for a period of similar length. In these rivers, H was in the range 0.8 (+/−0.15), which implies a stronger relation with groundwater during dry periods. In other river segments within the study region the probability of low flows appeared random, determined by H oscillating around the values for white noise (0.5 +/−0.15). The initial assessment of spatial clusters of the low flow parameters suggests no strict relationships, although a link to geologic characteristics and aquifer depth was noticed. At monthly scales, low flow occurrence followed precipitation patterns, with streamflow droughts first occurring in the Carolinas and along the Gulf Coast around May and then progressing upstream, reaching maxima around October for central parts of Mississippi, Alabama and Georgia. The relations for both annual and monthly scales are better represented with PACF, for which statistically significant lags were found in around 75% of stream nodes, while ACF explains on average only 20% of cases, indicating that streamflow droughts in the region occur in regular patterns (e.g., seasonal). This repeatability is of greater importance to defining patterns of extreme hydrologic events than the occurrence of high magnitude random events. The results of the research provide useful information about the spatial and temporal patterns of low flow occurrence across the Southeast US, and verify that the NWM retrospective data are able to differentiate the time processes for the occurrence of low flows.

Keywords: streamflow drought; low flows; national water model; multiannual patterns; autocorrelation; southeastern united states

 check for updates

Citation: Raczynski, K.; Dyer, J. Variability of Annual and Monthly Streamflow Droughts over the Southeastern United States. *Water* **2022**, *14*, 3848. https://doi.org/10.3390/w14233848

Academic Editor: Alina Barbulescu

Received: 7 November 2022
Accepted: 24 November 2022
Published: 26 November 2022

Publisher's Note: MDPI stays neutral with regard to jurisdictional claims in published maps and institutional affiliations.

Copyright: © 2022 by the authors. Licensee MDPI, Basel, Switzerland. This article is an open access article distributed under the terms and conditions of the Creative Commons Attribution (CC BY) license (https://creativecommons.org/licenses/by/4.0/).

1. Introduction

Due to increasing population stress on existing water resources and water quality, as well as uncertainty associated with current and future climate variability, understanding drought formation and evolution processes has become extremely important. As a result, the ability to accurately model and predict droughts is a critical step not only for maintaining the natural environment, but also for ensuring the needs for human water resources. The growing importance of droughts, especially in the light of the latest research showing the intensification of this phenomenon in the coming years [1–6]; Wang et al. [7], raises questions about forecast accuracy, and more importantly, the scale at which accurate drought predictions can be produced. The first, and perhaps the most important, problem in any forecast or analysis process for any natural phenomenon is data availability. For investigation of hydrological processes, river flow data are usually collected by governmental institutions, such as the United States (US) Geological Survey (USGS). Although the data collected by these institutions provide accurate and reliable observational datasets, the length of these datasets is related to the history and ability to perform continuous measurements at each gauge. This leads to limitations in the possibilities of spatial analysis of hydrological phenomena in areas with sparse and/or incomplete gauge information, and further hinders the development of modeling frameworks that allow for realistic simulations in places for which real data are not available.

Large-scale hydrologic models, in order to maintain computational feasibility and physical representativeness, must use spatial and temporal resolutions representative of the larger model domain; therefore, they often cannot accurately reflect local conditions related to geology, groundwater, heat fluxes, or evapotranspiration, which are important from the drought perspective [8–13]. On the other hand, complexity of local conditions makes it difficult to generate and maintain accurate local-scale models, especially when these conditions themselves evolve over space and time; therefore, some level of generalization must be introduced to produce a baseline simulation.

In 2016 the US National Oceanic and Atmospheric Administration (NOAA) contributed to the improvement of the accuracy and spatial coverage of data related to the observation, assessment, and prediction of hydrological extreme events over the continental United States (CONUS) by developing a hydrologic modelling framework called the National Water Model (NWM). The NWM is based on the Weather Research and Forecasting–Hydrological Modeling System (WRF-Hydro), and provides operational simulations of land surface and hydrologic conditions at a variety of time scales (e.g., short, medium, and long range). In addition to the operational version of the model, each major version of the simulation framework is used to produce a historical (also known as, retrospective) simulation for research and analysis purposes. These simulations are forced with the North American Land Data Assimilation (NLDAS) data sets for versions 1.2 and 2.0 and Office of Water Prediction Analysis of Record for Calibration (AORC) data for version 2.1. Thanks to this approach, the NWM has become the substrate for the large-scale distributed simulation of hydrological conditions across the US. Streamflow simulations are dependent on multiple surface and hydrological parameters, of which precipitation and snowmelt play major roles. The main limitations of the NWM include the inability to reproduce reservoir management flows [14,15], especially in previous versions of the model (i.e., v1.2 and v2.0) although the newer operational version of the model (v2.1) includes new reservoir treatment that leverages River Forecast Center (RFC), USGS and U.S. Army Corps of Engineers (USACE) data feeds. While this improves the model response for river segments located below reservoir, some artifacts might still exist [16]. The ability to simulate hydrological conditions at 2.7 million stream locations nationwide means not only better forecasts of water resources, but also improved safety and stability of communities, industry, and protection of life and property [17]. Additionally, retrospective simulation datasets provide continuous surface and hydrologic records for all computational nodes covered by the operational version of the NWM, which in turn allows for analysis of historical hydrological conditions without restrictions related to locations of river observation sites.

A primary motivation for the development of the NWM was the improvement of flood prediction information and dissemination, which are the costliest and deadliest type of natural disasters in the United States [18]. This approach contributed to an improved mathematical representation of the upper range of flows over a large spatial extent; however, due to the increasing importance of droughts in recent years [19], the ability to apply the

NWM to drought assessment has become an important topic [20,21]. This is especially true given that due to climate change water resources are more likely to behave in a non-stationary way [22], requiring the use of comprehensive physical models such as the NWM to provide meaningful predictions of hydrologic drought conditions. There is some evidence of NWM performance being linked to river basin size [23,24], or location, with streams underperforming in semiarid environments [25] or in rivers sensitive to snowmelt runoff [26]. As a result, the NWM has been shown to perform better in streams with a precipitation forced regime [21], and in general the NWM is able to capture major droughts [27] and general streamflow patterns in humid regions such as the Southeastern United States [20].

The aim of this work is to assess the variability of streamflow droughts at annual and monthly scales over the Southeastern United States, based on NWM retrospective v2.1 data, to quantify the spatial and temporal patterns of regional hydrologic drought. The study area is characterized by abundant water resources affected by stress due to industrial and agricultural water withdrawals. This stress is further exacerbated by advancing climate change resulting in changes of water resources and increasing drought risk [28–31]. The four main research questions posed in this paper are as follows: (i) can NWM retrospective data represent low flow occurrence patterns at different time scales, and differentiate regional dependencies, (ii) what are the patterns of hydrologic drought occurrence in terms of annual and monthly variability, (iii) what are the spatial patterns and associated physical drivers of streamflow drought generation and progression, and (iv) are these patterns reflected in the NWM retrospective data, such that machine learning-based occurrence models can be developed to predict future development of streamflow droughts? Autocorrelation and partial autocorrelation were used as the primary analysis tools within this study, providing information about statistically significant periods of streamflow droughts, as well as quantifying the significance of streamflow drought occurring as an extreme event, more or less randomly, versus reoccurrence in periods related to seasonality. The difference between the two occurrence patterns was further measured with the Hurst exponent statistic, which allows for assessment of so called "process memory". The analysis of variability in the hydrologic data will allow for the recognition of fluctuations in low flow events over time. This, in turn, will allow for the construction of models of the phenomenon, based on time dependencies, for example using machine learning approaches. Additionally analysis will provide critical information regarding not only the utility of the NWM retrospective data to define and represent low flow conditions and associated characteristics, but will allow for the generation of a baseline dataset that can be used for subsequent investigations of streamflow drought patterns and processes across the Southeast US for water resource assessments.

2. Study Area

The study area constitutes the southeastern part of the US, specifically identified in a hydrologic context as USGS Region 3: South Atlantic-Gulf Region (Figure 1). This region, which includes all rivers flowing to the Atlantic Ocean and the Gulf of Mexico between the James River catchment in Virginia and the Lower Mississippi River in Mississippi, comprises a total area of ~724,000 km². The area incorporates a diverse array of natural landscapes, with variable land use/cover, vegetation, meteorological, and geological characteristics that lead to a range of hydrologic conditions. The Coastal Plain, a major part of the region that represents around 60% of the total area, is composed mainly of soft unconsolidated sands, gravels, and clays or consolidated and semi-consolidated limestone. The northern regions within the Appalachian highlands contain mainly hard, consolidated rocks, indurated and metamorphosed sedimentary rocks, and crystallin igneous rocks. Groundwater is associated mainly with Cretaceous and Quaternary deltaic sand and gravel deposits, with daily groundwater discharge of around 0.3 km³ that moves seaward in a pattern reflective of the general layout of the regional river networks [32]. Elevation varies between −25–1589 m.a.s.l. over the area, and average stream density is 0.24 km/km² (total river length is ~175,000 km). Stream density has a high regional variability, with a denser network in the northern and northeastern areas and a sparse river network in the south. Approximately 60% of streamflow is contributed by baseflow and 40% by direct runoff [33].

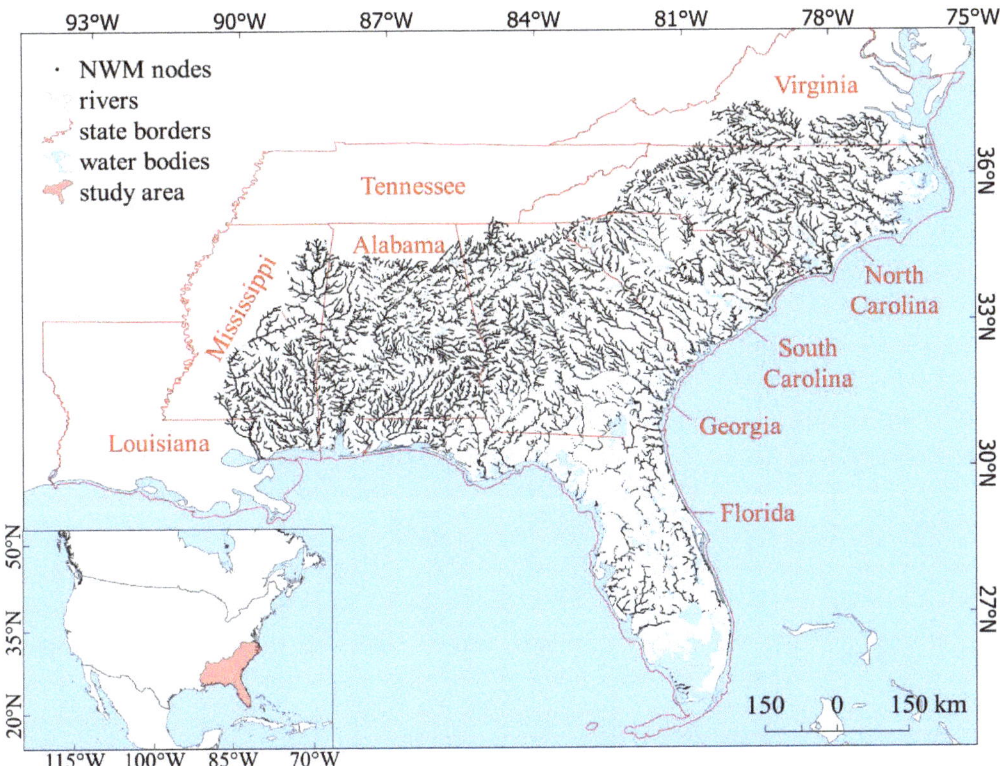

Figure 1. Southeast US study area with locations of NWM nodes.

The climate over the majority of the area is subtropical, with hot and humid summers. Mean annual temperature is between 14 °C–25 °C, depending on the region [25], and annual precipitation varies between 1000 to 2000 mm/year. The relation between water resources and climate variability is strong [34,35], and longer-term changes in precipitation, evaporation, and runoff are caused mainly by the El Niño-Southern Oscillation (ENSO), which affects terrestrial water storage and associated anomalies as well as streamflow discharges [35–38]. Occurrence of La Niña is connected with higher maximal temperatures and lower precipitation, mainly noticeable in June [29], with some evidence showing varying links to dry winters with La Niña in the past, that now depends purely on internal atmospheric variability [31,39]. Decreasing streamflow trends are observed during water-year and spring-summer periods, with strong evidence of abrupt step changes being of greater importance than gradual changes over past years [40]. Constant decrease in streamflow of rivers over the study area is linked to increasing sea surface temperatures [38,41]. Due to the specific environmental conditions of the region, increasing population, growing agriculture needs as well as changes in local climate patterns, Alabama, Mississippi and Florida are identified as regions facing water supply shortages in the future [42]. Further evidence also shows that the entire region faces a substantial drought threat both now and in the future, mainly due to increasing water demands, limited storage capacity, and agricultural dependance on precipitation [43–45], as local water supply regulations were often developed during wetter periods in the region [46].

3. Data and Methods

3.1. Hydrometeorological Data

Streamflow information for this study comes from the NWM retrospective v2.1 dataset, which contains 338,037 unique retrospective stream nodes (hereafter referred to as nodes) with a period of record of February 1979–December 2020. After data quality evaluation a decision was made to include only nodes of Strahler order 3 and higher, since lower orders contained over 80% zero values, which from the perspective of drought analysis introduces the risk of non-representative threshold levels and drought event statistics. This decision also relates to a general limitation of the NWM, such that the model underperforms in lower order streams [23,24]. Additional criteria of no more than 5% of zero or null data were introduced to avoid computational errors, and an additional 1198 nodes were characterized by almost unchanging minimum flow values, which led to the defined low flow threshold (see below for method details) being the same value as the minimum flow. These nodes were excluded from the study, leading to the inclusion of 60,750 nodes with hourly mean streamflow values, which were converted to daily mean flow values representing 00–00 UTC. The final dataset constituted daily flows for a period from 1 February 1979–31 December 2020, which equates to 2,551,500 stream years.

Precipitation data, which was used for analysis of low flow processes, was obtained from the U.S. Federal Government Climate Resilience Toolkit [47] for monthly and annual scales.

3.2. Low Flow Conditions Definition

In this study, low flow definition is based on the widely adopted threshold level method (TLM; [48]), whereby stream discharge is considered low flow if it is equal to or lower than a defined threshold level. There are many ways of calculating a threshold; however, this study adopts an objective breakpoint method to define unique threshold levels at each node. In this approach, the lower part of the flow duration curve (FDC) is considered as a series with a breakpoint that serves as the indicator of the moment of change from atmospheric supply to groundwater supply, which constitutes a natural marker for the beginning of low flow conditions. This method is described in detail by Raczynski and Dyer [49]. To accurately measure the seasonal and annual outflow deficits, no pooling method and no additional minimal time criteria were applied. In this study the term low flow refers to discharge values identified as lower or equal to a threshold discharge level, while low flow event/conditions are considered the same as streamflow drought—a series of low flows that occur over some period that lead to formation of hydrologic drought.

3.3. Statistical Analysis

A basic parameter used in this study is a low flow volume (V) which is calculated as a difference between the defined streamflow threshold and the flow hydrograph during drought episode:

$$V = \int_{t_1}^{t_2} (Q_t - Q)dt \tag{1}$$

where: V—volume (m^3), Q_t—threshold flow (m^3 s^{-1}), and Q—outflow (m^3 s^{-1}).

To describe changes in episodes occurrence, autocorrelation functions (ACF) were calculated for event volumes aggregated to monthly and annual scales. Autocorrelation allows for examination of seasonal or multi-annual variations in low flow conditions by estimating the degree of correlation between element with element shifted by k [50], where the length of this shift (also referred to as lag) may vary:

$$\rho_k = \frac{s_k}{s_0} \tag{2}$$

where s_0 is variance of the time series at t and s_k is covariance at k lag:

$$s_k = \frac{1}{n}\sum_{i=1}^{n-k}(y_i - \overline{y})(y_{i+k} - \overline{y}) = \frac{1}{n}\sum_{i=k+1}^{n}(y_i - \overline{y})(y_{i-k} - \overline{y}) \tag{3}$$

For this study, a time series constitutes of low flow volumes in a single node aggregated to monthly or annual scale.

In order to assess the possibility of applying data from NWM to occurrence models based on machine learning techniques, the number of lags needed to obtain statistically significant result should be estimated. Based on the information about the significance of the lags (q) of the autocorrelation functions, the usefulness of potential seasonal models can be estimated. The model based on the seasonality resulting from the autocorrelation relationship is the moving average (MA(q)) model, which is expressed by the number of statistically significant correlations of lags in ACF. For example, the MA(7) model means that the modeled dependence has statistically significant relationships up to the 7th lag (seven periods back—depending on the resolution of the tested series, e.g., months or years). In addition to ACF, partial autocorrelation functions (PACF) were calcualted to introduce the control of all lags. PACF explains partial correlation between the series and lags of itself. Significance of PACF lags (p) provides valuable information on potential lag steps in seasonal modeling using autoregressive models (AR(p)), as significant PACF lags (p) correspond to lags in AR(p) models on the same basis as ACF lags (q) are used for the MA(q) models [51,52]. Therefore, estimating whether and at which lag there are statistically significant relationships in ACF and PACF distributions constitute the basis for assessing whether the studied relationships can, at a later stage, be modeled using machine learning techniques, using the AR and/or MA seasonality models.

Summability or non-summability of the autocorrelation function is an indicator of the process memory length, which describes the tendency for grouping of natural extreme events into sequences. So-called "process memory" was first observed in hydrologic data by Hurst [53], which led to the development of the Hurst exponent (H) that describes the process memory length within a hydrologic data series. The detailed methodology for determining the value of the exponent was described by Koutsoyiannis [54]. Values of H close to 0.5 reflect white noise processes, where consecutive values are random, while H values closer to 1 reflect long process memory, understood as a tendency for similar events to group in longer sequences (large values are followed by large values, and vice versa). Although in natural processes the range of H is usually 0.5–1 [54], it is possible for H to range from 0–0.5, where values $H < 0.5$ means an anti-persistent series where high and low values appear alternately.

To classify spatial relations between groups of nodes with similar process memory, an unsupervised machine learning algorithm of K-means clustering was applied. The algorithm is used to group similar data into clusters by minimizing the objective function $J(z,A)$ with updating cluster centers [55]:

$$J(z, A) = \sum_{i=1}^{n}\sum_{k=1}^{c} z_{ik}||x_i - a_k||^2 \tag{4}$$

where c is number of clusters, x_i is the data point, a_k is cluster center, and z_{ik} is a binary variable that indicates if the data point is considered within the cluster.

All statistics were performed for two-tailed $\alpha = 0.05$.

4. Results and Discussion

4.1. Annual Distribution

To determine annual patterns of streamflow droughts, the daily water outflow deficiencies (volumes) were aggregated to the annual scale. It was expected that at least two different ACF shapes would be obtained as indicated in other works [56,57], and analysis confirmed that these patterns are present in the dataset. In the first case the pattern of low flow occurrence is close to random and the majority of the ACFs are statistically insignificant, with autocorrelation values shifting every year or so. *H* for these rivers is close to white noise with a mean of 0.59 (Figure 2), and 26% of the analyzed nodes were classified in this group. These patterns are usually described as related to meteorological conditions, and are easy to predict [10,58]. An example of this type of process is presented in Figure 2, where low flows occurred in around half of the 42-year study period (average of 18.5 years with low flow and 23.5 without low flow). On average there were 15 shifts between years with and without low flows, and episodes occur in two to three consecutive years and then disappear for a similar period.

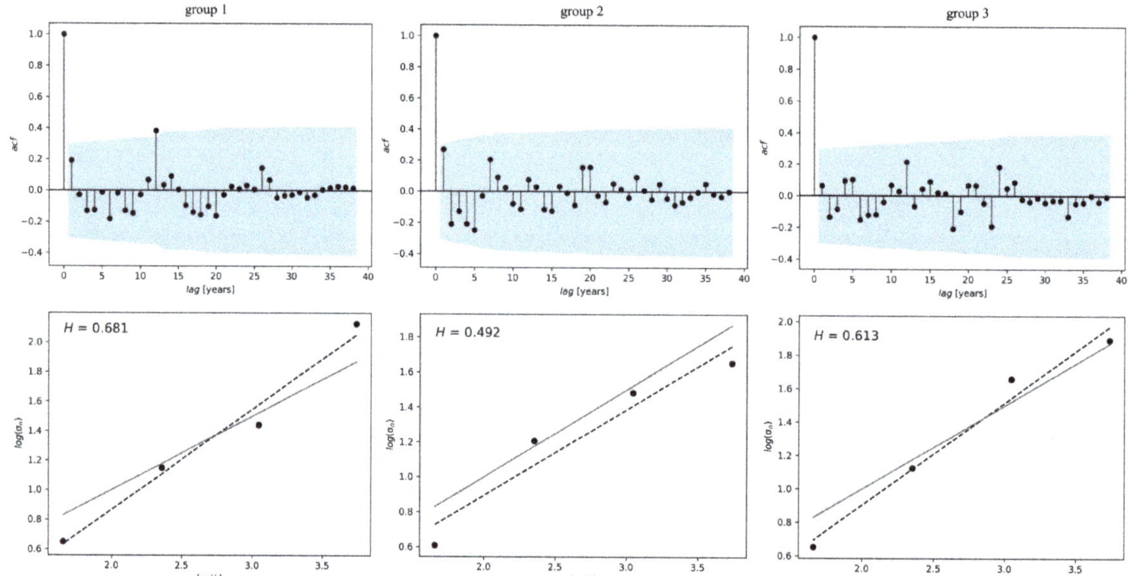

Figure 2. Examples of autocorrelation functions (ACF) (**top** row) and climacograms for Hurst exponent (**bottom** row) for the same, randomly selected nodes in each defined cluster; solid line–white noise process, dash line–Hurst-Kolmogorov process.

The second, and at the same time the largest group (65% of nodes), were rivers with low flows occurring in about 6–7 year intervals with 1–2 year breaks in occurrence (Figure 2). Similar to the first group, an average of 14 shifts were observed during the study period. In total, low flows occurred during 32 of the 42 years in the study period. *H* for this group is 0.70 and ACFs show higher repeatability, indicating that groundwater is of greater importance in these types of rivers, especially when multi-annual streamflow drought occurrence is present [11,59,60].

There is also a third group that constituted less than 9% of the analyzed nodes, and included rivers where low flow occurred almost annually (mean of 40 years with low flow over the study period; Figure 2). The average *H* is closer to representing white noise than for the second group, having a mean value of 0.63, which is most likely due to the high irregularity of shifts in occurrence along with the relatively short period (on average one year).

Analysis of the spatial distribution of the clusters showed no statistically significant correlation between the type of low flow occurrence and stream order, and in fact there are some rivers along which the cluster changes along the stream. Most nodes with low flow patterns that fall within the third group are located in North Carolina, especially in the Cape Fear and Pee Dee River systems, as well as in the upper Pearl River watershed in Mississippi. The lowest frequency of low flow occurs in central parts of Mississippi,

Alabama and Georgia, below the Appalachian piedmont, as well as in northern parts of the Carolinas in upstream river sections (Figures 3 and 4). The former relation might be explained by groundwater inflow to rivers located at the base of the Appalachian Mountains, while deeply allocated aquifers in the piedmont region of the Carolinas may explain the high magnitude of low flows in that area.

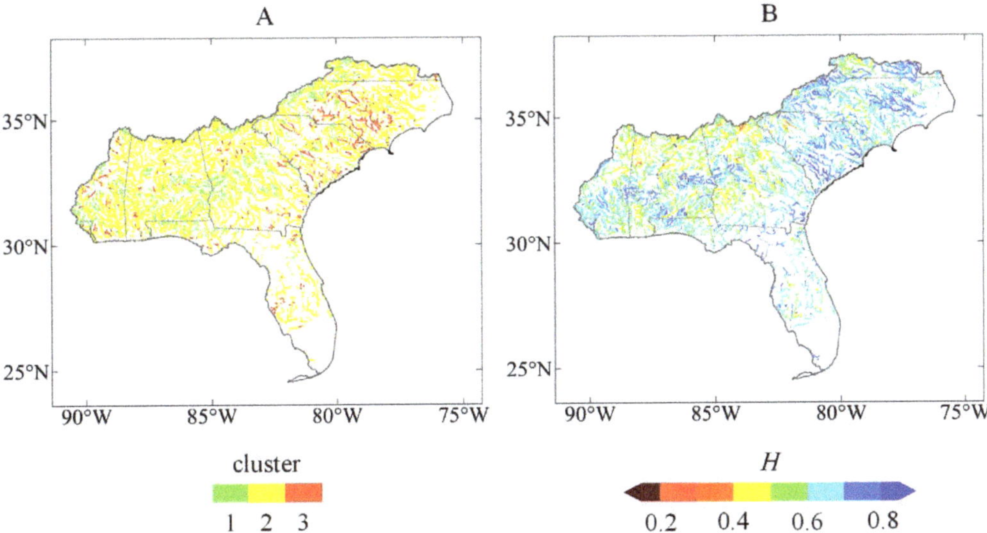

Figure 3. Annual low flow occurrence spatial clusters (**A**) and Hurst exponent values (**B**) in studied nodes.

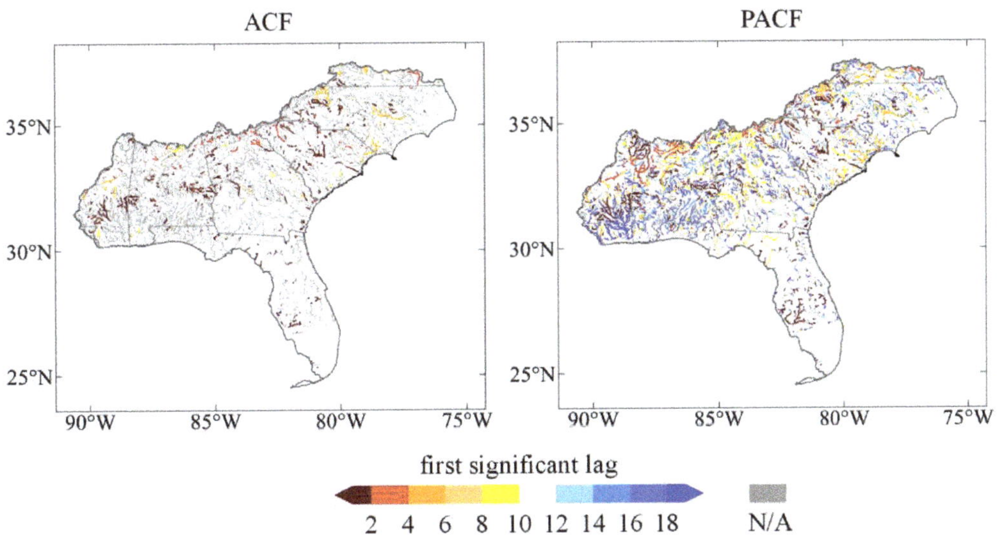

Figure 4. First significant lags for annual ACF and PACF in studied nodes.

High values of H are observed in both North and South Carolina, where over 80% of nodes are characterized by H higher than 0.6 and around half of nodes with H higher than 0.7. This suggests that longer process memory occurs in this region; however, defined ACF clusters do not accurately reflect spatial relations for part of study area. While Virginia and the Carolinas show relatively high values of *H*, suggesting strong seasonality, ACF clustering finds most of these rivers belong to the Cluster 1 (Figure 3). Additionally, statistically significant decreasing trends are observed in the Carolinas, meaning low flows are in general becoming more severe [61,62]. Multiyear droughts with high magnitudes were also identified in this region [62], which is confirmed by high *H* values found in this study. High magnitude, recurring streamflow droughts in the Carolinas with increasing trends are associated with decreased precipitation and increased potential evapotranspiration, especially in the July-September warm-season period, as the changes correspond to variations in meteorological factors [59,63]. This dependence is further intensified by long recovery times after dry conditions are gone [64] and agricultural practices such as irrigation that affect negatively water supplies in the region [65].

High values of *H* (>0.7) are also observed for central and southern parts of Alabama and Georgia (Figure 3B), where ACFs alter every 2–3 years. This is in contrast to clustering results, as *H* values suggest relatively easy to predict processes taking place in this region, while the cluster average was close to white noise. However, evidence of seasonality of low flows is seen from central Mississippi to north central Georgia, where increases in *H* are related to intervals of ACF Cluster 2. This might be due to high repeatability of 2–3-year patterns in these nodes while ACF functions were variable, resulting in a shape reminiscent of random processes. This relation is likely affected to some degree by ENSO, as during these conditions over the southeast US intense groundwater withdrawals for irrigation are observed that act to decrease baseflow and lower low flows [66]. Decreasing values of low flows in this region, however, might be also linked to additional human-induced influences [67], related mainly to land-use, population growth, and agriculture [66,68,69].

For less than half of the studied nodes in Alabama and Georgia, first, second and third lags were statistically significant. This relation is visible for other regions in the study area as well, where only a fraction of nodes with high *H* values yield significant autocorrelations. In total, 83.8% of all nodes do not have any significant lag (of the first 20 lags) and only 7.9% have the first lag significant (Table 1). These results indicate that there is no constant, univariate process that could be quantified by averaged seasonal models as residuals are not linearly dependent on current and past values. PACF distribution (Figure 4) in most nodes contains some statistically significant lags, and from all studied nodes, around 27% do not contain statistically significant *p* lags (Table 1). This dependence suggests that models based on seasonal repeatability defined by the autoregressive component (AR(*p*)) might reflect the actual changes in the annual occurrence of low flows better than moving average (MA(*q*)) models.

Table 1. Percent of nodes with statistically significant lags of annual ACF and PACF.

	No Lag	1st	2nd	3rd	4th	5th	6th	7th	8th	9th	10th	11th	12th	13th	14th	15th	16th	17th	18th	19th	20th
ACF	83.82	7.92	0.86	0.96	0.72	2.02	0.17	0.80	0.09	0.28	0.32	0.22	1.56	0.15	0.01	0.04	0.01	0.02	0.01	0.00	0.02
PACF	26.77	8.86	3.96	1.17	3.95	4.66	1.92	2.35	3.54	2.64	3.63	2.11	2.56	3.05	2.23	3.35	3.12	4.50	5.80	3.26	6.59

4.2. Monthly Distribution

Although annual distributions of low flow occurrences show some spatial dependencies, the monthly distribution provides a better understanding of the processes. In general, the distribution of monthly streamflow droughts follows precipitation patterns, which was also confirmed in other studies [5,63]. The relations are strongest along the Atlantic coast, where sandy soils lead to relatively rapid hydrologic response of river levels to rainfall. January low flow frequency is at or near zero in 64% of nodes over the study period, with mainly low magnitude events in Florida and southern Georgia; however, evidence suggests some relation to precipitation is also present in Virginia. In subsequent months, low flows continue to disappear in the majority of the study area except Florida, where winter precipitation is on average around 40 mm/month and there is a clear, strong relation with streamflow drought characteristics and precipitation through April, measured by the Spearman rank correlation (Figure 5). This pattern matches general climatic features, with subtropical regions north of Florida having a wet winter, while Florida has a relatively drier winter as it reflects a tropical climate.

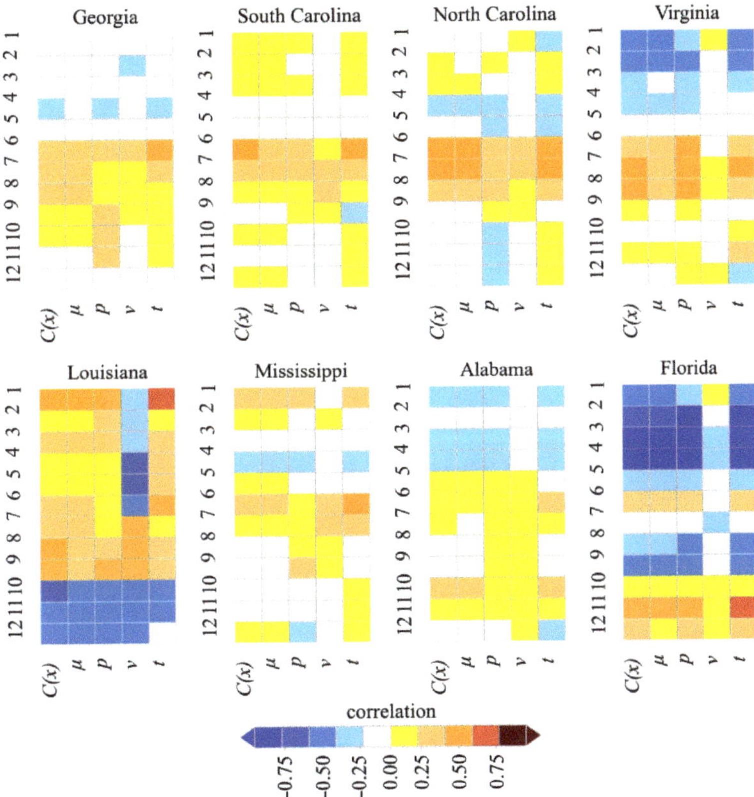

Figure 5. Spearman rank correlations between mean monthly precipitation and streamflow drought parameters, divided by state; $C(x)$—periodicity of streamflow drought occurrence, μ—mean length of process memory, p—number of periods with streamflow drought, v—mean monthly volume, t—mean duration.

During spring (April–June) there is also increased risk of flash droughts in northern and western Florida as well as south Georgia, which was observed by [70]. Starting from May the precipitation patterns begin to change due to the difference in climate patterns between Florida and the Gulf Coast and areas to the north, when sums of precipitation increase over Florida (tropical climate) and decrease through the central and northern study area (sub-tropical climate, Figure 6). In May and June low flows occur mainly over Florida and southern Georgia, while over the following months low flows continue to develop from the Gulf Coast toward central and western parts of the study region (Figure 6). During this time the relation of low flows to precipitation weakens (Figure 5), likely due to increased potential evaporation [59]. At the same time Florida's low flows disappear due to increased precipitation. This spatial direction is consistent with patterns in development of flash droughts found by Chen et al. [70]. In general, droughts related to climatic forcings are regionally specific, with a clear relation between increases in drought with precipitation decreases and potential evaporation increases, especially for north Florida, North Carolina, and Virginia [28,71].

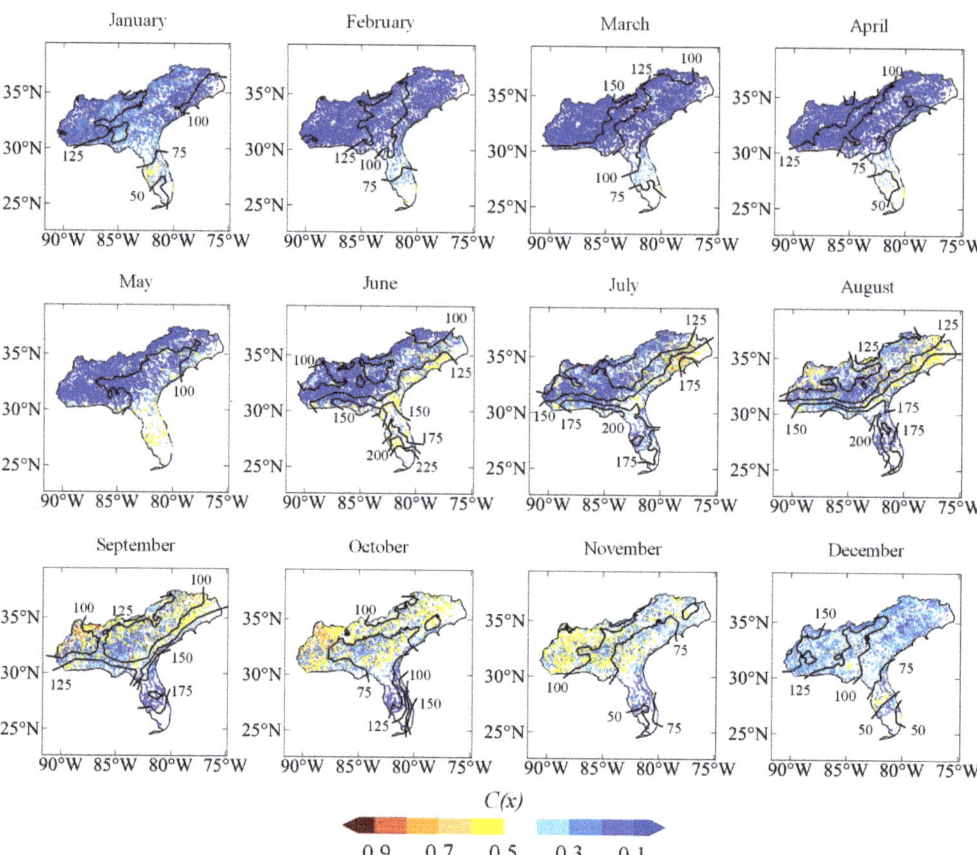

Figure 6. Periodicity ($C(x)$) of streamflow droughts in consecutive months with isohyets of average monthly precipitation (mm).

The highest periodicity of low flows changes by region, with the most intense low flows occurring in July in North Carolina (also annual maxima), in September and October in Mississippi and northern Alabama, and November in the piedmont region (Figure 6). The region of north-central Alabama was also found to be most prone to drought persistence within the study region [72]. The summer period (July–October) is also characterized by high periodicity, reaching 25–30 repeats with low flow each month over the study period, especially in the central and northern parts of the study region. This dependence is opposite to the precipitation distribution, where the Carolinas are characterized by high monthly mean precipitation reaching 175 mm in Coast area, while at the same time the periodicity is highest (Figure 6), emphasizing the role of lowering groundwater levels due to pumping presented by [59]. During late winter and spring a substantial number of nodes had no streamflow droughts during the entire study period, with the total percentage of nodes showing no low flows being 13% in February, 32% in March, 34% in April, and 16% in May.

Similar to annual observations, PACFs generally provide more information than ACFs. The latter on average were statistically significant for around 20% of nodes, with mostly insignificant functions in all 20 lags for April (85.3%) and the lowest values over the winter period (December–February, on average 79%). In most cases, significant ACF lags were related to lag 1 or 12, which follows seasonal and annual patterns. When considering PACFs, on average 60% of nodes contain statistically significant lags, with the highest number for November (79%) and lowest for March (33%). In general, ACFs and PACFs provide the same pattern as described before, with the lowest explanatory power during March and April due to lowest number of drought episodes, and then increasing in lag significance from May over the Carolinas and Gulf Coast before progressing inland. Around June and July the highest concentration of significant lags is found in eastern parts of the study region, while during late summer and fall the western regions are better explained by both ACFs and PACFs (Figures 7 and 8). This pattern is confirmed in monthly precipitation distributions as well as *H* values (Figure 9). Florida is characterized by the lowest number of significant ACF and PACF lags and varying H values, which could be attributable to the high number of drought episodes interrupted by precipitation from tropical cyclones, which for Florida is the case for over 30% of episodes [73]. This might also explain the relations for south Georgia and the Carolinas, where tropical cyclone precipitation accounts for the cessation of between 20–30% of droughts, while for the Appalachian region this is less than 10% [73].

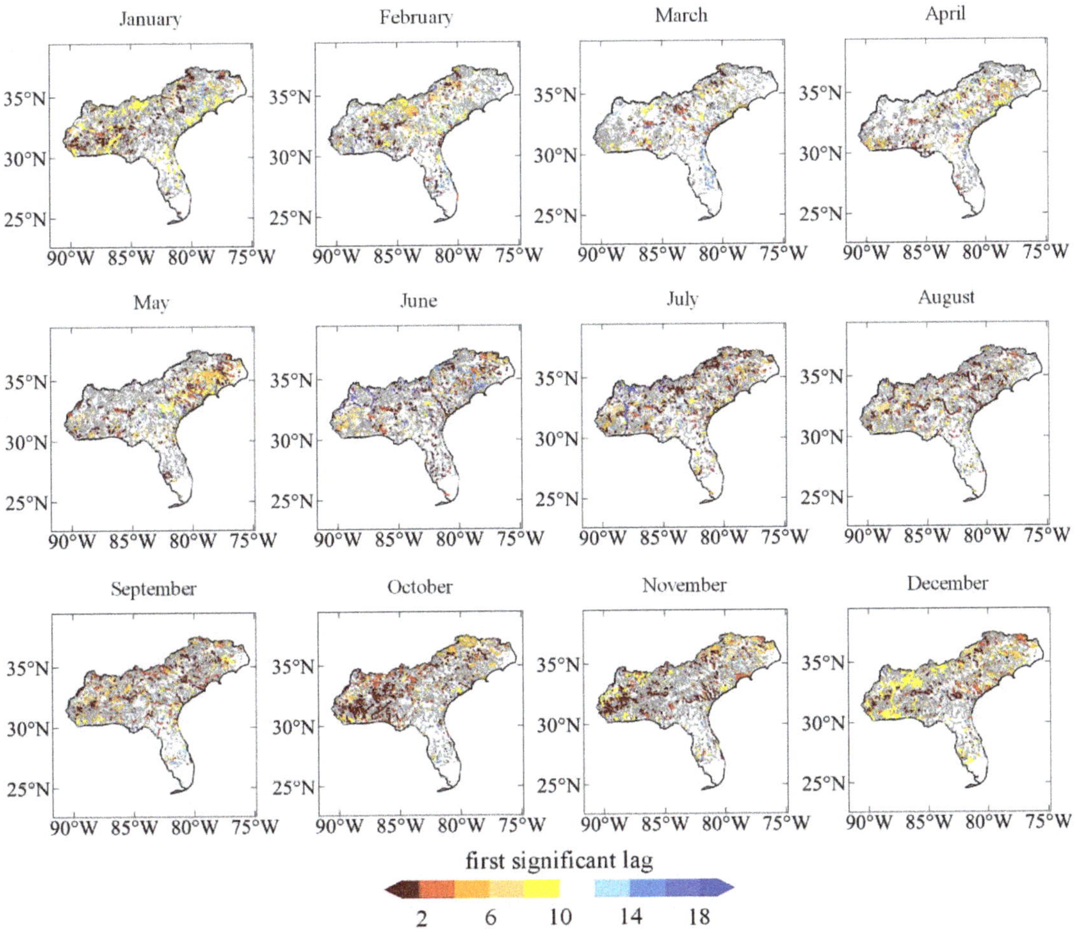

Figure 7. First significant lag in monthly ACF distributions.

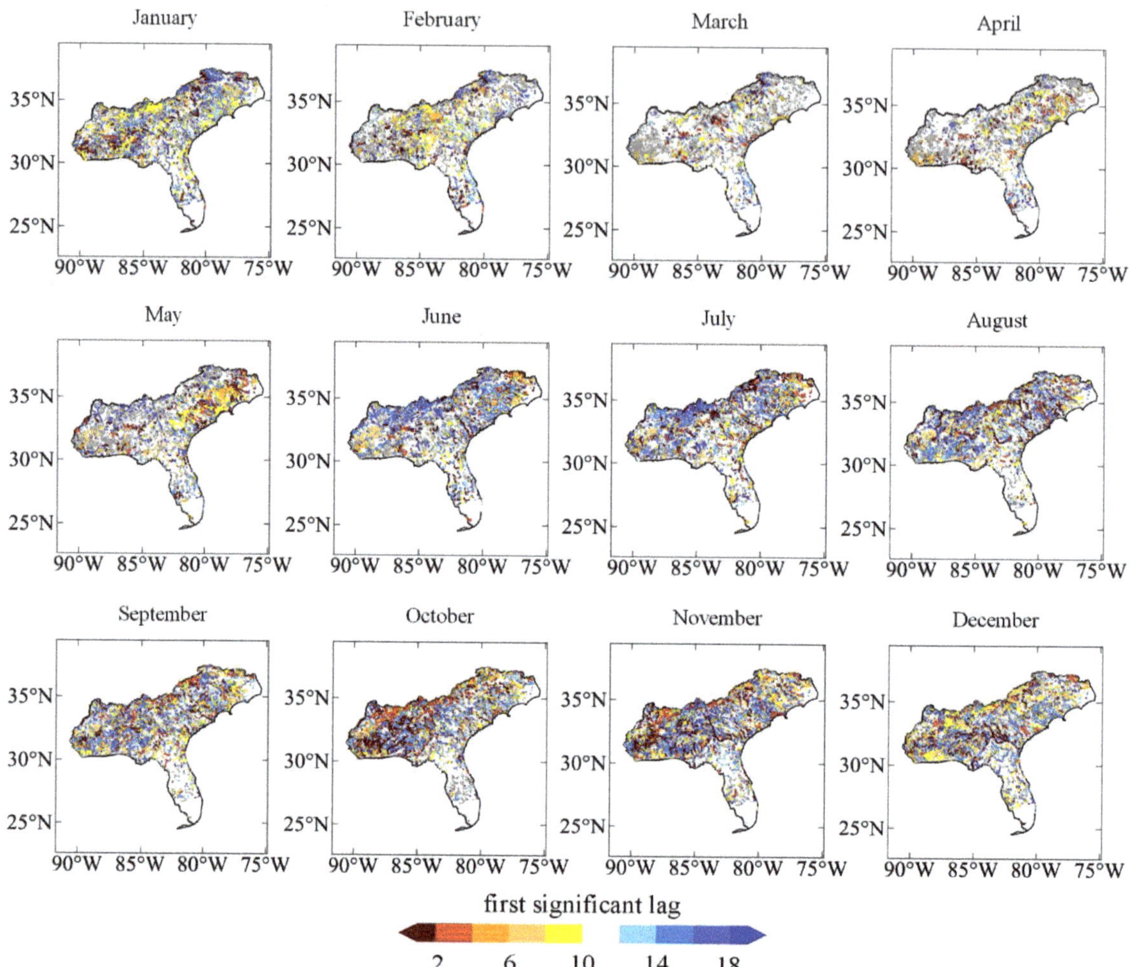

Figure 8. First significant lag in monthly PACF distributions.

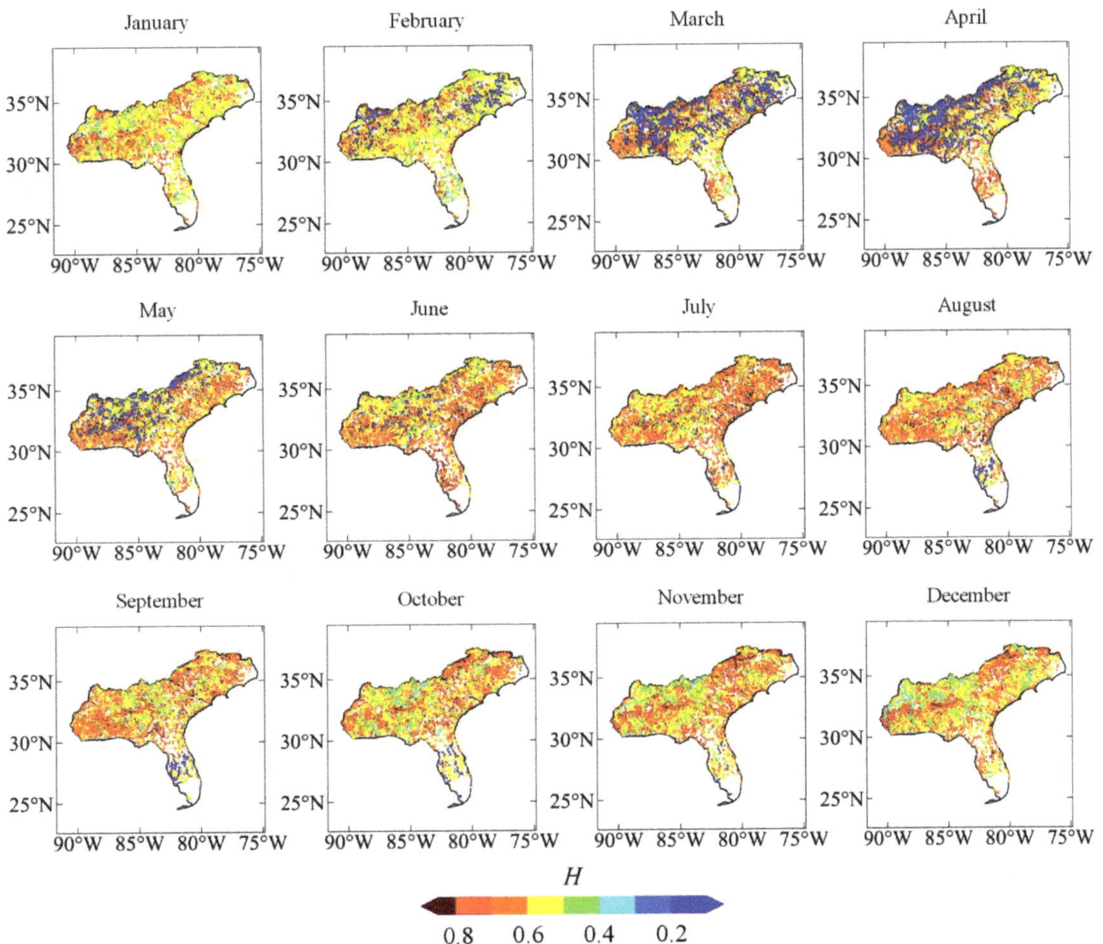

Figure 9. Hurst exponent values for monthly series.

5. Conclusions

This article assesses the variability of streamflow droughts at annual and monthly scales over the Southeastern United States and quantifies temporal and spatial patterns of hydrologic droughts in the region. As hydrologic input data the NWM retrospective v2.1 daily flows for period February 1979–December 2020 for 60,750 nodes was used. Streamflow droughts were identified using an objective threshold approach [49] and the ACF and PACF were calculated based on aggregated annual and monthly flow series.

At annual scales the Carolinas are characterized by a high periodicity of streamflow droughts with occurrence almost every year. The presence of high process memory is further confirmed by high H values both over the Carolinas as well as central parts of Alabama and Georgia. ACFs values, however, are mostly insignificant over the study region, with around 80% of nodes showing no significant relationship. At the same time, PACFs explained around 75% of temporal relations, with monthly aggregated data showing clearer spatial patterns. Except for Florida, which exhibits a tropical climate pattern with a dry winter, streamflow droughts rarely occur during spring and then begin to increase in frequency around May over the Carolinas and Gulf of Mexico regions before progression inland which reflects general precipitation patterns. Eastern parts of the study area are characterized by droughts during late spring/early summer, with western parts showing increased drought by late summer/early fall. This coincides with the progression of decreased warm-season rainfall over the study area north of Florida, reflecting the subtropical climate patterns, while Florida streamflow droughts occur mainly during winter, as reflected by the drier winter representative of the tropical climate. Monthly ACF and PACF dependences are confirmed by H values, with highest values (reaching $H = 0.8$) for June-July over the Carolinas and September for the Gulf Coast area. Major parts of Mississippi, Alabama and Georgia have H close to 0.2 for the March-April period, which suggests an alternating character of events. This is also found in ACF annual functions, where only around half of studied years had low flow episodes. Overall, PACFs are better adjusted to spatio-temporal relations, and yield more statistically significant results than ACFs.

Since PACF yields more statistically significant results than ACF over the study area for both annual and monthly series, autoregressive models (AR(p)) will be better adjusted to capture seasonality, than moving average (MA(q)) based models. This in turn implies that repeatability (represented by AR models) is of greater importance in the region with respect to drought occurrence than extreme events occurrence (represented by MA models).

The results of this study are refl"ctiv' of the NWM retrospective dataset (v2.1.); however, this study does not assess the accuracy of the model data against observations. Some artifacts and/or differences in model performance over the study region may be present in the results; therefore further research should focus on exploring spatial patterns and tendencies in extreme hydrologic events using available observed data over the Southeastern US study region. Using the results from this work as a baseline, a comparison of results using similar methods applied to observed data will help to determine whether the NWM retrospective dataset sufficiently reflects patterns and trends in extreme hydrologic events. Additionally, as this work indicates the potential usefulness in application of AR(p) machine learning models to quantify schemes and future predictions based on detected significant lags, such an approach could be considered future applications using either simulated or measured datasets.

Author Contributions: Conceptualization K.R. and J.D.; methodology K.R.; validation K.R. and J.D.; formal analysis K.R.; investigation K.R.; resources J.D.; data curation K.R. and J.D.; writing—original draft preparation K.R. and J.D.; writing—review and editing K.R. and J.D.; visualization K.R.; supervision J.D.; project administration J.D.; funding acquisition J.D. All authors have read and agreed to the published version of the manuscript.

Funding: This research was funded by the National Oceanic and Atmospheric Administration (NOAA), grant number NA19OAR4590411.

Data Availability Statement: Publicly available datasets of NOAA National Water Model Retrospective v2.1 were analyzed in this study. This data can be found here: https://registry.opendata.aws/nwm-archive/.

Conflicts of Interest: The authors declare no conflict of interest. The funders had no role in the design of the study; in the collection, analyses, or interpretation of data; in the writing of the manuscript, or in the decision to publish the results.

References

1. Cook, B.I.; Ault, T.R.; Smerdon, J.E. Unprecedented 21st century drought risk in the American Southwest and Central Plains. *Sci. Adv.* **2015**, *1*, e1400082. [CrossRef] [PubMed]
2. Marengo, J.A.; Torres, R.R.; Alves, L.M. Drought in Northeast Brazil—Past, present, and future. *Theor. Appl. Climatol.* **2017**, *129*, 1189–1200. [CrossRef]
3. Papadimitriou, L.V.; Koutroulis, A.G.; Grillakis, M.G.; Tsanis, I.K. High-end climate change impact on European runoff and low flows—Exploring the effects of forcing biases. *Hydrol. Earth Syst. Sci.* **2016**, *20*, 1785–1808. [CrossRef]
4. Spinoni, J.; Naumann, G.; Carrao, H.; Barbosa, P.; Vogt, J. World drought frequency, duration, and severity for 1951–2010: World Drought Climatologies for 1951–2010. *Int. J. Clim.* **2014**, *34*, 2792–2804. [CrossRef]
5. Touma, D.; Ashfaq, M.; Nayak, M.A.; Kao, S.-C.; Diffenbaugh, N.S. A multi-model and multi-index evaluation of drought characteristics in the 21st century. *J. Hydrol.* **2015**, *526*, 196–207. [CrossRef]

6. Vicente-Serrano, S.M.; Lopez-Moreno, I.; Beguería, S.; Lorenzo-Lacruz, J.; Sanchez-Lorenzo, A.; García-Ruiz, J.M.; Azorin-Molina, C.; Morán-Tejeda, E.; Revuelto, J.; Trigo, R.; et al. Evidence of increasing drought severity caused by temperature rise in southern Europe. *Environ. Res. Lett.* **2014**, *9*, 044001. [CrossRef]
7. Wang, Z.; Zhong, R.; Lai, C.; Zeng, Z.; Lian, Y.; Bai, X. Climate change enhances the severity and variability of drought in the Pearl River Basin in South China in the 21st century. *Agric. For. Meteorol.* **2018**, *249*, 149–162. [CrossRef]
8. Bormann, H.; Pinter, N. Trends in low flows of German rivers since 1950: Comparability of different low-flow indicators and their spatial patterns. *River Res. Appl.* **2017**, *33*, 1191–1204. [CrossRef]
9. Guzha, A.; Rufino, M.; Okoth, S.; Jacobs, S.; Nóbrega, R. Impacts of land use and land cover change on surface runoff, discharge and low flows: Evidence from East Africa. *J. Hydrol. Reg. Stud.* **2018**, *15*, 49–67. [CrossRef]
10. Haslinger, K.; Koffler, D.; Schöner, W.; Laaha, G. Exploring the link between meteorological drought and streamflow: Effects of climate-catchment interaction. *Water Resour. Res.* **2014**, *50*, 2468–2487. [CrossRef]
11. Stoelzle, M.; Stahl, K.; Morhard, A.; Weiler, M. Streamflow sensitivity to drought scenarios in catchments with different geology. *Geophys. Res. Lett.* **2014**, *41*, 6174–6183. [CrossRef]
12. Van Loon, A.F.; Rangecroft, S.; Coxon, G.; Naranjo, J.A.B.; Van Ogtrop, F.; Van Lanen, H.A.J. Using paired catchments to quantify the human influence on hydrological droughts. *Hydrol. Earth Syst. Sci.* **2019**, *23*, 1725–1739. [CrossRef]
13. Van Loon, A.; Laaha, G. Hydrological drought severity explained by climate and catchment characteristics. *J. Hydrol.* **2015**, *526*, 3–14. [CrossRef]
14. Kim, J.; Read, L.; Johnson, L.E.; Gochis, D.; Cifelli, R.; Han, H. An experiment on reservoir representation schemes to improve hydrologic prediction: Coupling the national water model with the HEC-ResSim. *Hydrol. Sci. J.* **2020**, *65*, 1652–1666. [CrossRef]
15. Viterbo, F.; Read, L.; Nowak, K.; Wood, A.; Gochis, D.; Cifelli, R.; Hughes, M. General Assessment of the Operational Utility of National Water Model Reservoir Inflows for the Bureau of Reclamation Facilities. *Water* **2020**, *12*, 2897. [CrossRef]
16. Khazaei, B.; Read, L.K.; Casali, M.; Sampson, K.M.; Yates, D. Improvement of Lake and Reservoir Parameterization in the NOAA National Water Model. In *World Environmental and Water Resources Congress 2021: Planning a Resilient Future along America's Freshwaters*; American Society of Civil Engineers: Reston, VA, USA, 2021; pp. 552–560.
17. Hooper, R.; Nearing, G.; Condon, L. Using the National Water Model as a Hypothesis-Testing Tool. *Open Water J.* **2017**, *4*.
18. National Weather Service (NWS). US Flood Loss Report, Hydrologic Services [Online]. Available online: https://www.weather.gov/water/ (accessed on 12 June 2022).
19. Forzieri, G.; Feyen, L.; Russo, S.; Vousdoukas, M.; Alfieri, L.; Outten, S.; Migliavacca, M.; Bianchi, A.; Rojas, R.; Cid, A. Multi-hazard assessment in Europe under climate change. *Clim. Chang.* **2016**, *137*, 105–119. [CrossRef]
20. Dyer, J.; Mercer, A.; Raczyński, K. Identifying Spatial Patterns of Hydrologic Drought over the Southeast US Using Retrospective National Water Model Simulations. *Water* **2022**, *14*, 1525. [CrossRef]
21. Hansen, C.; Shiva, J.S.; McDonald, S.; Nabors, A. Assessing Retrospective National Water Model Streamflow with Respect to Droughts and Low Flows in the Colorado River Basin. *JAWRA J. Am. Water Resour. Assoc.* **2019**, *55*, 964–975. [CrossRef]
22. Milly, P.C.D.; Betancourt, J.; Falkenmark, M.; Hirsch, R.M.; Kundzewicz, Z.W.; Lettenmaier, D.P.; Stouffer, R.J. Climate change. Stationarity Is Dead: Whither Water Management? *Science* **2008**, *319*, 573–574. [CrossRef]
23. Rojas, M.; Quintero, F.; Krajewski, W.F. Performance of the National Water Model in Iowa Using Independent Observations. *JAWRA J. Am. Water Resour. Assoc.* **2020**, *56*, 568–585. [CrossRef]
24. Johnson, J.M.; Munasinghe, D.; Eyelade, D.; Cohen, S. An integrated evaluation of the National Water Model (NWM)–Height Above Nearest Drainage (HAND) flood mapping methodology. *Nat. Hazards Earth Syst. Sci.* **2019**, *19*, 2405–2420. [CrossRef]
25. Lahmers, T.M.; Hazenberg, P.; Gupta, H.; Castro, C.; Gochis, D.; Dugger, A.; Yates, D.; Read, L.; Karsten, L.; Wang, Y.-H. Evaluation of NOAA National Water Model Parameter Calibration in Semi-Arid Environments Prone to Channel Infiltration. *J. Hydrometeorol.* **2021**, *22*, 2939–2969. [CrossRef]
26. Garousi-Nejad, I.; Tarboton, D.G. A comparison of National Water Model retrospective analysis snow outputs at snow telemetry sites across the Western United States. *Hydrol. Process.* **2022**, *36*, e14469. [CrossRef]
27. Xu, L.; Wang, H.; Chelliah, M.; Dewitt, D. National Water Model for Drought Monitoring: A Preliminary Evaluation, Science and Technology Infusion Climate Bulletin. In Proceedings of the 45th NOAA Annual Climate Diagnostics and Prediction Workshop, Virtual Workshop, 20–22 October 2020; pp. 97–100.
28. Ficklin, D.L.; Maxwell, J.T.; Letsinger, S.L.; Gholizadeh, H. A climatic deconstruction of recent drought trends in the United States. *Environ. Res. Lett.* **2015**, *10*, 044009. [CrossRef]
29. Mourtzinis, S.; Ortiz, B.V.; Damianidis, D. Climate Change and ENSO Effects on Southeastern US Climate Patterns and Maize Yield. *Sci. Rep.* **2016**, *6*, 29777. [CrossRef] [PubMed]
30. Mulholland, P.J.; Best, G.R.; Coutant, C.C.; Hornberger, G.M.; Meyer, J.L.; Robinson, P.J.; Stenberg, J.R.; Turner, R.E.; Vera-Herrera, F.; Wetzel, R.G. Effects of Climate Change on Freshwater Ecosystems of the South-Eastern United States and the Gulf Coast of Mexico. *Hydrol. Process.* **1997**, *11*, 949–970. [CrossRef]
31. Seager, R.; Tzanova, A.; Nakamura, J. Drought in the Southeastern United States: Causes, Variability over the Last Millennium, and the Potential for Future Hydroclimate Change*. *J. Clim.* **2009**, *22*, 5021–5045. [CrossRef]
32. Cederstrom, D.J.; Boswell, E.H.; Tarver, G.R. *Summary Appraisals of the Nation's Ground-Water Resources, South Atlantic-Gulf Region*; U.S. Government Printing Office: Washington, DC, USA, 1979.

33. Chen, H.; Teegavarapu, R.S.V. Comparative Analysis of Four Baseflow Separation Methods in the South Atlantic-Gulf Region of the U.S. *Water* **2019**, *12*, 120. [CrossRef]

34. Joshi, N.; Kalra, A.; Lamb, K.W. Land–Ocean–Atmosphere Influences on Groundwater Variability in the South Atlantic–Gulf Region. *Hydrology* **2020**, *7*, 71. [CrossRef]

35. Sagarika, S.; Kalra, A.; Ahmad, S. Interconnections between oceanic–atmospheric indices and variability in the U.S. streamflow. *J. Hydrol.* **2015**, *525*, 724–736. [CrossRef]

36. Abtew, W.; Trimble, P. El Niño–Southern Oscillation Link to South Florida Hydrology and Water Management Applications. *Water Resour. Manag.* **2010**, *24*, 4255–4271. [CrossRef]

37. Adusumilli, S.; Borsa, A.A.; Fish, M.A.; McMillan, H.K.; Silverii, F. A Decade of Water Storage Changes Across the Contiguous United States From GPS and Satellite Gravity. *Geophys. Res. Lett.* **2019**, *46*, 13006–13015. [CrossRef]

38. Clark, C.; Nnaji, G.A.; Huang, W. Effects of El-Niño and La-Niña Sea Surface Temperature Anomalies on Annual Precipitations and Streamflow Discharges in Southeastern United States. *J. Coast. Res.* **2014**, *68*, 113–120. [CrossRef]

39. Williams, A.P.; Cook, B.I.; Smerdon, J.E.; Bishop, D.A.; Seager, R.; Mankin, J.S. The 2016 Southeastern U.S. Drought: An Extreme Departure From Centennial Wetting and Cooling. *J. Geophys. Res. Atmos.* **2017**, *122*, 10888–10905. [CrossRef]

40. Kalra, A.; Piechota, T.C.; Davies, R.; Tootle, G.A. Changes in U.S. Streamflow and Western U.S. Snowpack. *J. Hydrol. Eng.* **2008**, *13*, 156–163. [CrossRef]

41. Sadeghi, S.; Tootle, G.; Elliott, E.; Lakshmi, V.; Therrell, M.; Kam, J.; Bearden, B. Atlantic Ocean Sea Surface Temperatures and Southeast United States streamflow variability: Associations with the recent multi-decadal decline. *J. Hydrol.* **2019**, *576*, 422–429. [CrossRef]

42. Keellings, D.; Engström, J. The Future of Drought in the Southeastern U.S.: Projections from Downscaled CMIP5 Models. *Water* **2019**, *11*, 259. [CrossRef]

43. Apurv, T.; Cai, X. Drought Propagation in Contiguous U.S. Watersheds: A Process-Based Understanding of the Role of Climate and Watershed Properties. *Water Resour. Res.* **2020**, *56*, e2020WR027755. [CrossRef]

44. Cook, B.I.; Mankin, J.S.; Marvel, K.; Williams, A.P.; Smerdon, J.E.; Anchukaitis, K.J. Twenty-First Century Drought Projections in the CMIP6 Forcing Scenarios. *Earths Futur.* **2020**, *8*, e2019EF001461. [CrossRef]

45. Dai, A. Drought under global warming: A review. *WIREs Clim. Chang.* **2011**, *2*, 45–65. [CrossRef]

46. Pederson, N.; Bell, A.; Knight, T.A.; Leland, C.; Malcomb, N.; Anchukaitis, K.; Tackett, K.; Scheff, J.; Brice, A.; Catron, B.; et al. A long-term perspective on a modern drought in the American Southeast. *Environ. Res. Lett.* **2012**, *7*, 014034. [CrossRef]

47. U.S. Federal Government. Climate Resilience Toolkit. [Online]. 2014. Available online: http://climate.gov (accessed on 11 June 2022).

48. Yevjevich, V.; An Objective Approach to Definitions and Investigations of Continental Hydrologic Droughts. Colorado State University. 1967. Available online: https://linkinghub.elsevier.com/retrieve/pii/0022169469901103 (accessed on 12 June 2022).

49. Raczyński, K.; Dyer, J. Development of an Objective Low Flow Identification Method Using Breakpoint Analysis. *Water* **2022**, *14*, 2212. [CrossRef]

50. Box, G.E.P.; Jenkins, G.M.; Reinsel, G.C.; Ljung, G.M. *Time Series Analysis: Forecasting and Control*; John Wiley & Sons: Hoboken, NJ, USA, 2015.

51. Brockwell, P.J.; Davis, R.A. *Time Series: Theory and Methods*; Springer Science & Business Media: Berlin, Germany, 2009.

52. Box, G.E.; Jenkins, G.M. *Time Series Analysis: Forecasting and Control*, 3rd ed.; Prentice Hall: Englewood Cliffs, NJ, USA, 1994.

53. Hurst, H.E. Long-Term Storage Capacity of Reservoirs. *Trans. Am. Soc. Civ. Eng.* **1951**, *116*, 770–808. [CrossRef]

54. Koutsoyiannis, D. Hydrology and change. *Hydrol. Sci. J.* **2013**, *58*, 1177–1197. [CrossRef]

55. Sinaga, K.P.; Yang, M.-S. Unsupervised K-Means Clustering Algorithm. *IEEE Access* **2020**, *8*, 80716–80727. [CrossRef]

56. Raczyński, K.; Dyer, J. Multi-annual and seasonal variability of low-flow river conditions in southeastern Poland. *Hydrol. Sci. J.* **2020**, *65*, 2561–2576. [CrossRef]

57. Raczyński, K.; Dyer, J. Simulating low flows over a heterogeneous landscape in southeastern Poland. *Hydrol. Process.* **2021**, *35*, e14322. [CrossRef]

58. Modarres, R. Streamflow drought time series forecasting. *Stoch. Hydrol. Hydraul.* **2007**, *21*, 223–233. [CrossRef]

59. Sadri, S.; Kam, J.; Sheffield, J. Nonstationarity of low flows and their timing in the eastern United States. *Hydrol. Earth Syst. Sci.* **2016**, *20*, 633–649. [CrossRef]

60. Van Loon, A.F.; Van Lanen, H.A.J. A process-based typology of hydrological drought. *Hydrol. Earth Syst. Sci.* **2012**, *16*, 1915–1946. [CrossRef]

61. Stephens, T.A.; Bledsoe, B.P. Low-Flow Trends at Southeast United States Streamflow Gauges. *J. Water Resour. Plan. Manag.* **2020**, *146*, 04020032. [CrossRef]

62. Patterson, L.A.; Lutz, B.D.; Doyle, M.W. Characterization of Drought in the South Atlantic, United States. *JAWRA J. Am. Water Resour. Assoc.* **2013**, *49*, 1385–1397. [CrossRef]

63. Kam, J.; Sheffield, J. Changes in the low flow regime over the eastern United States (1962–2011): Variability, trends, and attributions. *Clim. Chang.* **2016**, *135*, 639–653. [CrossRef]

64. Ahmadi, B.; Ahmadalipour, A.; Moradkhani, H. Hydrological drought persistence and recovery over the CONUS: A multi-stage framework considering water quantity and quality. *Water Res.* **2019**, *150*, 97–110. [CrossRef] [PubMed]

65. Poshtiri, M.P.; Pal, I. Patterns of hydrological drought indicators in major U.S. River basins. *Clim. Chang.* **2016**, *134*, 549–563. [CrossRef]
66. Singh, S.; Mitra, S.; Srivastava, P.; Abebe, A.; Torak, L. Evaluation of water-use policies for baseflow recovery during droughts in an agricultural intensive karst watershed: Case study of the lower Apalachicola-Chattahoochee-Flint River Basin, southeastern United States. *Hydrol. Process.* **2017**, *31*, 3628–3644. [CrossRef]
67. Dudley, R.; Hirsch, R.; Archfield, S.; Blum, A.; Renard, B. Low streamflow trends at human-impacted and reference basins in the United States. *J. Hydrol.* **2020**, *580*, 124254. [CrossRef]
68. Cruise, J.F.; Laymon, C.A.; Al-Hamdan, O. Impact of 20 Years of Land-Cover Change on the Hydrology of Streams in the Southeastern United States1: Impact of 20 Years of Land-Cover Change on the Hydrology of Streams in the Southeastern United States. *JAWRA J. Am. Water Resour. Assoc.* **2010**, *46*, 1159–1170. [CrossRef]
69. Nagy, R.C.; Lockaby, B.G.; Helms, B.; Kalin, L.; Stoeckel, D. Water Resources and Land Use and Cover in a Humid Region: The Southeastern United States. *J. Environ. Qual.* **2011**, *40*, 867–878. [CrossRef]
70. Chen, L.G.; Gottschalck, J.; Hartman, A.; Miskus, D.; Tinker, R.; Artusa, A. Flash Drought Characteristics Based on U.S. Drought Monitor. *Atmosphere* **2019**, *10*, 498. [CrossRef]
71. Sehgal, V.; Sridhar, V. Watershed-scale retrospective drought analysis and seasonal forecasting using multi-layer, high-resolution simulated soil moisture for Southeastern U.S. *Weather Clim. Extremes* **2019**, *23*, 100191. [CrossRef]
72. Ford, T.; Labosier, C.F. Spatial patterns of drought persistence in the Southeastern United States. *Int. J. Clim.* **2014**, *34*, 2229–2240. [CrossRef]
73. Maxwell, J.T.; Ortegren, J.T.; Knapp, P.A.; Soulé, P.T. Tropical Cyclones and Drought Amelioration in the Gulf and Southeastern Coastal United States. *J. Clim.* **2013**, *26*, 8440–8452. [CrossRef]

Case Report

Floods Simulation on the Vedea River (Romania) Using Hydraulic Modeling and GIS Software: A Case Study

Cristian Popescu [1] and Alina Bărbulescu [1,2,*]

[1] Doctoral School, Technical University of Civil Engineering of Bucharest, 122-124, Lacul Tei Av., 020396 Bucharest, Romania

[2] Department of Civil Engineering, Transilvania University of Brasov, 5, Turnului Str., 900152 Brasov, Romania

* Correspondence: alina.barbulescu@unitbv.ro or alina.barbulescu@utcb.ro

Abstract: Extreme hydro-meteorological phenomena have become more frequent in recent years compared to the year 2000 in Europe, including Romania. Flooding occurs from heavy rainfalls favored by natural and anthropogenic factors such as the valley's flat slope or settlements situated near the river. Ţigăneşti and Brânceni villages (from southern Romania) are no exception and have been affected by floods many times. One of these events is that from 2005, when the flow reached $676 \text{ m}^3/\text{s}$ (a value 80 times higher than the normal flow of the Vedea River) in Brânceni. This paper aims to present a simulation of the flood that occurred during 3–6 July 2005 and its impact on the settlements, roads, and land, using field observation (including some from 2005), GIS software (ArcGIS), software for flood simulations (HEC-RAS—Hydrologic Engineering Center River Analysis System), and flow data from the Romanian National Institute of Hydrology. Simulations were run in HEC-RAS. The obtained flooded areas imported back into GIS (Geographic Information System) were used to determine the area covered by water and the length of affected roads. The surface and number of flooded buildings were calculated using different tools from ArcMap. Results were interpreted, commented on, and compared with data and maps provided by the Romanian Water National Administration. The simulation shows that the villages would be protected from the flood by building a levee along the Vedea River. Significant losses can be prevented, and money can be saved.

Keywords: flood; ArcGIS; simulation; hydraulic modeling; catchment; flow

Citation: Popescu, C.; Bărbulescu, A. Floods Simulation on the Vedea River (Romania) Using Hydraulic Modeling and GIS Software: A Case Study. *Water* **2023**, *15*, 483. https://doi.org/10.3390/w15030483

Academic Editors: Yijun Xu and Enrico Creaco

Received: 23 November 2022
Revised: 11 January 2023
Accepted: 20 January 2023
Published: 25 January 2023

Copyright: © 2023 by the authors. Licensee MDPI, Basel, Switzerland. This article is an open access article distributed under the terms and conditions of the Creative Commons Attribution (CC BY) license (https://creativecommons.org/licenses/by/4.0/).

1. Introduction

Natural hazards are sources of potential harm and dangers. They include phenomena that significantly impact the natural and human environment, with destructive consequences and high material losses. These include earthquakes, volcanic eruptions, gravitational processes such as landslides, rock falls, or avalanches, and hydrological phenomena such as floods, flash floods, forest fires, desertification, storms, and hurricanes. The hazards' impact can be so severe that the entire inhabited zones are destroyed while relief appearance, vegetation, fauna, and soils are modified. Even if the natural phenomena have a smaller impact, they can still affect the natural environment and inhabited areas. For example, part of the population must be relocated until the phenomena end or, if the impact is too severe, the inhabitants need to be permanently relocated [1].

Central Europe is affected by flooding when high precipitations occur (in May, June, or July) and sometimes when the snow melts very fast in the areas situated close to the mountains (subalpine zones) [2]. Western countries such as the United Kingdom, The Netherlands, France, and Belgium can be affected by heavy rainfalls and storms because of the oceanic influence [3]. Landslides and avalanches can occur in mountainous regions such as the Alps, Carpathians, or the Caucasus due to global warming and human intervention [4]. Summers can be arid in the far eastern parts of Europe. Due to continental influences, drought is frequent. Moreover, the rain quantity is small throughout the year, while the annual thermic amplitude between summer and winter is high. The winters are harsh in countries in eastern and northern Europe due to freezing weather, blizzard, and heavy snowfall. Forest fires can be encountered in Mediterranean areas due to intense heat [3].

Romania is located at the junction between the Mediterranean, continental, and oceanic influences. Thus, different extreme phenomena can occur, such as drought in the southeast, extreme cold, strong winds, and heavy snow in northeastern parts of the country due to continental influences and in eastern parts of Transylvania due to thermic inversions in the mountain depressions [5]. Even if desertification is not highly encountered, it occurs in some small areas in the southwest and southeast. Vegetation fires can appear in the summer due to intense heat, while flooding and flash flooding appear in May, June, and July due to heavy rainfalls or fast snow melting in the spring [4]. The precipitation, lower than in western Europe, is unequally distributed in time and space in Romania. Its volume is sometimes concentrated in one or two months, while in the rest of the year, weeks or months without precipitation are recorded, leading to floods followed by the soil deprivation of water and further erosion and desertification [3].

The impact of such extreme phenomena is so severe sometimes that agricultural production is compromised, houses are destroyed (and people must leave them due to flash floods), or the roads are closed by heavy snow or destroyed by floods [6]. Therefore, identifying the areas with risk is essential to take measures and exclude or at least reduce the human and material losses.

Hydrological phenomena are common in Romania on catchments such as Siret, Timiş, Ialomiţa, or Mureş. Different studies [1,6–9] have been dedicated to flash flood analysis to forecast and avoid their negative impact. Most research focused on atmospheric circulation and aimed at determining the mechanism of the heavy rain apparition. The results give an overview of the negative impact of such phenomena and offer a clear perspective to the authorities based on which they must take measures, such as relocating the inhabitants susceptible to flooding and building river dams or water reservoirs for flow regulation.

This paper aims to evaluate the impact of floods in the study area, considering the terrain slope, land use, shape of the river catchment, riverbed slope, or proximity of inhabited areas. Simulations have been performed using HEC-RAS, and comparisons with the available data have been made. The simulation shows that the villages are protected from flooding by building a levee along the Vedea River. The study must be extended to other regions since it provides information to the authorities. Taking into account their results, emergency institutions must be prepared with human and material resources (vehicles, shelters, food, or drinks) when instant actions are needed [10,11].

2. Study Area and Methodology

2.1. Study Area

The study area—Ţigăneşti and Brânceni villages—is situated in southern Romania, in Teleorman county (Figure 1), 100 km southwest of Bucharest, 10 km south of the county capital, Alexandria [12] situated in Teleorman County, in the catchment of Vedea River. In 2005, the population of Ţigăneşti was over 5000 inhabitants, while Brânceni had a population of 2900 inhabitants [12].

The terrain is relatively flat in the interfluve zones, with valleys that are medium in size downstream for the rivers in Teleorman. Because of steep slopes, narrower meadows, and hilly areas, flash floods can appear, favoring water accumulation downstream in Teleorman [1]. The valleys in Teleorman can be situated at least 20 m below the altitude of the interfluve, thus creating a favorable factor for water accumulation [6]. According to field observation and maps, the catchments of the Vedea River and its tributaries are elongated, with an almost flat riverbed, with a medium width, bordered by slopes with medium inclination. These characteristics, combined with showers lasting for many days, lead to severe flooding [13].

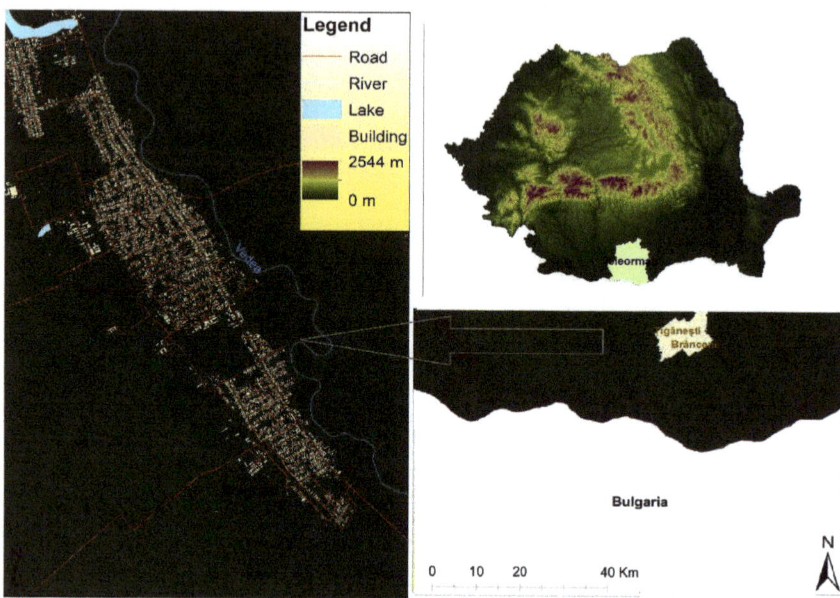

Figure 1. Location of Ţigăneşti and Brânceni.

The last significant flood on the Vedea River happened in 2014, but the most important one in recent times took place in July 2005, with a severe impact on Ţigăneşti and Brânceni villages. In the 1990s, floods occurred annually, with minor consequences on settlements but affecting the agricultural terrains. The flood analyzed in the present research occurred during 3–6 July 2005, affecting a significant number of buildings, roads, and terrains, and the first author was present there when it took place.

Ţigăneşti and Brânceni villages are situated at an altitude of about 35–37 m, close to the Vedea River. The slopes bordering the valleys can climb to altitudes higher than 60 m. The climate is temperate continental (transition influences) with cold winters (average of −3 degrees Celsius in January) and warm summers (average of 22 degrees Celsius in July), high-temperature amplitudes (over 25 degrees between January and July) with precipitations between 500 and 600 mm/m^2/year, important quantities falling in May, June, and July [14].

Meadow vegetation (close to Vedea), patches of deciduous forests, and steppe are found in the analyzed zone. Most of the former steppe and forest were replaced with agricultural terrains, settlements, and roads [14]. The absence of extensive forests on the slopes can increase the chances of water runoff into the river [13]. Meadow and chernozem soils are specific to Ţigăneşti and Brânceni. Flooding apparition is favored if the infiltration capacity is exceeded and the soils are saturated with water [15].

2.2. Methodology

The present research is based on field observations, personal observations from the flood that occurred in 2005, and hydrological data from the National Institute of Hydrology. GIS (Geographic Information Software) computing and HEC-RAS 6.2 2D version simulation have been used to evaluate the flood impact on the settlements, land, and roads in the two villages. To process data for HEC-RAS in ArcMap, the HEC-GeoRAS extension of ArcGIS was installed.

The GIS software used to process the data is ArcMap (ArcGIS) from ESRI. GIS database [16] was imported into ArcMap to obtain maps of the terrain, settlements, river, and roads [17], while HEC-RAS was used to simulate floods [18].

The following databases were used: Digital Terrain Model (DTM) of Romania at 30 m, road network, buildings, rivers, and territorial administrative units (TAUs). The coordinate system used was Stereo 1970 (31700).

First, it was necessary to process the large data mentioned above to obtain the information necessary for the simulation in the two villages. The shape of the Vedea valley containing the two villages was obtained by extracting the TAUs and cutting polygons. For example, in ArcGIS, the TAUs or buildings are polygon shapefiles, while rivers are polylines. Moreover, points are equivalent to points of interest [19].

The DTM of the inhabited areas and surroundings of Ţigăneşti and Brânceni was obtained using "Raster Clip" from Raster Processing (Data Management Tools).

A raster image file is a rectangular array of regularly sampled values, known as pixels. In GIS, a raster can be converted into a shapefile.

To create a more realistic appearance, "Hillshade" was applied from the Raster Surface extension of the 3D Analyst Tools alongside an increase of contrast in the display properties of the newly obtained raster layer. Then, transparency was set up for the raster layer obtained from DTM and the "Hillshade" raster layer. The two were overlayed, obtaining maps such as those in Figure 2. The roads and buildings were added after using "Clip" from Geoprocessing.

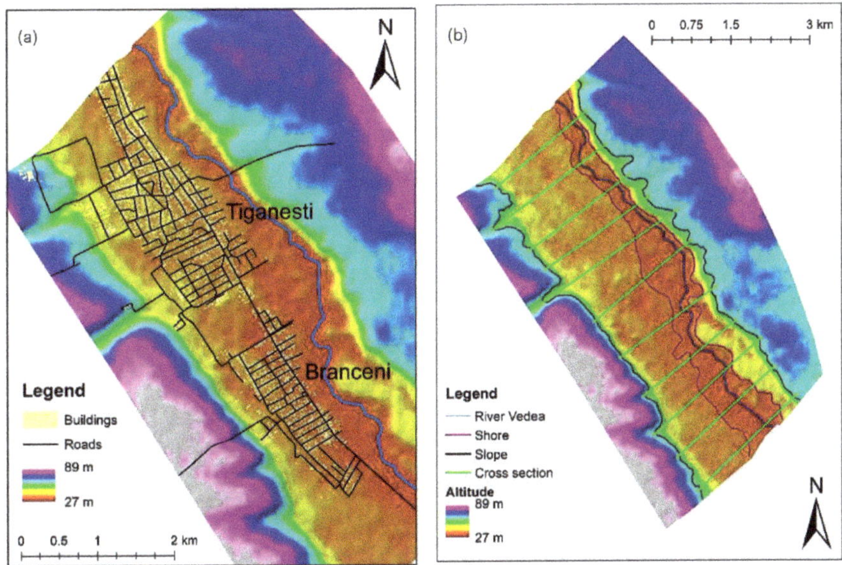

Figure 2. (**a**) DTM of Ţigăneşti and Brânceni; (**b**) RAS layers, including 11 cross-sections.

The next step was to activate the HEC-GeoRAS option and prepare the terrain for flood simulations in HEC-RAS. The RAS layers were manually created, such as stream centerline (or thalweg, the lowest point of a river), bank lines (shore), flow paths (slope), and cross-sections lines (transversal lines from an interfluve or hill to another where the depth of the flood can be observed after stimulation in HEC-RAS) [20].

In Figure 2b, the four elements necessary for flood simulation can be seen. After processing, data were exported for further processing in HEC-RAS.

From HEC-RAS, the obtained data in ArcGIS were added. Before running the simulation, several actions are required. In the "Steady Flow Data" under Edit, the flow value in cubic meters/second (m^3/s) was added, while in the Boundary Conditions, the "Critical Depth" was set up. In the present situation, the river selected for analysis is Vedea [18].

In "Geometric Data in Tables", the "Manning's n or k values" tab is accessed to set up the roughness coefficient (n). Therefore, it is essential to specify the correct values. An increase of n will cause a decrease in the water flow velocity across a surface [21]. High values mean an increased favorability to water accumulation. Table 1 contains the values of n for each cross-section. The value $n\#1$ represents the coefficient for the left bank, $n\#2$ for the main channel, and $n\#3$ for the right bank. The values for the main channel are smaller because it is smoother (based on field observations the surface is earth channel, weedy, $n = 0.03$), while the banks are rougher (pasture, agricultural land, $n = 0.035$). The numbers in Table 1 are based on a list of roughness coefficient values for different surfaces [21].

Table 1. Manning's roughness coefficient for the Vedea River.

River Station (Equivalent to Cross-Section)	$n\#1$	$n\#2$	$n\#3$
1	0.035	0.03	0.035
2	0.035	0.03	0.035
3	0.035	0.03	0.035
4	0.035	0.03	0.035
5	0.035	0.03	0.035
6	0.035	0.03	0.035
7	0.035	0.03	0.035
8	0.035	0.03	0.035
9	0.035	0.03	0.035
10	0.035	0.03	0.035
11	0.035	0.03	0.035

Manning's equation, used in HEC-RAS for steady flows gives the flow rate as a function of the channel velocity, flow area, and channel slope [21].

$$Q = VA = (1/n)AR^{2/3}\sqrt{S} \tag{1}$$

where:

Q = Flow rate (m^3/s)
V = Velocity (m/s)
A = Flow area (m^2)
n = Manning's roughness coefficient—setup by the HEC-RAS user (Table 1)
R = Hydraulic radius, (m)
S = Channel slope, (m/m)

After setting the variables, the simulation can be run by accessing the "Steady Flow Analysis". After processing, the level of flooding for each cross-section and the flooded surface can be observed [22].

According to the documentation from HEC-RAS, the energy Equation (2) is used for the profile calculations:

$$Z_2 + Y_2 + a_2 V_2^2/(2g) = Z_1 + Y_1 + a_1 V_1^2/(2g) + h_e \tag{2}$$

where:

Z_1, Z_2—elevation of the main channel inverts (m)
Y_1, Y_2—depth of water at cross-sections (m)
V_1, V_2—average velocities (total discharge/total flow area) (m/s^2)
a_1, a_2—velocity weighting coefficients
g—gravitational acceleration (m/s^2)
h_e—energy head loss (m)

HEC-RAS simulates floods based on the information specified by the user and mentioned above in the methodology: stream centerline, bank lines, flow paths, cross-sections, digital terrain model, roughness coefficient, flow value (example: 676 m^3/s), and specifying the information under "steady flow data" as critical depth [22].

After obtaining the flooded areas (2D), the accumulated water level for each cross-section can be visualized. In the present research, 11 cross-sections can be accessed [17]. Figure 3 shows the valley shape, the flooded areas, and the water depth for two cross-sections. The user can view the flooded surface by accessing "X-Y-Z perspective plots" from the View tab (Figure 4). The length and depth of the water for the entire valley differ for each cross-section.

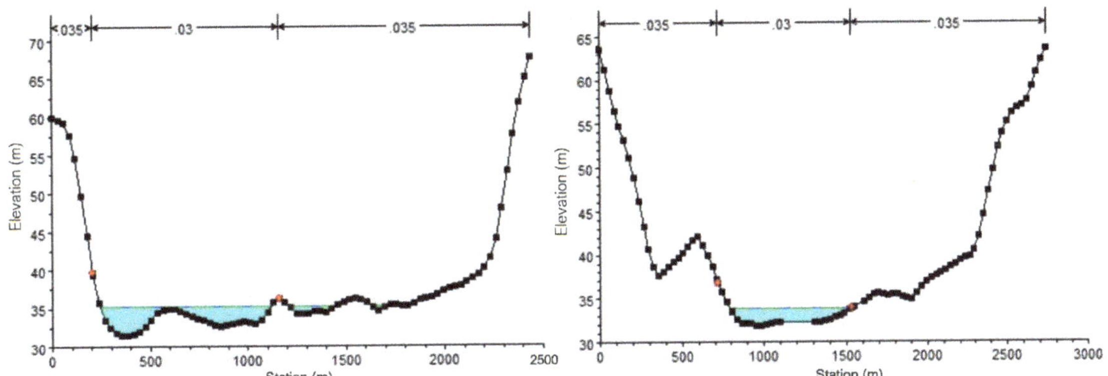

Figure 3. The Valley cross-sections (water volume represented in blue and river banks in red).

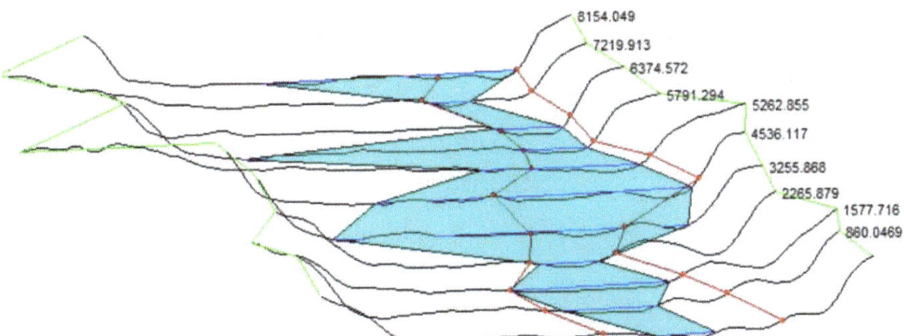

Figure 4. Flooded surface in Ţigăneşti and Brânceni area (the water volume represented in blue).

After verifying the correctness of the obtained data, the flooded area was exported from HEC-RAS as a raster layer using RAS Mapper to be imported into ArcMap. Figure 5 shows a shape of a flooded area. The blue area represents the flooded surface, the green lines the cross sections, while the color palette from green to gray represents altitudes. In ArcMap, the imported raster of the flood was reclassified from multiple values (multiple values come from the depth of the flooded area) into a single value using "Reclassify" from the Raster Reclass extension (3D Analyst Tools). It is essential to have a single value for the raster because, in the following steps, it is easier to quantify the affected buildings and roads and calculate the flooded area.

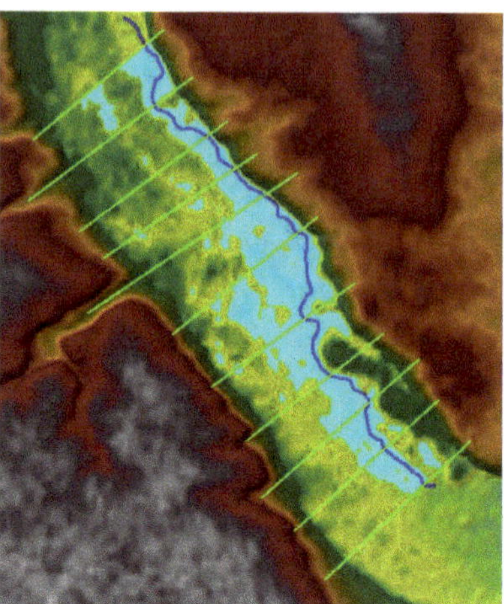

Figure 5. The flooded area visible in RAS Mapper (HEC-RAS).

Further, the raster was converted in GIS into a shapefile using Conversion Tools (from Raster to Polygon shapefile) [23].

For computing the flooded area, a new column was created ("surface") in the attributes of the flooded area polygon shapefile. Using "calculate geometry" in the attribute table, the flooded surface in "km^2" was estimated.

Then, the surface of the total surface of the flooded buildings was computed. First, with the help of "Clip" from Geoprocessing, the affected buildings were obtained based on the flood polygon shapefile [23]. In the attribute table of the affected buildings, the surface of each item can be observed. To calculate the total surface of the buildings in "m^2", "Summary Statistics" from Statistics extension (Analysis Tools) was used. A table under "List by source" (Table of Contents) with the total surface of the buildings was obtained.

When using ArcMap, one can also determine the number of flooded buildings. First, polygons corresponding to each structure were transformed into points using "Feature to Point" from the Features extension (Data Management Tools). Further, the points were summarized using "Spatial Join" from the Overlay extension (Analysis Tools) based on the flooded area polygon shapefile [24]. In the attribute table, under the new column called "Join_Count" of the newly created polygon, one can see the number of affected buildings [24].

The length of the flooded roads can also be calculated in ArcMap. Using "Clip" from the Geoprocessing tab, the flooded roads were obtained based on the flooded area polygon shapefile [25]. In the attribute table, the distance of each flooded road in meters was computed using "calculate geometry" under a newly created column. The next step was to determine the total flooded roads. Using "Summarize Statistics" from the Statistics extension (Analyst Tools), the total length of flooded roads was calculated. The new table is also available under "List by source" (Table of Contents) [25].

The methodology used to calculate the surface of the flooded area can be applied to any flooded area polygon shapefile obtained based on flow values (m^3/s). Of course, the surface of the affected buildings and the length of flooded roads are different depending on the flow values used to calculate them. The flow values used in this study were 200, 400, 676, and 800 m^3/s.

Even though the values in July 2005 did not exceed 676 m^3/s based on the data from the National Institute of Hydrology, it was interesting to see the flooded surface at a bigger value, 800 m^3/s, because after discussing with the villagers that lived for decades in Țigănești and Brânceni, it was learned that larger floods occurred in the 1970s and 1980s.

For Brânceni, the DTM was modified in HEC-RAS by adding a levee to see how it protects the village from flooding at different flow values.

The losses were also calculated based on the Ministry of Environment and national road company data. The model was validated based on the observations from 2005 (similarities in how the flood area appeared in reality and the simulation), the configuration of the Vedea Valley, and values from the National Institute of Hydrology. Moreover, the observations from 2014, when important flow values occurred, showed that the village could be protected even at high values.

In the HEC-RAS simulation, the water level did not exceed the dike of approximately 2 m at flows up to 800 m^3/s. Based on a document from the Romanian Waters, the levee was projected to withstand flows of 813 m^3/s; it is one of the reasons why the value of 800 cubic meters was also chosen in the analysis. A similar situation was observed in 2014 when even though the flow was very high, the water level did not exceed the newly constructed levee (the difference between the water level and the top of the dike was a little over half a meter).

The study limitation comes from the data available from the National Institute of Hydrology that does not include the water depth measured at the hydrometric station, but only the flow rate. Thus, the matching between the simulation and the event on the site had to be done using the above methods.

3. Results and Discussion

The flow registered for river Vedea in Ţigăneşti and Brânceni during 3–6 July 2005 was significantly high to affect the two villages. Usually, the river flow is around 8.36 m^3/s. Thus, the flow exceeded more than 80 times when the flood occurred. On the 3 July at noon, river Vedea had a flow of abound 200 m^3/s. The next day, it was at its maximum of 676 m^3/s at 11 a.m. and 12 p.m. On the 6 July at 6 a.m., the flow value decreased to 133 m^3/s. The water depth increased by almost 5 m from the thalweg. The water level reached up to 1–2 m in the inhabited area.

According to Figure 6a, at a flow of 200 m^3/s, most of the areas affected included crops, significant parts of the meadow located in the proximity of the river, pasture areas, but also some roads and houses in the northern and central part of Brânceni and several settlements, roads in the southeastern and northern parts of Ţigăneşti and a bridge in the same village. Because of the impressive quantity of precipitation felt in that period (2005 was the rainiest year in recent times), several ponds formed in the two villages, flooding roads and buildings.

Figure 6b shows that at a flow of 400 m^3/s, the flooded area increased compared to that for 200 m^3/s, covering important surfaces with houses and roads in Brânceni (the northern and central parts and smaller numbers in the south). Ţigăneşti was less affected by comparison with Brânceni. Agricultural terrains and pastures were flooded as well.

When the highest flow rate of 676 m^3/s was recorded, according to Figure 7a, the areas affected were significant parts of Brânceni (north, center, and south). By comparison, a smaller surface Ţigăneşti was affected (in the north and southeast). The flooded area was higher compared to that at 400 m^3/s. Ponds that flood houses and roads were also formed. In Brânceni, the access (National Road 51) to Alexandria (in the north) and Zimnicea was cut because of the flooding, making the authorities' intervention difficult. The bridge affected in Ţigăneşti had only local importance, connecting the village with the pasture land located in the east.

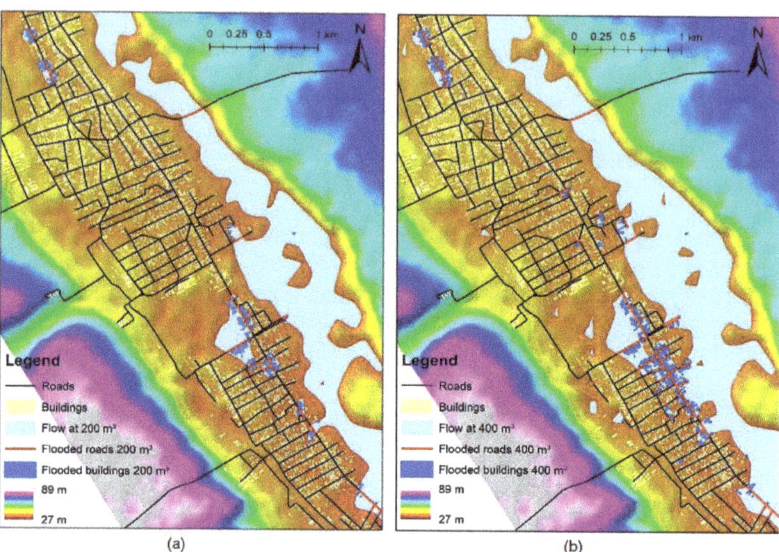

Figure 6. Flooding at (a) 200 m^3/s, (b) 400 m^3/s.

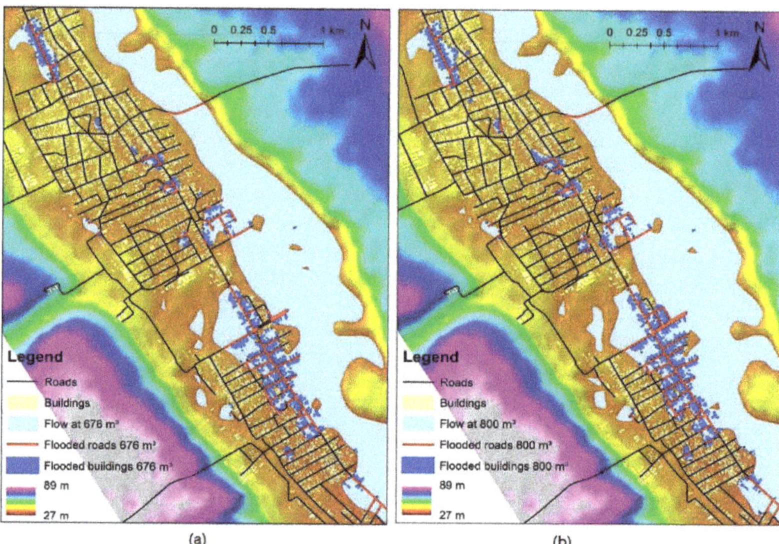

Figure 7. Flooding at (a) 676 m^3/s, (b) 800 m^3/s.

Even though the flow rate of 800 m^3/s is hypothetical, it was probably recorded in the 1970s when bigger floods occurred. A larger surface was affected compared to the one at 676 m^3/s. One-third of the settlements and roads in Brânceni is entirely covered by water, according to Figure 7b.

The most critical factors that favored the settlements' flooding are their location near the river, the terrain elevation [26], very high precipitations that lasted for days, almost flat terrain in the river valley, the elongated shape of the catchment, absence of large forested areas, and precarious hydro-technical works. The water depth in the villages exceeded in some parts 1.5 m. Figures 6 and 7 show that Brânceni was more affected because the settlements are located at a lower elevation compared to Ţigăneşti.

According to Figure 8, in Branceni, almost the entire section of the main road was flooded. The village supply of goods was affected for a couple of days, while the electricity was stopped for four days. The village was accessible only by large 4 × 4 vehicles through its edge from the interfluve (hill).

Figure 8. Flooded areas in Brânceni at 676 m^3/s.

The village hall, school, kindergarten, church, monastery, dispensary, police station, and several shops were completely flooded. Here, the water levels exceeded 1 m. After the flooding, one house was destroyed, the structure of several settlements was affected, animals drowned, household crops were compromised that year, and water from the fountains was no longer safe to drink. In addition, after the waters' withdrawal, areas covered with dirt and waste remained.

Figure 9 contains photos from that period. The left one presents a flooded house on a lateral street. The second shows National Road 51 covered by water, while the third is from the center of Brânceni.

Figure 9. Photos from the 2005 flooding in Brânceni.

Table 2 contains the estimated loss by flooding in the four scenarios. At 200 m^3/s, the flooded area was 2.7 km^2, and 306 buildings were affected, representing 23,314 m^2 (3.39% of the total surface of the buildings). The length of the flooded roads was 3212 m (4.84% of the total length of the roads).

Table 2. List of values for flooded areas, buildings, and roads at different flow values.

Flow Rate	200 m^3/s	400 m^3/s	676 m^3/s	800 m^3/s	Damages
Flooded area (km^2)	2.7	3.46	4.16	4.45	Total surface of buildings (m^2)
Number of flooded buildings	306	674	1117	1380	687,718
Surface of flooded buildings (m^2)	23,314	52,278	99,693	108,469	Total length of roads (m)
Length of flooded roads (m)	3212	6483	10,287	12,125	66,238

At 400 m^3/s, the flooded surface was 3.46 km^2 and 674 buildings were affected, representing 52,278 m^2 (7.6% of the buildings' surface). The distance of the flooded roads was 6483 m (9.78% of the total length of the roads).

At a flow of 676 m^3/s, the flooded area was 4.16 km^2, and 1117 buildings were flooded, representing 14.49% of the buildings' total surface. The length of the flooded roads was 15.53% of the total length of the roads.

Regarding the hypothetical flow of 800 m^3/s, the flooded surface would be 4.45 km^2, and 1380 buildings and 12.125 km would be affected.

Table 3 contains the estimated flood losses at a flow of 676 m^3/s in both villages. The building unit value is a national estimation from the Ministry of Environment from 2005. The current value from 2022 was calculated by multiplying the 2005 value with the annual inflation from 2006 to 2022. The cost for the complete reconstruction of one kilometer of the national road that crosses the villages in the flat area, such as the one studied in this article, comes from the contracts of the national road company (CNAIR). The unit value loss in the case of buildings is 14,959 lei, while the total loss is more than 3 million euros. In the case of the national road, the unit value (1 km) is 12,223,600 lei, while the estimated loss is more than 8 million euros if the road should be rebuilt. Thus, the losses can be very high.

Table 3. Estimated losses in Romanian currency (lei) and euros for year 2022.

Infrastrucure Elements	Number of Houses/km	Unit Value (RON)	Estimated Losses (RON)	Estimated Losses (Euros)
Houses	1117	14,959	16,709,203	3,389,433
Length of national roads (km)	3.306 km	12,223,600	40,411,221.6	8,197,704

In 2005, the flood impact was very high, comparable with a hydrological phenomenon with a probability of 1%. Figure 10a is the result of our simulation—the flood at 676 m^3/s—while Figure 10b is the map obtained from the National Administration Romanian Water, showing the flood with a probability of 1% in the same zone [27,28]. One can find in the two images similarities regarding the flooded surfaces. For example, the most affected zones are the northern part of Brânceni and the pasture areas between Ţigăneşti and Brânceni. Moreover, in both representations, Ţigăneşti village was less affected.

After 2005, a tall river levee of around two meters was built in Brânceni to defend the village from flooding. The levee proved to be useful in 2014 when the water level of the Vedea River augmented again due to heavy rainfall [29]. In the simulation from Figure 11 at (a) 200 and (b) 400 m^3/s the village is not flooded because the levee protects it. This applies also to the higher flows of 676 and 800 m^3/s (Figure 12).

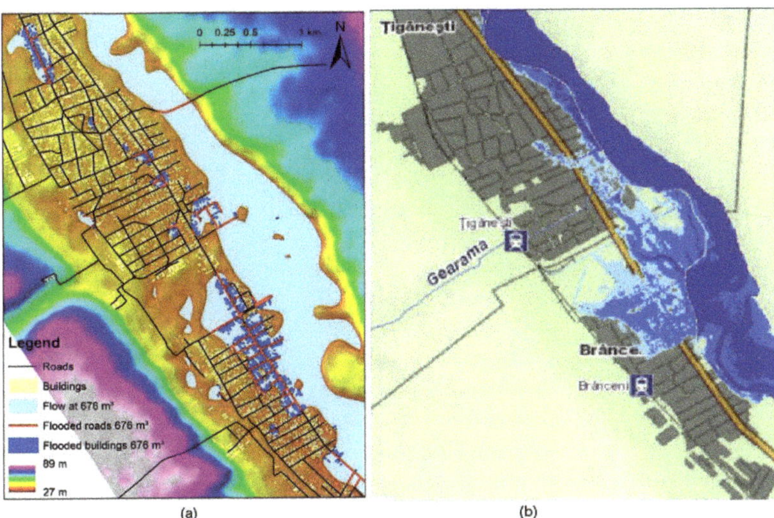

Figure 10. (**a**) Flood at 676 m^3/s and (**b**) flood with a probability of 1% (image from 2014, RoWater) [27].

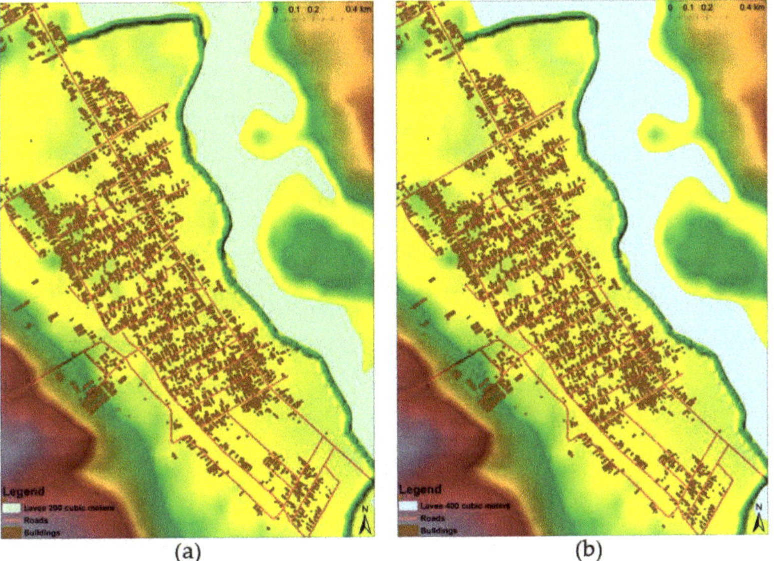

Figure 11. Flooding simulation (**a**) 200 m^3/s, (**b**) 400 m^3/s after building the levee.

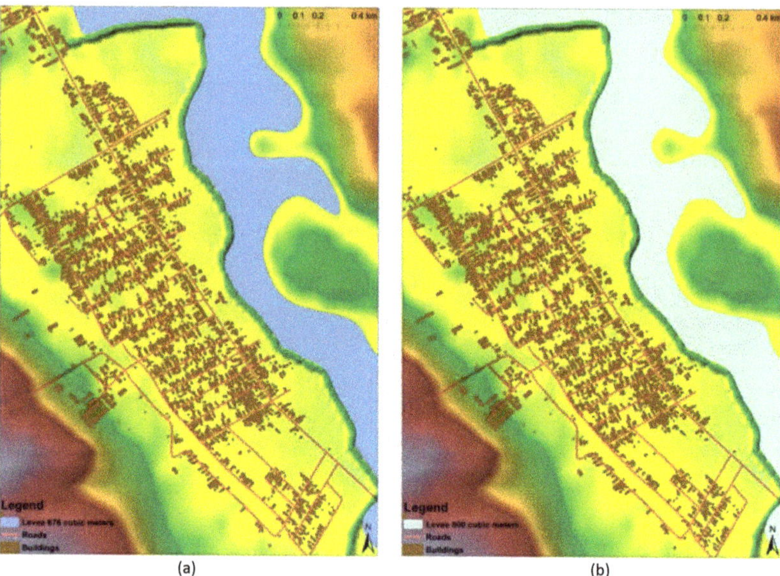

(a)   (b)

Figure 12. Flooding at (**a**) 676 m^3/s, (**b**) 800 m^3/s after building the levee.

Based on the information from the European climate change website (climate.ec.europa.eu), the cost for lower river levees is 3 million euros/km. The combined distance of the dikes in Brânceni and Ţigăneşti is 7 km. Thus, the implementation cost is 21 million euros. Because the levee prevents floods, the potential loss becomes, in this case, a benefit (avoid the houses and roads losses). The benefit-cost is 11.58 million euros if the flood occurs only once. In the past 30 years, there have been two major floods (2005 and 2014) and other small ones. It is hard to determine the difference between the implementation and benefit for 100 years, for example. However, if the flow of 676 m^3/s occurred twice in 100 years, then the benefit would be 11.52 million euros multiplied by 2 (a total of 23.16 million euros). Therefore, it is worth building levees in this area because the difference between implementation and benefit is 23.16 − 21 = 2.16 million euros.

4. Conclusions

In this paper, we presented of case study on the floods of the Vedea River, Romania. The location was selected because it is prone to flooding when large quantities of precipitations fall due to the elongated catchment shape, the location of the settlements close to the Vedea River, and the almost horizontal slope of the valley.

At the beginning of July 2005, the maximum recorded value of the flow registered was 676 m^3/s (80 times higher than the average flow), affecting 1117 houses, administrative buildings, 12.287 km of roads, crops, and pasture areas, especially in the northern and central parts of the Brânceni village. The water level exceeded 1 m in the central area of Brânceni. Ţigăneşti village was less affected. The flooding that occurred in July 2005 brought attention from the national media. The emergency authorities used special 4 × 4 vehicles, inflatable boats, and water pumps. The intervention was difficult, especially in Brânceni, because the paved road was completely covered with water, and the access from the two cities, Alexandria and Zimnicea, was cut off.

Extreme flooding phenomena are hard to predict. They frequently occur in May, June, and July, or spring (after heavy winters, when the snow melts rapidly due to the abrupt temperature change). The region of Ţigăneşti and Brânceni is no exception.

The simulation of the flood impact was performed at flow rates of 200, 400, 676 m^3/s, and 800 m^3/s, indicating an increase in the number of the flooded buildings from 306 to

1380, of the surface of flooded buildings from 23,314 m^2 to 108,469 m^2, and the number of kilometers of flooded roads from 3.212 to 12.125, in the worst case (800 m^3/s) compared to the best one (200 m^3/s). Losses were very high at a flow of 676 m^3/s (11.58 million euros) before the levee construction.

The situation of the two villages improved after the flooding in 2005 since, in Brânceni, a high and solid levee was built on the right bank, while in Ţigăneşti, protection works against floods with a probability of occurrence of 5% were carried out [28]. The new levee in Brânceni protects the village from floods even at flow values of 800 m3/s, as results from the performed simulation.

Similar studies will be carried out for other zones where flooding is expected and will be publicly available to provide the authorities with solid documentation for making decisions for population protection.

This paper is unique for the Brânceni and Ţigăneşti villages quantifying the flooded roads and buildings (including building surface). It determined the flooded areas at different flow values, the potential losses, and the economic loss. It also showed the usefulness of the levee. So far, no such research has been carried out for the discussed area.

Author Contributions: Conceptualization, C.P. and A.B.; methodology, C.P.; software, C.P.; validation, C.P. and A.B.; formal analysis, A.B.; investigation, C.P.; resources, C.P. and A.B.; data curation, C.P.; writing—original draft preparation, C.P.; writing—review and editing, A.B.; visualization, C.P. and A.B.; supervision, A.B.; project administration, A.B.; funding acquisition, C.P. and A.B. All authors have read and agreed to the published version of the manuscript.

Funding: This research received no external funding.

Institutional Review Board Statement: Not applicable.

Informed Consent Statement: Not applicable.

Data Availability Statement: Data will be available on request from the authors.

Conflicts of Interest: The authors declare no conflict of interest.

References

1. Roşu, A. *The Physical Geography of Romania*; Editura Didactică şi Pedagogică: Bucharest, Romania, 1980. (In Romanian)
2. Gaume, E.; Bain, V.; Bernardara, P.; Newinger, O.; Barbuc, M.; Bateman, A.; Blaškovičová, L.; Blöschl, G.; Borga, M.; Dumitrescu, A.; et al. A compilation of data on European flash floods. *J. Hydrol.* **2009**, *367*, 70–78. [CrossRef]
3. Ciulache, S.; Ionac, N. *Essential in Meteorology and Climatology*; Editura Universitară: Bucharest, Romania, 2007. (In Romanian)
4. Ielenicz, M.; Comanescu, L. *General Physical Geography with Elements of Cosmology*; Editura Universitară: Bucharest, Romania, 2009. (In Romanian)
5. Velcea, V.; Savu, A. *Geography of the Romanian Carpathians and Subcarpathians*; Editura Didactică şi Pedagogică: Bucharest, Romania, 1982. (In Romanian)
6. Costache, R.; Zaharia, L. Flash-flood potential assessment and mapping by integrating the weights-of-evidence and frequency ratio statistical methods in GIS environment—Case study: Bâsca Chiojdului River catchment (Romania). *J. Earth Syst. Sci.* **2017**, *126*, 59. [CrossRef]
7. Stancalie, G.; Antonescu, B.; Oprea, C.; Irimescu, A.; Catana, S.; Dumitrescu, A.; Barbuc, M.; Matreata, S. Representative flash flood events in Romania Case studies. In *Flood Risk Management: Research and Practice*; Samuels, P., Huntington, S., Allsop, W., Harrop, J., Eds.; Taylor & Francis: London, UK, 2008; pp. 1587–1597.
8. Stefanescu, V.; Stefan, S. Synoptic context of floods and major flashfloods in Romania during 1948–1995. *Rom. Rep. Phys.* **2011**, *63*, 1083–1098.
9. Necula, M.F. Observations on the September 2007 flash flood in the Tecucel River Basin. *Present Environ. Sustain. Develop.* **2010**, *4*, 363–376.
10. Aivaz, K.-A.; Florea Munteanu, I.; Stan, M.-I.; Chiriac, A. A Multivariate Analysis on the Links Between Transport Noncompliance and Financial Uncertainty in Times of COVID-19 Pandemics and War. *Sustainability* **2022**, *14*, 10040. [CrossRef]
11. Aivaz, K.-A. Correlations Between Infrastructure, Medical Staff and Financial Indicators of Companies Operating in the Field of Health and Social Care Services. The Case of Constanta County, Romania. In *Under the Pressure of Digitalization: Challenges and Solutions at Organizational and Industrial Level*, 1st ed.; Edu, T., Schipor, G.L., Vancea, D.P.C., Zaharia, R.M., Eds.; Filodiritto Editor Proceedings: Bologna, Italy, 2021; pp. 17–25.
12. Ghinea, D. *Geographic Encyclopedia of Romania*; Editura Enciclopedica: Bucharest, Romania, 2000. (In Romanian)

13. Prăvălie, R.; Costache, R. The analysis of the susceptibility of the flash-floods genesis in the area of the hydrographical basin of Basca Chiojdului River. *Forum Geogr.* **2014**, *XIII*, 39–49. [CrossRef]
14. Ielenicz, M. *Physical Geography of Romania, Vol. II—Climate, Waters, Vegetation, Soils, Environment*; Editura Universitară: Bucharest, Romania, 2007. (In Romanian)
15. Drobot, R. *Hydrology and Hydrogeology Lessons*; Editura Didactică şi Pedagogică: Bucharest, Romania, 2020. (In Romanian)
16. European Environment Agency. GIS Data. Available online: https://www.eea.europa.eu/data-and-maps/data/eea-coastline-for-analysis-2/gis-data (accessed on 13 April 2022).
17. Costache, R. Flash-flood Potential Index mapping using weights of evidence, decision trees models and their novel hybrid integration. *Stoch. Environ. Res. Risk Assess.* **2019**, *33*, 1375–1402. [CrossRef]
18. Costache, R.; Prăvălie, R.; Mitof, I.; Popescu, C. Flood vulnerability assessment in the low sector of Sărăţel catchment. Case study: Joseni village. *Carpathian J. Earth Environ. Sci.* **2015**, *10*, 161–169.
19. Salvati, L.; Zitti, M. Assessing the impact of ecological and economic factors on land degradation vulnerability through multiway analysis. *Ecol. Indic.* **2009**, *9*, 357–363. [CrossRef]
20. Tehrany, M.S.; Pradhan, B.; Jebur, M.N. Flood susceptibility mapping using a novel ensemble weights-of-evidence and support vector machine models in GIS. *J. Hydrol.* **2014**, *512*, 332–343. [CrossRef]
21. Zhe, L.; Juntao, Z. Calculation of Field Manning's Roughness Coefficient. *Agric. Water Manag.* **2001**, *49*, 153–161.
22. Khosravi, K.; Shahabi, H.; Pham, B.T.; Adamowski, J.; Shirzadi, A.; Pradhan, B.; Dou, J.; Ly, H.-B.; Gróf, G.; Ho, H.L.; et al. A comparative assessment of flood susceptibility modeling using Multi-Criteria Decision-Making Analysis and Machine Learning Methods. *J. Hydrol.* **2019**, *573*, 311–323. [CrossRef]
23. Gatto, M.P.A.; Misiano, S.; Montrasio, L. On the Use of MATLAB to Import and Manipulate Geographic Data: A Tool for Landslide Susceptibility Assessment. *Geographies* **2022**, *2*, 341–353. [CrossRef]
24. Guiwen, L.; Zhang, Q.; Yang, Z.; Li, T. Optimizing Multi-Way Spatial Joins of Web Feature Services. *ISPRS Int. J. Geo-Inf.* **2017**, *6*, 123. [CrossRef]
25. Bulai, M.; Ursu, A. Creating, testing and applying a GIS road travel cost model for Romania. *Geogr. Tech.* **2012**, *15*, 8–18.
26. Armenakis, C.; Du, E.X.; Natesan, S.; Persad, R.A.; Zhang, Y. Flood Risk Assessment in Urban Areas Based on Spatial Analytics and Social Factors. *Geosciences* **2017**, *7*, 123. [CrossRef]
27. National Administration of Romanian Waters. Available online: https://rowater.ro (accessed on 10 April 2022).
28. Planul de Management al Riscului la Inundaţii, Administraţia Bazinală de Apă Argeş—Vedea (The Plan of Risk Management in Floods, Water Basin Administration Argeş—Vedea). Available online: https://rowater.ro/despre-noi/descrierea-activitatii/managementul-situatiilor-de-urgenta/directiva-inundatii-2007-60-ce/harti-de-hazard-si-risc-la-inundatii/ (accessed on 21 May 2022).
29. Kryzanowski, A.; Brilly, M.; Rusjan, S.; Schnabl, S. Structural flood-protection measures referring to several European case studies. *Nat. Hazards Earth Syst. Sci. Discuss.* **2013**, *1*, 247–274. [CrossRef]

Disclaimer/Publisher's Note: The statements, opinions and data contained in all publications are solely those of the individual author(s) and contributor(s) and not of MDPI and/or the editor(s). MDPI and/or the editor(s) disclaim responsibility for any injury to people or property resulting from any ideas, methods, instructions or products referred to in the content.

 *water*

Article

Changes in the Stability Landscape of a River Basin by Anthropogenic Droughts

Laura E. Garza-Díaz * and **Samuel Sandoval-Solis**

Department of Land, Air, and Water Resources, University of California, Davis, CA 95616, USA
* Correspondence: legarza@ucdavis.edu

Abstract: As water resources enter the era of the Anthropocene, the process of anthropogenic droughts arises as the interplay between climate cycles and human-centered water management in rivers. In their natural conditions, rivers exhibit a natural hydrologic variability, wet and dry cycles, that are a vital property for promoting ecological resilience. Human activities alter the temporal variability of streamflow, a resilience property of river systems. We argue that anthropogenic droughts in river basins can lead to changes in the resilience properties of the system depicted in stability landscapes. This study aims to analyze anthropogenic droughts and the changes provoked to the stability landscapes of the streamflow system of a river basin. We use 110 years of regulated and naturalized streamflow data to analyze the hydrologic variability (wet periods and droughts) of a river system. First, we determined the streamflow drought index (SDI), and the results were assessed using probability distribution functions to construct stability landscapes and explore the resilience properties of the system. The transboundary basin of the Rio Grande/Rio Bravo (RGB) is used as a case study. Our main findings include evidence of resilience erosion and alterations to the properties of the stability landscape by the human-induced megadrought in the RGB, which resulted from extensive anthropogenic alteration and fragmentation of the river system. The novelty of this research is to provide a baseline and move forward into quantifying ecological resilience attributes of river basins in water resources planning and management.

Keywords: anthropogenic drought; ecological resilience; river basin; stability landscape

 **check for updates**

Citation: Garza-Díaz, L.E.; Sandoval-Solis, S. Changes in the Stability Landscape of a River Basin by Anthropogenic Droughts. *Water* **2022**, *14*, 2835. https://doi.org/10.3390/w14182835

Academic Editor: Alina Barbulescu

Received: 6 August 2022
Accepted: 6 September 2022
Published: 12 September 2022

Publisher's Note: MDPI stays neutral with regard to jurisdictional claims in published maps and institutional affiliations.

Copyright: © 2022 by the authors. Licensee MDPI, Basel, Switzerland. This article is an open access article distributed under the terms and conditions of the Creative Commons Attribution (CC BY) license (https://creativecommons.org/licenses/by/4.0/).

1. Introduction

As social-ecological systems (SES), river basins are inherently bound to a fundamental property of ecological resilience: dynamism, expressed by the temporal variability of the natural flow regime. Historic cycles of flooding and drought in the natural flow regime are integral components of most intact running water ecosystems [1] as these exert dominant controls on ecosystem structure and function [2]. As water resources are well into the era of the Anthropocene, climate change and human dominance pose pressing challenges to the hydrologic cycle and its components, putting the integrity and resilience of river basins at higher risk. The human influence on the global hydrological cycle is now the dominant force behind changes in water variability across the world and in regulating and triggering hydrologic resilience changes in the Earth system. Globally, extreme weather or climate events are expected to become more frequent and increase in intensity and duration, due to climate change and are largely exacerbated by the persistent pressures of human water demands in creating such extreme environmental conditions. The complex and interrelated processes between natural and human-induced changes drive the development of anthropogenic droughts [3–6]; a compound multidimensional and multiscale phenomenon governed by the combination of natural water variability, climate change, human decisions and activities, and altered micro-climate conditions due to changes in land and water management [3]. The growing frequency of precipitation extremes, especially droughts, will have profound consequences on the hydrologic variability of the streamflow systems and the natural flow regime, creating selective pressures in the environment and society. In return, this will affect the resilience of river basins and the capacity of systems to withstand shocks and perturbations without modifying their functional identity and adapting to changing conditions [7].

Resilience theory applied to water systems can offer a perspective on the understanding of anthropogenic droughts as one of the central disturbances of streamflow dynamics and the potential changes in hydrological resilience across all scales, from local watersheds to regional and transboundary basins. Catastrophic disturbances such as anthropogenic megadroughts can cause shifts in ecosystems into alternative states, through which many ecosystems can lose their functionality and identity. This phenomenon can be assessed by determining the relationships between natural drivers and processes that allow for ecosystem functioning (e.g., streamflow) and anthropogenic pressures (e.g., water use, land use change, and management practices). To see how resilience is affected by changes in hydrologic conditions, we may construct stability landscapes [8] which are good approximations for understanding resilience concepts [9]. The metaphor of stability landscapes in resilience theory depicts the various stable states of a system as a series of "basins of attraction," which are regions in state space in which a system tends to remain (Figure 1—retrieved from Dakos and Kefi, 2002 [10]) and have been used to explain the dynamics of several ecosystems and the components of resilience including resistance, latitude, precariousness, and panarchy [11]. Stability landscapes help understand the properties of dynamical systems and have been used to represent resilience characteristics of shallow lakes [12], urban water systems [13], tropical forest and savanna [14], climate states [15], plant patterns in drylands [16,17], and river management [18].

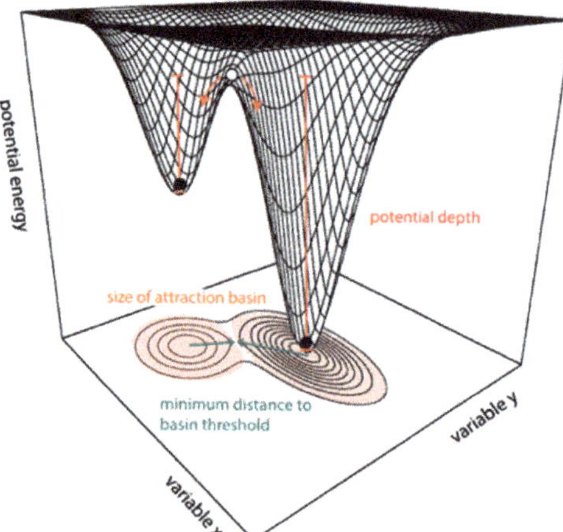

Figure 1. A (hypothetical) stability landscape of a two-dimensional system with hilltops and valleys, also known as a marble-in-a-cup or balls-and-cups landscape. Black balls are found at the bottom of the valley and represent stable states. Retrieved from Dakos and Kefi, 2020 [10].

A stability landscape with several basins of attraction corresponds to the various stable states in which a system will exist. As streamflow in river basins is modified by exogenous drivers (precipitation, exchange rates) and endogenous processes (infrastructure, management practices), the streamflow system may move from one basin of attraction to another when substantial disturbances occur (e.g., hurricanes, dry spells, ENSO patterns, management practices) and affect the state variables. State variables include temporal or spatial characteristics, and when these occur, the set of variables will persist in one of many possible configurations, which may shift to a different configuration or equilibrium after a perturbation [9]. However, changes in environmental conditions that affect processes between state variables, such as river fragmentation or changes in the natural flow regime, will alter the shape of the stability landscape as these pressures directly affect state variables.

We argue that the evidence of anthropogenic drought in river basins can lead to changes in the stability landscape, such as changes in position, width, depth, and configuration of the basins of attraction. The problem is that, for most systems, its quantification is challenging to operationalize, and we usually do not know the shape of the stability landscape, but in principle, we could derive their shapes [10,19] to assess the resilience properties of resistance, latitude, precariousness, and panarchy. The objective of this study is to assess anthropogenic droughts and the changes provoked to the stability landscapes of the streamflow system of a river basin. This study assessment is twofold: (1) analyze the hydrologic variability (floods and droughts) of a river system by comparing the natural and regulated flow regimes using long-term streamflow data, and (2) construct stability landscapes and explore properties of resilience in terms of changes in the basins of attraction of the natural and anthropogenic state. The transboundary basin of the Rio Grande/Rio Bravo, located half in the United States (U.S.) and the other half in Mexico, will be used as a case study given its arid, water-limited, and drought-prone landscape, its binational context, and its long history of human manipulation. This research identifies the current anthropogenic state of a transboundary basin in comparison to its natural state and approximates the metaphor of stability landscapes and basins of attraction using streamflow as a representation of the resilience conditions of river basins which can be used in any local, regional, or international scale worldwide.

2. Materials and Methods

To analyze the basin-wide dynamics, this study uses 110 years of monthly streamflow from 1900 to 2010 at eight control points (i.e., hydrologic gauge stations) to portray the natural and anthropogenic states of the RGB. Four control points are selected in the mainstem of the river basin: San Marcial, El Paso, Above Amistad Dam, and Anzalduas. And four at the outlet of the main sub-basins: Rio Conchos, Pecos River, Rio Salado, and Rio San Juan (Figure 2). The overall methodology includes (1) data collection of historical streamflow data, including inflows and outflows of the river system; (2) converting gaged or observed flows to naturalized flows using a water mass balance; (3) performing a hydrologic drought assessment for the observed and naturalized flows to observe the hydrologic variability of the river basin; and (4) developing of stability landscapes to compare resilience attributes between the naturalized and anthropogenic states of the river basin.

2.1. Case Study

The transboundary Rio Grande/Bravo (RGB) basin is a water-scarce basin full of extreme climate conditions, from heavy snowfall and tropical storms to prolonged minimal precipitation, which ranges from 190 to 2250 mm per year and an average temperature range of $-2\,^\circ$C to 25 $^\circ$C. As one of the largest drainage basins in North America, the Rio Grande-Rio Bravo (RGB) extends approximately 557,000 km^2 between the United States of America (U.S.) and Mexico. The RGB provides water to eight states, three in the U.S. (Colorado, New Mexico, and Texas) and five in Mexico (Chihuahua, Coahuila, Durango, Nuevo León, and Tamaulipas). Snowmelt from the Rocky Mountains and monsoon runoff from the Sierra Madre Occidental flows mostly through arid regions, including the Chihuahuan Desert, North America's largest desert. The northern branch of the RGB joins the Rio Conchos at La Junta de los Rios near Ojinaga (Chihuahua)/Presidio (Texas) to form the mainstem river. Several other tributaries contribute to streamflow, including but not limited to the Pecos River, which originates in New Mexico and flows through Texas until the mainstem, and other Mexican tributaries such as the Rio Salado and the Rio San Juan, which originate in the states of Coahuila and Nuevo León, respectively. The annual average natural supply of the Rio Grande delivered to the Gulf of Mexico was between 10 and 12 km^3 [20].

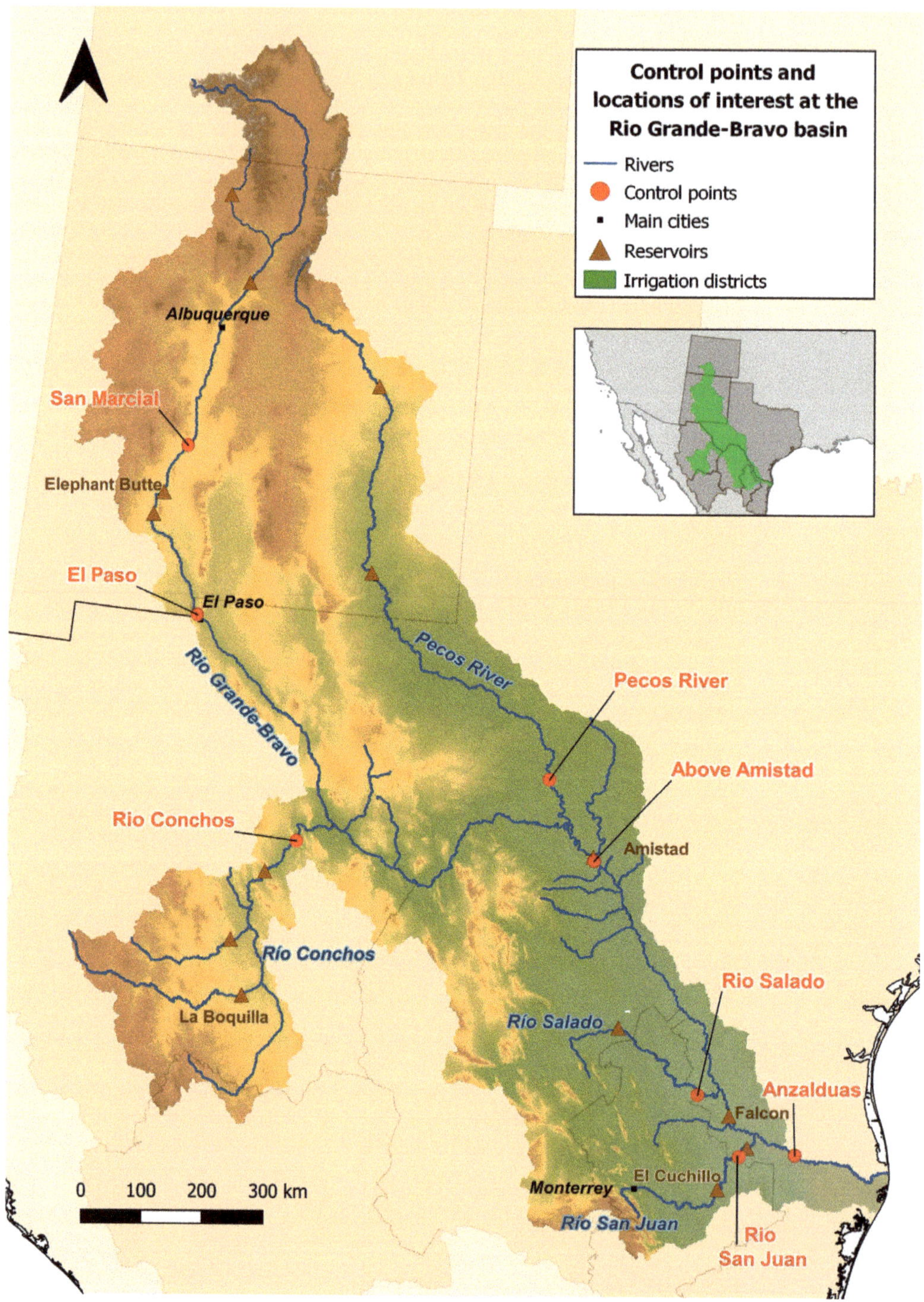

Figure 2. Control points and locations of interest at the Rio Grande-Bravo Basin.

2.2. Data Collection

Long-term streamflow data are required to represent specific conditions of river basins, including the dynamics and behaviors of hydrologic, climatic, anthropogenic, and seasonal variables over extended periods in a river basin. This analysis requires two streamflow datasets: (1) Observed flow regimes, which represent a clear manifestation of the Anthropocene, including water diversions, withdrawals, and reservoir operations, among others. Observed flow data were obtained from the Mexican National Water Commission (Comisión Nacional del Agua [CONAGUA]), the International Boundary and Water Commission (IBWC), and the U.S. Geological Survey (USGS). (2) Natural flow regimes represent streamflow without anthropogenic impacts, removing the impacts of reservoirs, diversions, return flows, groundwater sources, and any other water management practice and assuming to capture the relevant characteristics of climate and natural river basin hydrology [21]. Naturalized streamflow data sources were retrieved from previous studies, including the Upper RGB at Rio Grande Del Norte, Colorado, to the Rio Grande Above Presidio, Texas [22]. Then for the Lower RGB, daily and monthly naturalized data was retrieved from below Presidio/Ojinaga to Anzalduas, Tamaulipas from 1900–1943 [23,24], and from 1950 to 2008 [25]. Data gaps were calculated using streamflow naturalization.

2.3. Streamflow Naturalization

Streamflow naturalization is used in observed flow regimes for removing anthropogenic influence disturbances such as impoundments of rivers, land-use changes, water extractions, return flows, and other factors from streamflow time series. As the influence of humans continues to have a direct impact on river flows, the natural and anthropogenic parts of observed flows need to be distinguished [26,27]. The method used to naturalize flow is the water balance, which is the most widely used, despite the fact that it is primarily governed by data availability. This approach consists of decomposing flow into a natural part and an influenced part by removing the volume variation induced by the source of influence (e.g., reservoirs) [28] by accounting for the system's gains and losses for the desired time frame [21]. The mass water balance equation (Equation (1)) is the following:

$$Q_t^{nat} = GF_t + O_t - I_t + \Delta S_t \tag{1}$$

where Q_t^{nat} is the natural flow, G_t is the observed/gauged flows, O_t is the outflows, I_t is the inflows, and ΔS_t is the change of reservoir storage at a given daily time step t.

Outflows include evaporation losses from the reservoir and streamflow losses, obtained from the Mexican National Data Bank for Superficial Waters (Banco Nacional de Datos de Aguas Superficiales [BANDAS]) and IBWC. Moreover, any consumptive use, including agriculture diversions retrieved by the Agricultural Statistics of the Irrigation Districts in Mexico (Estadísticas Agrícolas de los Distritos de Riego), domestic and industrial water uses obtained by CONAGUA. Inflow data include agriculture and urban returns, flows, precipitation in the reservoir, and streamflow gains obtained by BANDAS and CONAGUA. Furthermore, the change of storage was obtained from BANDAS and IBWC. Lastly, to validate our results, we performed a statistical analysis comparison between our results and available research including the studies of Orive de Alba [29] and Blythe and Schmidt [22]. The goodness of fit criteria used from Moriasi et al. [30] were the coefficient of determination (R2), index of agreement (d), Nash-Sutcliffe efficiency (NSE), and percent bias (PBIAS).

2.4. Streamflow Drought Index

The streamflow drought index (SDI) developed by Nalbantis and Tsakiris [31] is used to characterize the severity of hydrological droughts. To capture decadal changes and long-term droughts in the basin for each control point. First, the cumulative streamflow of the naturalized streamflow data was estimated in a time window of 120 months. Then, the aggregated time series were fitted to probability distribution functions (normal, log-normal, and gamma) using the Kolmogorov–Smirnov (K-S) test; the log-normal distribution function (*p*-value less than 0.5) was selected based on the goodness of fit at a 95 percent confidence level and the least sum squared error between each probability distribution function (see Supplemental Materials). The software used to test and select the best probability distribution function was the Python package: fitter [32]. At last, the estimation of the cumulative probability is transformed into a standard normal random variable with a mean zero and standard deviation of one, resulting in the values of the naturalized SDI.

$$SDI_{i,k} = \frac{V_{i,k} - \overline{V}_k}{S_k} \tag{2}$$

where $SDI_{i,k}$ is the standard drought index value, $V_{i,k}$ is the cumulative streamflow volume, $\overline{V}_k$ is the mean, and S_k is the standard deviation of the cumulative streamflow volume for an *i-th* hydrological year with a period length of *k*. Consecutively, the observed streamflow data is evaluated by correlating the cumulative observed streamflow volumes with the closest aggregated naturalized volume; then, its corresponding SDI value is assigned. Hydrologic wet states are values between 0 and 3, and dry states between 0 and −3. For this study, eight states of hydrological droughts representing different severities are used (Table 1), which is the criterion of Nalbantis and Tsakiris [31] modified by Garza-Díaz and Sandoval-Solis [33].

Table 1. Description of hydrologic stated based on a modified streamflow drought index (SDI) criterion by Garza-Díaz and Sandoval-Solis [33].

Description of State	Criterion
Extremely dry	$-2 < SDI \leq -3$
Severely dry	$-1 < SDI < -2$
Dry	$-0.5 < SDI < -1$
Moderately dry	$0 < SDI < -0.5$
Moderately wet	$0 < SDI < 0.5$
Wet	$0.5 < SDI < 1$
Severely wet	$1 < SDI < 2$
Extremely wet	$2 < SDI \leq 3$

2.5. Computation of Stability Landscapes

Properties of the stability landscape in environmental systems are commonly linked to the geometric properties of a potential function [10]. Where minima and maxima respectively correspond to stable and unstable equilibria of the basins of attraction, the slopes of the potential surface are proportional to the rates of change in the system [10]. Even if this method is widely used, finding a potential function for systems with more than one dimension can be difficult [34]. Alternative measures have been applied to other systems, including the use of probability distribution functions (pdf) as it is closely related to the potential function where local minima of the potential function correspond to local maxima in the pdf [35]. Hypothetical three-dimensional stability landscapes for the river basin were computed directly from the pdf of the natural and regulated SDI values. These figures depict the conditional probability of a given SDI value (SDI_t) given a previous SDI value (SDI_{t-1}). For instance, given that the system had an SDI of −3 in the previous year ($SDI_{t-1} = -3$), what is the probability of having an SDI value of X in the present year. The pdfs dominant modes serve as proxies of the shape of the basins of attraction and are used to reflect the stability landscape properties and how they change over time.

3. Results and Discussion

3.1. Data Validation

Results of the analysis comparison between the streamflow estimations from the period of record of 1900–1943 from Orive de Alba [29] were $R^2 = 0.9$, d = 0.9, NSE = 0.9, and PBIAS = 3.6. In addition, the comparison between Blythe and Schmidt [22] with a period of record is 1900–2010 are R^2 = 0.9, d = 0.9, NSE = 0.9, and PBIAS = 1.8. The statistical performance for both comparisons was very good according to the criteria of da Silva et al. [36].

3.2. Hydrologic Variability of the Natural State of a River Basin

The RGB basin spans a climatic gradient from semi-arid to subhumid; its environment is vulnerable to extreme hydroclimatic events [37]; and to investigate its dichotomy, the hydrologic variability of the natural state of the RGB is depicted in a 120-month SDI analysis (Figure 3) which allowed identification of hydrologic drought and flood events.

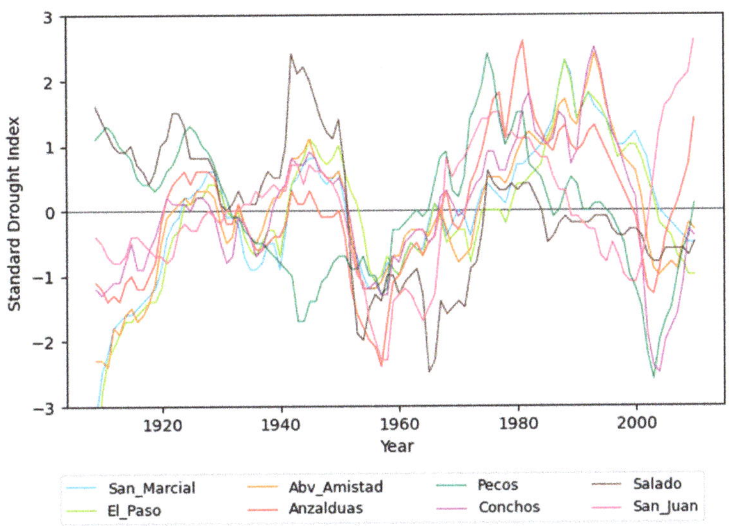

Figure 3. Streamflow drought index of the naturalized control points of the Rio Grande-Bravo Basin.

Overall, the hydrological behavior of the basin indicates recurrent periods of water stress (Table 2). Droughts in this basin are common and, on average, can span from 10 to 25 years, including consecutive extremely and severely dry periods ranging from 5 to 9 years. In contrast, wet periods tend to be shorter, from 11 to 16 years; extremely and severely wet periods could typically last from 2 to 4 years. Alternating dry and wet cycles could last 24 years in the mainstem of the RGB; these cycles are correlated with ocean-atmosphere climate variability [38].

Table 2. Hydrologic periods of the Rio Grande–Bravo basin. Each hydrologic period is the average of the consecutive number of years that ranges from specific SDI values.

Control Point	Hydrologic Period (Average of Consecutive Years)			
	Dry (−3 to −0.5) [1]	Extremely Dry (−3 to −2) [1]	Wet (0.5 to 4) [1]	Extremely Wet (2 to 4) [1]
San Marcial	10	8	11	2
El Paso	13	8	15	2
Above Amistad	13	9	16	3
Anzalduas	13	6	12	2
Rio Conchos	12	6	16	3
Pecos River	25	2	16	4
Rio Salado	23	5	13	2
Rio San Juan	18	4	15	3

Table 2. *Cont.*

Control Point	Hydrologic Period (Average of Consecutive Years)			
	Dry (−3 to −0.5) [1]	Extremely Dry (−3 to −2) [1]	Wet (0.5 to 4) [1]	Extremely Wet (2 to 4) [1]
Average	16	6	14	3
Median	13	6	15	3

Note: [1] streamflow drought index (SDI) values.

3.2.1. Synchronous and Asynchronous Wet and Dry Periods

Synchronous and asynchronous wet and dry periods occurred along the RGB mainstem due to the difference in physiographic and climatic main controls in the RGB, snowmelt runoff in the headwaters of the San Juan Mountains, and the strong influence of the North American monsoon gives rise to two different hydroclimate regions: the hydroclimatic snowmelt variability in the headwaters of the RGB (the northern branch, including: San Marcial, El Paso, and Pecos River) and the North American monsoon variability experienced downstream of its confluence with the RGB (the southern branch, including Above Amistad, Anzalduas, Rio Conchos, Rio Salado, and Rio San Juan). This can be shown in the overlap and out of phase of droughts and wet periods that are concurrent in specific decades and regions, and other times are out of phase and independent. For example: synchronous wet periods occurred in the late 70s and the 80s, which were the wettest of the century, and matching droughts years include 1909–1920, the 1930s, 1950s, and 2005–2010. Although in some of these periods, the severity was not as extreme as in other regions. For example, the drought experienced in 1910 by the Rio Conchos was less severe than those in San Marcial or El Paso, or the wettest period was more severe for Anzalduas than El Paso. On the contrary, asynchronous wet and dry periods can also occur; for example: the beginning of the twentieth century was particularly wet for the Pecos River and the Rio Salado, which showed positive SDI values from 1900–1930. After this wet period, these rivers exhibit contrasting dry/wet periods from 1940 to 1950, where the Pecos River has the second driest period on record while the Rio Salado shows its wettest period. In addition, all control points exhibit differences in severities and durations, even if these overlap, indicating that one or more underlying circulation mechanisms influence the entire basin [39].

3.2.2. Occurrence of Droughts

The RGB is vulnerable to extreme hydroclimatic events, especially droughts, which are expected to become more severe in this region by the end of the 21st century. Paleoclimate reconstructions using tree rings have been used in the RGB to reconstruct streamflow. For the Pecos River, a 700-year paleoclimate reconstruction estimated streamflow declines in a multi-century context, setting the drought of 1950–1957 as one of the highest ranked based on magnitude and intensity, slightly less severe as the 11-year drought of 1772–1782 [40]. For the RGB near Del Norte [39] and the Rio Conchos [41], a 344-year (1749–1933) reconstruction of seasonal precipitation and a 243-year (1775–2015) reconstruction of streamflow volume reported an extraordinary drought from 1950 to 1957 and from 1948 to 1958, respectively. These studies coincide with our research where the severely dry period for the natural streamflow system is estimated, from 1950 to 1965, for several control points, including Pecos River and Rio San Juan. The drought of the 1950s has been well documented in rainfall, discharge, and dendro-chronological data and is consistent with drought spells in northern Mexico [42]. However, in our records, the most severe drought in the Rio Conchos was in 2005, and the second driest in the 1950s. Nonetheless, the study of Ortega-Gaucin [43], reports from 1997 to 2008 as an extraordinarily hydrological dry period for the portion of the RGB located in Mexican territory, specifically the severe and extremely dry period from 2000–2008 in the control points of Rio Conchos. Moreover, San Marcial and El

Paso experienced extreme and severe drought in the early 1900s, a decade distinguished by predominantly below-average flows in the northern branch of the RGB [39].

3.2.3. Occurrence of Snowfall and Hurricanes

Snowfall and hurricanes significantly affect the water availability throughout the basin. The RGB (San Marcial and El Paso) and Rio Conchos showed an exceptionally wet decade between the 1980s and 1990s, as reported by the northern branch using a 445-year streamflow reconstruction forecast [44] and streamflow data along the RGB mainstem (at Johnson Ranch) and the Rio Conchos [45]. The Rio Salado shows its wettest period in the 1970s, which coincides with estimates of Ortiz-Aguilar [46]. Then the 1900s was extraordinarily wet within the context of the Pecos basin, only broken by the widespread 1950s drought, which was ended by the 1980s wet event. In addition, the 20th century was the wettest in the Pecos basin over the past 700 years [40]. Heavy rains, influenced by tropical storms and hurricanes that hit the RGB from the Pacific and Atlantic Oceans, have increased in frequency. These storms, concentrated in short periods, are responsible for high annual discharge in the RGB. In the Rio Salado and Rio San Juan, the hurricanes Beulah, 1967; Allen, 1980; Barry, 1983; and Gilbert, 1988 [47] resulted in an extremely wet and wet period, respectively. In the 2000s, hurricanes Emily, 2005; Dean, 2007; Dolly, 2008; and Alex, 2010 [47] resulted in a severely wet period for the Rio San Juan basin and in Anzalduas, the outlet of the RGB.

3.2.4. Impacts of Climate Change

Effects of climate change are already altering the RGB streamflow timing and volume through changes in rainfall, snowfall and snowpack, and increased temperatures and evapotranspiration rates [48]. Despite that this study did not distinguish the effects of climate change and human impacts separately; and climate and hydrologic forced models (e.g., rainfall-runoff models) are needed as additional research to distinguish the impact of climate change on the natural streamflow. The intensity and frequency of dry and wet conditions for the natural system in Figure 3 have increased since 1950. Extreme hydroclimatic events, such as intense precipitation and drought, are expected to increase in this region by the end of the 21st century [49,50]. For example, streamflow declines are occurring in tributaries upstream of Albuquerque between 1980 and 2016 [51]. In addition, in the past 40 years, snow drought has impacted the RGB headwaters in Colorado and New Mexico [52]. Moreover, elevated evapotranspiration rates since 1980 in the Rio Conchos, Rio Salado, and Rio San Juan are affecting crop production [53] and changes in air temperature exacerbate water quality issues in border cities of the southern branch of the RGB [54]. Furthermore, there has been an increase in the frequency of tropical cyclones and hurricanes since 1950 generated in the Pacific Ocean [55] resulting in economic losses by flooding and crop destruction.

3.3. The Modern Hydrology: A Perennial Human-Induced Extreme Drought

A comparison between the natural and modern streamflow variability in the mainstem of the RGB is shown in Figure 4 and the subbasin control points in Figure 5. The natural hydrology of the RGB exhibits a strong hydrologic variability with alternating dry and wet periods. In contrast, the regulated hydrology lacks the cyclical periods of wet and dry periods highlighted in the natural system; it shows a permanent state of human-induced extreme drought in the basin. The lack of hydrologic variability intensifies the dry states' severity and frequency, shifting from a possible wet or moderately wet to a dry, moderately dry, or even extremely dry period that could last several years. The loss of this dynamism puts the system in a perennial and extreme dry state for most of the sites for decades, in some regions more severe than others, yet the magnitude and extent of the dry state permeate all regions of the RGB. In the RGB mainstem, perennial extreme dry periods started in San Marcial and El Paso in 1920 (for 90 years), above Amistad in 1939 (for 71 years), and in Anzalduas since the beginning of the 20th century (for 110 years).

Anzalduas represents the response of the entire RGB basin given its location near the outlet; it shows that since the early 1900s, water diversions and flow regimes modified the basin as if it was in a perennial drought. For the main tributaries of the RGB, perennial extreme dry periods started in the Pecos River in 1945 (for 65 years) and the Rio Conchos in 1960 (50 years). In the San Juan and Rio Salado basins, they appear to have periods of extreme drought that are separated by periods of dry and moderately dry periods; these can be explained by the1980s wet period in the San Juan and the severely and extremely wet period in the 1970s in the Salado basin.

3.3.1. Causes of the Perennial Human-Induced Drought

At the core of this permanent state of human-induced drought is the interplay of human development and climate. Since the 1870s, the RGB has been subject to a long history of human manipulation [20]. The present perennial drought state is the result of increased water demands (for agriculture, municipal, and industrial), water agreements (at the international, interstate, regional and local scales), water overallocation, and the construction of large water infrastructure (reservoirs, canals, levees) [33,37]. Water resources are often insufficient to meet human and environmental requirements due to the natural water scarcity in the basin and the increased human water demand. The RBG basin provides water for more than 10.4 million inhabitants. Moreover, the basin supports extensive irrigated agriculture, comprising approximately 780 thousand hectares of irrigated land [33] and accounting for 83% of water withdrawals in the RGB [37]. In the U.S., the extent of irrigation activities expanded during the 19th century after the Desert Land Act of 1877 [56,57], prompting a disproportionate expansion of agricultural land, water diversions for irrigation, and water consumption. In the U.S., irrigated agriculture accounts for 80 to 90% of the overall water use. The main crops are forage, cotton, pecans, and vegetables [58]. In contrast, as a result of the Mexican Revolution in 1917, the Mexican Agrarian Reform implemented a prolonged distribution of land, where more than half of the Mexican territory was assigned to farmers [59]. A total of 11 irrigation districts were created, totaling 458 thousand hectares of irrigated land [33], where the states of Chihuahua and Tamaulipas account for 87% of the total irrigated areas. In both countries, the large-scale farming systems require large reservoir projects and extensive channelization, which started in 1916 with Elephant Butte in New Mexico and La Boquilla in Chihuahua. Since then, 27 large dams (greater than 16,000 Mm^3 of storage capacity) have been built in the basin, including two international dams: Amistad and Falcon.

As streamflow is reduced by overconsumption and climate change, access to water is becoming a looming crisis, and droughts have become more devastating due to increased use of water resources for human purposes, changes in regulations for water allocation between users, states, or countries. Management actions for concealing water shortages and increasing water supply through more river engineering in one area certainly affect downstream communities. For example, the construction of El Cuchillo Dam in Rio San Juan during the drought of 1990 aimed to supply water for the city of Monterrey in Nuevo Leon. However, this action led to a diminishing water supply for farmers in Tamaulipas. Droughts have also triggered a change in regulations for water allocation, whether in international agreements or state water allocation systems [37]. For instance, the Pecos River Compact [60] between New Mexico and Texas promotes collaboration and sharing of water resources. However, constraining surface water use created an increase in groundwater use, that ultimately ended up in groundwater overdraft that diminished baseflows that downstream users depended on. Droughts have also triggered conflicts among water users, states, and countries. For example, the drought in the late 1990s triggered disputes between farmers and the federal government in Mexico. From 1997 to 2002, Mexico incurred a substantial water debt to the U.S. The Rio Conchos basin was not able to deliver water to U.S. and Mexican downstream water users due to drought and increased water use in the Rio Conchos basin. At that time, the Mexican government resolved this conflict by delivering water to the U.S. from other tributaries and from

Mexican water stored in the international reservoirs, leaving without downstream water users in Tamaulipas. The imbalance between supply and demand creates a complex web of governance structure, infrastructure, and user conflicts, which translate into compounding effects for anthropogenic droughts.

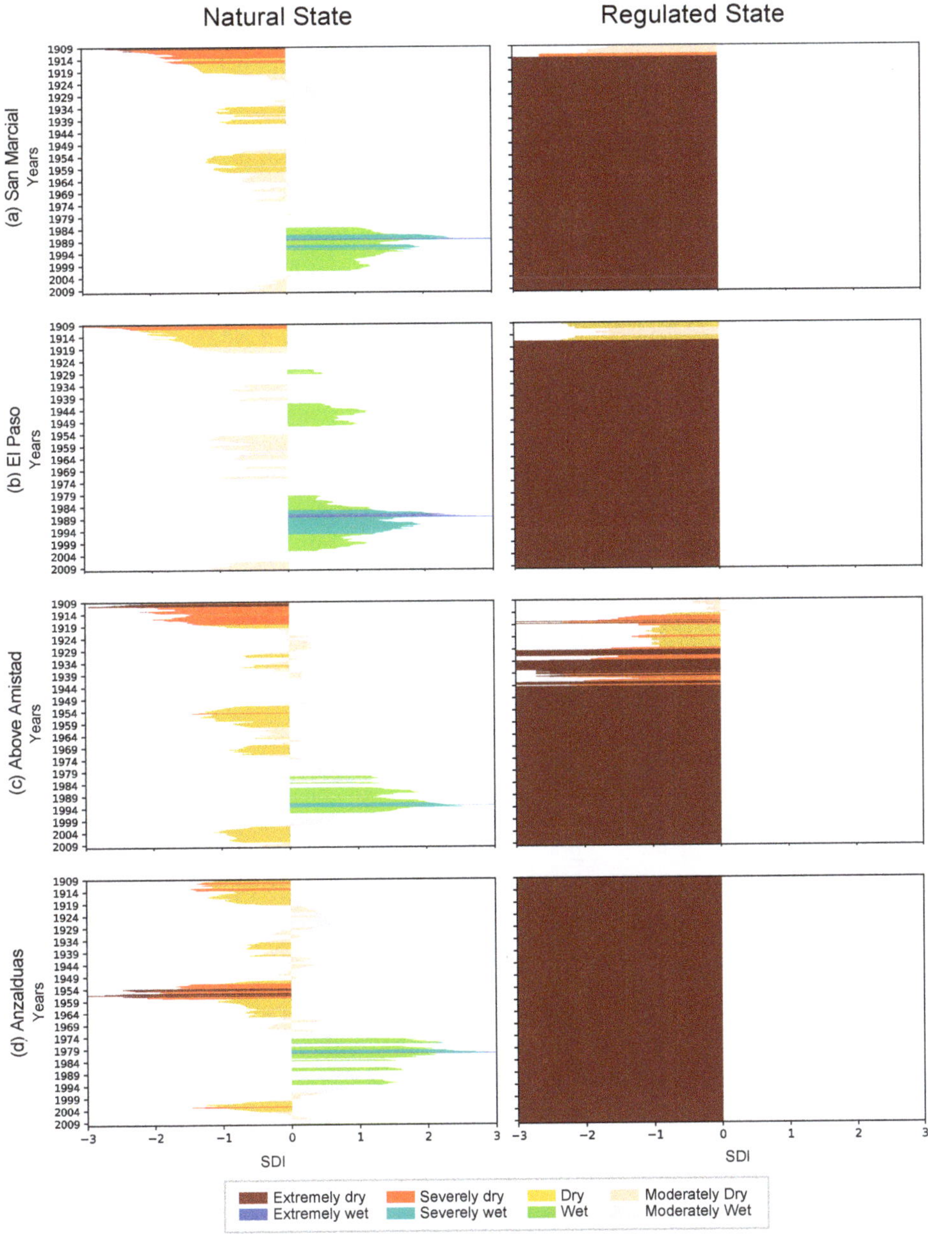

Figure 4. Streamflow drought index (SDI) indicating the hydrologic variability of the natural (**left**) and the regulated (**right**) state of four mainstem control points of the Rio Grande-Bravo Basin at (**a**) San Marcial, (**b**) El Paso, (**c**) Above Amistad, and (**d**) Anzalduas.

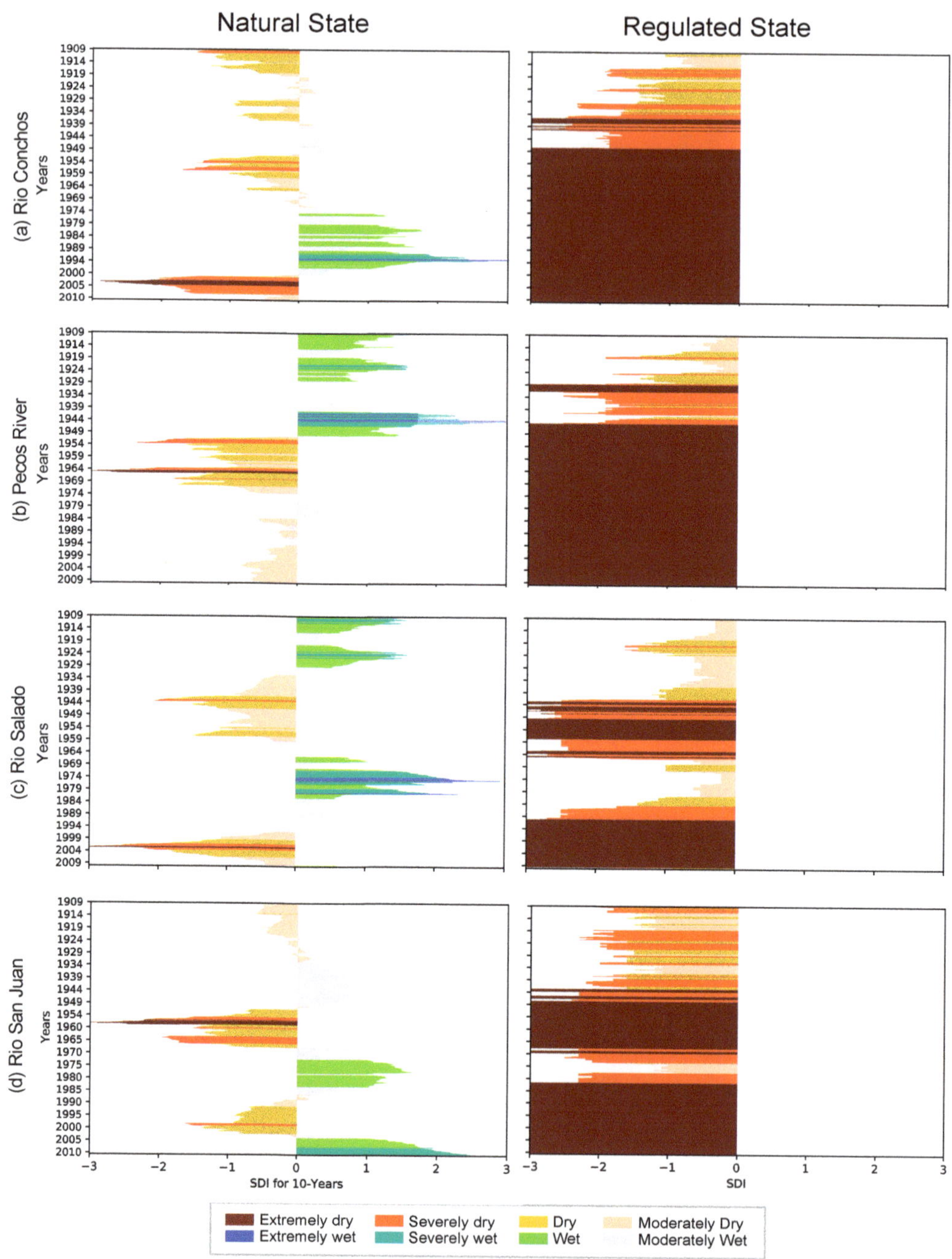

Figure 5. Streamflow drought index (SDI) indicating the hydrologic variability of the natural (**left**) and the regulated (**right**) state of four subbasin control points of the Rio Grande-Bravo Basin at (**a**) Rio Conchos, (**b**) Pecos River, (**c**) Rio Salado, and (**d**) Rio San Juan.

3.3.2. The Degradation Toll of the Environment Due to Human Activities

Land use change, reservoir development, straightening of the main river, and over-extraction of water have a high degradation toll on ecosystems by altering the river's natural flow pattern, timing, temperature, and quantity of river flows. By changing the temporal variation of streamflow in river basins, assemblages of riparian species are profoundly transformed because their life cycle is synchronized with the timing, magnitude, duration, and rate of change, of the natural flow regime. For example, lack of fall monsoonal flooding facilitates the invasion by non-native organisms by shifting regionally endemic species (e.g., generalist red shiner; *Cyprinella lutrensis*) to dominant generalist fish species (e.g., endemic Tamaulipas shiner; *Notropis braytoni*) [61]. In addition, other native species have gone locally extinct in some areas of the RGB (e.g., the Rio Grande Monkeyface mollusk; *Quadruka couchiana*), while others have been listed as endangered species (e.g., the Rio Grande silvery minnow; *Hybognathus amarus*). In addition, reduced flood flow frequency has enhanced invasive vegetation encroachment and caused channel incision and narrowing [37]. Native ecosystems are adapted to droughts; however, the level and persistence of the current human-induced drought are severely affecting river ecosystems and species throughout the basin. In the 20th century, the flow of the RGB had been reduced by nearly 95% of its natural flow [22,23], and at least 30 springs have gone dry in the states of Chihuahua and Coahuila [62,63].

3.3.3. The Human-Induced Megadrought

The perennial drought state of the RGB can be better described as an anthropogenic megadrought; a compound multidimensional and multiscale phenomenon governed by the combination of natural water variability, human decisions, increased water use for human activities, climate change, and altered microclimate conditions due to changes in land and water management [3]. Since the early 2000s, the Rio Grande/Bravo has been listed among the most at-risk rivers in the world [64]. Other regions in the world are experiencing anthropogenic megadrought, for instance, across Canada, the United States, and Mexico [5], and in South America, a multi-year dry spell has been referred to as the Central Chile Mega Drought [65]. These examples point out that anthropogenic forcing is critical to explain the perennial dry states of regions, given its capability of transforming a dry spell into a full-blown multiyear megadrought [4]. The regulated state in Figures 4 and 5 show that the human-induced megadrought has become the new normal in the RGB, posing environmental and socioeconomic hardship, including the unwanted anthropogenic consequences of altering natural systems beyond their resilience carrying capacity. Prolonged droughts cause major fluctuations in the structure and functioning of the RGB; resilience erosion can trigger changes in the stability landscape of the system or even changes in regimes.

3.4. Stability Landscape Metaphor: Resistance, Latitude, Precariousness, and Panarchy

The resilience of a system can be described using the stability landscape metaphor [11] by characterizing the components that govern a system's dynamics: resistance, latitude, precariousness, and panarchy. A three-dimensional stability landscapes is used to estimate, visualize, and compare the resilience attributes of the natural and regulated flow regimes (Figure 6). The topology of the stability landscape is portrayed by the occurring valleys and hilltops [11] that delineate the boundaries between the basins of attraction and represent the states where the system exists for a determined period of time. The resistance indicates how easy or difficult the system can be changed between states; it is expressed by the depth of the basin. The latitude is the maximum amount the system can be changed and is depicted as the width of the basin of attraction. Wide basins mean a greater number of system states can be experienced without crossing a threshold, while deep basins indicate greater perturbations are required to change the current state of the system away from the attractor [66]. The precariousness indicates the trajectory of the system at a given time within the stability landscape and how close it is to crossing it. Finally, panarchy acknowledges that systems are dynamic and continually passing through "adaptive cycles" at various scales [67]. Like any metaphor, there are limitations to using stability landscapes as a decision-making tool. Nonetheless, it is a valuable resilience concept that helps us to think about ecosystem dynamics and how human management might affect resilience properties.

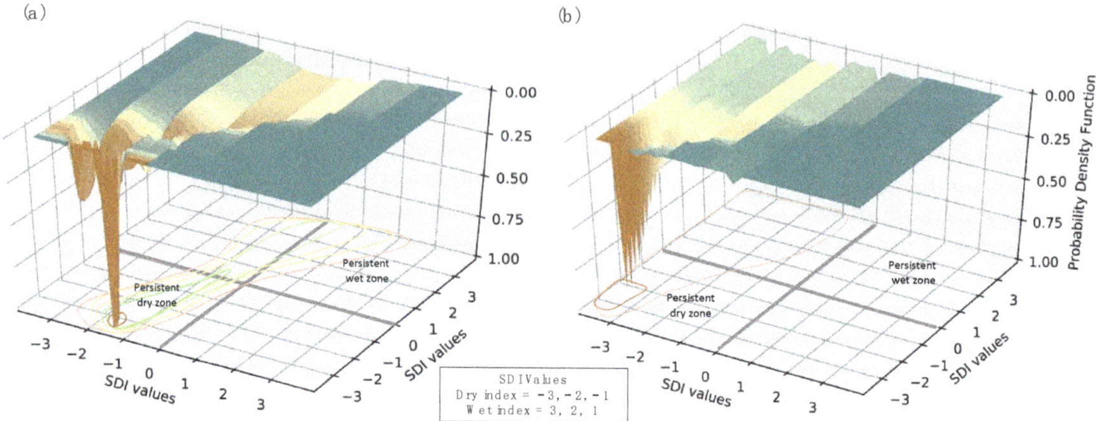

Figure 6. Stability landscapes of the Rio Grande-Rio Bravo at Above Amistad control point. (**a**) Natural flow regime (Figure 5A), two states are identified: (1) a persistent dry zone, characterized by a constricted-deep basin of attraction; and (2) a persistent wet zone, portrayed as one shallow-wide basin. (**b**) The regulated flow regime shows a single wide-deep basin of attraction pertaining to the persistent dry zone.

3.4.1. The Dynamic RGB Natural Stability Landscape

In the natural flow regime (Figure 6A), two states are identified: (1) a dry state portrayed as a constricted-deep basin of attraction located in the persistent dry zone; and (2) a wet state located portrayed as a shallow-wide basin located in the persistent wet zone. Valley bottoms correspond to the highest likelihood value of the system to remain in a given state; they are the modes of the probability density distribution [10]. At a given time, if the system is in a dry state, the system will remain in this state between 15 and 20 years, or if the system has transitioned to a wet state, it will remain in this state between 2 and 5 years. Based on the duration and frequency of both states, the basins of attraction differ in width, depth, and the number of valleys. In general, the RGB basin will tend to remain in a dry state, and greater perturbations are needed to move the system out of the persistent dry zone. In contrast, the RGB basin will remain less time in a wet state, and smaller perturbations will likely move the system away from the persistent wet zone. In essence, the stability landscape of the natural flow regime incorporates a diverse topography with different shapes and valleys where environmental stochasticity in the form of perturbations, such as hurricanes, droughts, tropical depressions, ENSO events, among others, will expose the system to a wide range of dynamics under the two stable states: dry and wet.

3.4.2. The Precarious RGB Regulated Stability Landscape

In contrast, the regulated flow regime (Figure 6B) has only a dry state depicted as a single wide-deep basin of attraction located in the persistent dry zone. Anthropogenic forcing (e.g., increased water use for agriculture) has altered the dynamics of river basins, changing the behavior and functionality of the natural ecosystem and causing alterations in the topology of the stability landscape. In the absence of environmental stochasticity due to the water regulations and streamflow diminishment, the resilience of the natural system erodes, and precariousness increases, moving the system closer to crossing a threshold. Precariousness is the result of management actions under historical conditions that have transformed the system and as a result, the number of states [68]. The anthropogenic megadrought in the RGB is likely the driver that transformed the stability landscape, reducing and shrinking the two states (dry and wet) of the natural stability landscape into the one state (dry) of the regulated system.

The intrinsic nature of coupled human–environmental systems and the adaptive cycles of panarchy in the RGB basin modified the stability landscape eroding its resilience. There is a higher resistance (depth of the basin) in the regulated system (Figure 6B) in comparison with the natural system (Figure 6A), indicating that greater forces and perturbations are required to move the system out of the current dry state. Even if the critically endangered RGB is trapped in an undesirable and unsustainable human-induced megadrought state, our society has the ability to modify the current stability landscape through transformability—the capacity to create a fundamentally new system when ecological, economic, or social structures make the existing system untenable [11]. The challenge is to reduce or avoid the human activities (e.g., modification of flow regime due to water storage in reservoirs that only meet human water needs) that create undesirable basins of attractions and move toward stability landscapes that resemble the natural state (e.g., implementation of environmental flows through dem releases mimicking the natural flow regime). "Different management actions would be required to initiate a transformative change that envisions and restores natural dynamic processes. Reservoir re-operation and environmental flows are strategies targeted to minimize hydrologic alteration by incorporating water releases that include functional flow metrics such as timing, frequency, magnitude, duration, and rate of change of the natural flow regime. Other management actions include environmentally, socially, and climate responsible agriculture, such as adequate selection of crops, deficit irrigation, and the implementation of cover crops which are measures to reduce consumptive water use."

4. Conclusions

Natural hydrologic variability is vital for promoting ecological resilience, as it governs the water quantity, quality, habitat, and health of riverine ecosystems. In the Anthropocene, the alteration of natural flow variation by human-induced changes is the dominant force in social-ecological systems, causing changes in flow regimes and the resilience properties of river basins. This study demonstrates how human development and human-centered water management regulations are the main drivers of the anthropogenic megadrought in the Rio Grande. In addition, we demonstrate how this process has produced changes in the stability landscape of these river basins, including changes in the topology (resistance and latitude), the trajectory (precariousness), and the dynamic processes of a natural system (panarchy). The stability landscape alteration is depicted as the modification of two basins of attraction, which represent the natural wet and dry hydrologic states, into a single basin of attraction representing a permanent dry state. The implication of the resilience erosion in the RGB indicates that streamflow conditions have changed sufficiently to provide early warning signals of crossing a resilience threshold, meaning that the system could suffer consequences. As a society, we are already experiencing the effects of a water crisis, and current management practices and policies are beginning to migrate into placing aspects of social-ecological resilience analysis at the core of integrated water resources management. Aside from the limitations to operationalizing the concept of stability landscapes, the broader impact of this study is that it sheds light on quantifying ecological resilience attributes in river basins. We believe that a shift toward addressing resilience in river basins is a prerequisite to understanding current systems and reconnecting our societies with adaptable strategies aimed to be in sync with the dynamics of natural resources. Scenario planning and adaptive management are also necessary to overcome undesirable systems and foster flexibility and adaptability. Our ability to understand the dynamic processes of the natural system and modify our outdated vision of highly manipulated systems to obtain maximum yields is the most effective way to manage sustainable, resilient river basins in the face of increasing environmental and social change.

Supplementary Materials: The following supporting information can be downloaded at: https://www.mdpi.com/article/10.3390/w14182835/s1, Information about testing different probability distribution functions for the aggregated naturalized time series data is included.

Author Contributions: L.E.G.-D. and S.S.-S. conceptualized the study; L.E.G.-D. performed the formal analysis and wrote the original paper; S.S.-S. reviewed and edited the paper and supervised the project. All authors have read and agreed to the published version of the manuscript.

Funding: This research was financially supported by The National Council of Science and Technology of Mexico (CONACYT) through the Doctoral Fellowship Grant awarded to the first author Grant Number: CVU-598945.

Institutional Review Board Statement: Not applicable.

Informed Consent Statement: Not applicable.

Data Availability Statement: The data supporting this study's findings are openly available in HydroShare at: Garza-Diaz, L. E., S. Sandoval Solis (2022). Natural and Regulated Monthly Streamflow Data for the Rio Grande/Rio Bravo Basin, HydroShare, http://www.hydroshare.org/resource/89728c8779c644d7a6ce110406516849, accessed on 1 August 2022.

Conflicts of Interest: The authors declare no conflict of interest.

References

1. Lytle, D.A.; Poff, N.L. Adaptation to Natural Flow Regimes. *Trends Ecol. Evol.* **2004**, *19*, 94–100. [CrossRef] [PubMed]
2. Grantham, T.E.; Mount, J.F.; Stein, E.D.; Yarnell, S.M. *Making the Most of Water for the Environment: A Functional Flows Approach for California Rivers*; Public Policy Institute of California: San Francisco, CA, USA, 2020.
3. AghaKouchak, A.; Mirchi, A.; Madani, K.; Di Baldassarre, G.; Nazemi, A.; Alborzi, A.; Anjileli, H.; Azarderakhsh, M.; Chiang, F.; Hassanzadeh, E.; et al. Anthropogenic Drought: Definition, Challenges, and Opportunities. *Rev. Geophys.* **2021**, *59*, e2019RG000683. [CrossRef]
4. Sthale, D. Anthropogenic Megadrought. *Science* **2020**, *368*, 238–239. [CrossRef]
5. Williams, A.P.; Cook, E.R.; Smerdon, J.E.; Cook, B.I.; Abatzoglou, J.T.; Bolles, K.; Baek, S.H.; Badger, A.M.; Livneh, B. Large Contribution from Anthropogenic Warming to an Emerging North American Megadrought. *Science* **2020**, *368*, 314–318. [CrossRef] [PubMed]
6. Ashraf, S.; Nazemi, A.; AghaKouchak, A. Anthropogenic Drought Dominates Groundwater Depletion in Iran. *Sci. Rep.* **2021**, *11*, 9135. [CrossRef]
7. Holling, C.S. Resilience and Stability of Ecological Systems. *Annu. Rev. Ecol. Syst.* **1973**, *4*, 1–23. [CrossRef]
8. Scheffer, M.; Carpenter, S.R.; Dakos, V.; van Nes, E.H. Generic Indicators of Ecological Resilience: Inferring the Chance of a Critical Transition. *Annu. Rev. Ecol. Evol. Syst.* **2015**, *46*, 145–167. [CrossRef]
9. Beisner, B.E.; Haydon, D.T.; Cuddington, K. Alternative Stable States in Ecology. *Front. Ecol. Environ.* **2003**, *1*, 376–382. [CrossRef]
10. Dakos, V.; Kéfi, S. Ecological Resilience: What to Measure and How. *Environ. Res. Lett.* **2022**, *17*, 043003. [CrossRef]
11. Walker, B.; Holling, C.S.; Carpenter, S.R.; Kinzig, A. Resilience, Adaptability and Transformability in Social-Ecological Systems. *Ecol. Soc.* **2004**, *9*, 5. [CrossRef]
12. Scheffer, M.; Bascompte, J.; Brock, W.A.; Brovkin, V.; Carpenter, S.R.; Dakos, V.; Held, H.; van Nes, E.H.; Rietkerk, M.; Sugihara, G. Early-Warning Signals for Critical Transitions. *Nature* **2009**, *461*, 53–59. [CrossRef] [PubMed]
13. Blackmore, J.; Plant, R. Risk and Resilience to Enhance Sustainability with Application to Urban Water Systems. *J. Water Resour. Plan. Manag. ASCE* **2008**, *134*, 224–233. [CrossRef]
14. Hirota, M.; Holmgren, M.; Van Nes, E.H.; Scheffer, M. Global Resilience of Tropical Forest and Savanna to Critical Transitions. *Science* **2011**, *334*, 232–235. [CrossRef] [PubMed]
15. Livina, V.N.; Kwasniok, F.; Lenton, T.M. Potential Analysis Reveals Changing Number of Climate States during the Last 60 Kyr. *Clim. Past* **2010**, *6*, 77–82. [CrossRef]
16. Berdugo, M.; Kéfi, S.; Soliveres, S.; Maestre, F. Plant Spatial Patterns Identify Alternative Ecosystem Multifunctionality States in Global Drylands. *Nat. Ecol. Evol.* **2017**, *1*, 3. [CrossRef]
17. Berdugo, M.; Vidiella, B.; Solé, R.V.; Maestre, F.T. Ecological Mechanisms Underlying Aridity Thresholds in Global Drylands. *Funct. Ecol.* **2022**, *36*, 4–23. [CrossRef]
18. Sendzimir, J.; Magnuszewski, P.; Flachner, Z.; Balogh, P.; Molnar, G.; Sarvari, A.; Nagy, Z. Assessing the Resilience of a River Management Regime: Informal Learning in a Shadow Network in the Tisza River Basin. *Ecol. Soc.* **2008**, *13*, 11. [CrossRef]
19. Pillar, V.D.; Blanco, C.C.; Müller, S.C.; Sosinski, E.E.; Joner, F.; Duarte, L.D.S. Functional Redundancy and Stability in Plant Communities. *J. Veg. Sci.* **2013**, *24*, 963–974. [CrossRef]
20. Enríquez-Coyro, E. *El Tratado Entre México y Los Estados Unidos de América Sobre Ríos Internacionales: Una Lucha Nacional de Noventa Años*; Tomos I y II, Segunda Edición; Comisión Nacional del Agua: Mexico City, Mexico, 1976.

21. Wurbs, R.A. Methods for Developing Naturalized Monthly Flows at Gaged and Ungaged Sites. *J. Hydrol. Eng.* **2006**, *11*, 55–64. [CrossRef]
22. Blythe, T.L.; Schmidt, J.C. Estimating the Natural Flow Regime of Rivers with Long-Standing Development: The Northern Branch of the Rio Grande. *Water Resour. Res.* **2018**, *54*, 1212–1236. [CrossRef]
23. Gonzalez-Escorcia, Y.A. Determinación Del Caudal Natural En La Cuenca Transfronteriza Del Río Bravo/Grande. Master's Thesis, Instituto Politécnico Nacional, Mexico City, Mexico, 2016.
24. Loredo-Rasgado, J. Determinación y Análisis de Los Valores de Huella Hídrica En La Región Hidrológico Administrativa VI/Río Bravo. Master's Thesis, Instituto Politécnico Nacional, Mexico City, Mexico, 2018.
25. Silva Hidalgo, H. Modelo Matemático Para La Distribución De Agua Superficial En Cuencas Hidrológicas. Ph.D. Thesis, Centro de Investigación en Materiales Avanzados, S.C., Chihuahua, Mexico, 2010.
26. Montanari, A.; Young, G.; Savenije, H.H.G.; Hughes, D.; Wagener, T.; Ren, L.L.; Koutsoyiannis, D.; Cudennec, C.; Toth, E.; Grimaldi, S.; et al. "Panta Rhei—Everything Flows": Change in Hydrology and Society—The IAHS Scientific Decade 2013–2022. *Hydrol. Sci. J.* **2013**, *58*, 1256–1275. [CrossRef]
27. Littlewood, I.G.; Marsh, T.J. Re-Assessment of the Monthly Naturalized Flow Record for the River Thames at Kingston since 1883, and the Implications for the Relative Severity of Historical Droughts. *Regul. Rivers: Res. Manag.* **1996**, *12*, 13–26. [CrossRef]
28. Terrier, M.; Perrin, C.; de Lavenne, A.; Andréassian, V.; Lerat, J.; Vaze, J. Streamflow Naturalization Methods: A Review. *Hydrol. Sci. J.* **2021**, *66*, 12–36. [CrossRef]
29. Orive-Alba, A. *Informe Técnico Sobre El Tratado International de Aguas Irrigación En México*; Comisión Nacional de Irrigación: Mexico City, Mexico, 1945.
30. Moriasi, D.N.; Arnold, J.G.; van Liew, M.W.; Bingner, R.L.; Harmel, R.D.; Veith, T.L. Model Evaluation Guidelines for Systematic Quantification of Accuracy in Watershed Simulations. *Trans. ASABE* **2007**, *50*, 885–900. [CrossRef]
31. Nalbantis, I.; Tsakiris, G. Assessment of Hydrological Drought Revisited. *Water Resour. Manag.* **2009**, *23*, 881–897. [CrossRef]
32. Cokelaer, T. Fitter: A Tool to Fit Data to Many Distributions and Best One(s), GitHub Repository 2014. Available online: https://github.com/cokelaer/fitter (accessed on 10 August 2022).
33. Garza-Díaz, L.E.; Sandoval-Solis, S. An Ecological Resilience Assessment to Detect Hydrologic Regime Shifts and Thresholds in River Basins. *Ecol. Soc.* 2022; *in review*.
34. Rodríguez-Sánchez, P.; van Nes, E.H.; Scheffer, M. Climbing Escher's Stairs: A Way to Approximate Stability Landscapes in Multidimensional Systems. *PLoS Comput. Biol.* **2020**, *16*, e1007788. [CrossRef]
35. Srinivasan, V.; Kumar, P. Emergent and Divergent Resilience Behavior in Catastrophic Shift Systems. *Ecol. Model.* **2015**, *298*, 87–105. [CrossRef]
36. Da Silva, M.G.; de Aguiar Netto, A.d.O.; de Jesus Neves, R.J.; Do Vasco, A.N.; Almeida, C.; Faccioli, G.G. Sensitivity Analysis and Calibration of Hydrological Modeling of the Watershed Northeast Brazil. *J. Environ. Prot.* **2015**, *6*, 837. [CrossRef]
37. Sandoval-Solis, S.; Paladino, S.; Garza-Diaz, L.E.; Nava, L.F.; Friedman, J.R.; Ortiz-Partida, J.P.; Plassin, S.; Gomez-Quiroga, G.; Koch, J.; Fleming, J.; et al. Environmental Flows in the Rio Grande—Rio Bravo Basin. *Ecol. Soc.* **2022**, *27*, e20. [CrossRef]
38. Ingol-Blanco, E.M.; McKinney, D.C. *Modeling Climate Change Impacts on Hydrology and Water Resources: Case Study Rio Conchos Basin*; Center for Research in Water Resources, University of Texas: Austin, TX, USA, 2011.
39. Woodhouse, C.A.; Stahle, D.W.; Díaz, J.V. Rio Grande and Rio Conchos Water Supply Variability over the Past 500 Years. *Clim. Res.* **2012**, *51*, 147–158. [CrossRef]
40. Harley, G.L.; Maxwell, J.T. Current Declines of Pecos River (New Mexico, USA) Streamflow in a 700-Year Context. *Holocene* **2018**, *28*, 767–777. [CrossRef]
41. Sifuentes-Martínez, A.R.; Villanueva, J.; Teodoro, D.; Allende, C.; Estrada, J. 243 Years of Reconstructed Streamflow Volume and Identification of Extreme Hydroclimatic Events in the Conchos River Basin, Chihuahua, Mexico. *Trees* **2020**, *34*, 1347–1361. [CrossRef]
42. Návar-Cháidez, J.d.J. Water Scarcity and Degradation in the Rio San Juan Watershed of Northeastern Mexico. *Front. Norte* **2011**, *23*, 125–150.
43. Ortega-Gaucin, D.; Ortega-Gaucin, D. Caracterización de las sequías hidrológicas en la cuenca del río Bravo, México. *Terra Latinoam.* **2013**, *31*, 167–180.
44. Lehner, F.; Wahl, E.R.; Wood, A.W.; Blatchford, D.B.; Llewellyn, D. Assessing Recent Declines in Upper Rio Grande Runoff Efficiency from a Paleoclimate Perspective. *Geophys. Res. Lett.* **2017**, *44*, 4124–4133. [CrossRef]
45. Sandoval-Solis, S.; McKinney, D.C. Integrated Water Management for Environmental Flows in the Rio Grande. *J. Water Resour. Plan. Manag.* **2014**, *140*, 355–364. [CrossRef]
46. Aguilar Ortiz, T.L.O. *Análisis Histórico De La Sequía Hidrológica En La Cuenca Mexicana Del Río Salado*; Instituto Mexicano de Tecnología del Agua: Mexico City, Mexico, 2018; p. 2.
47. *CONAGUA Caracterización Fluvial E Hidráulica De Las Inundaciones En México*; Comisión Nacional del Agua: Mexico City, Mexico, 2014; p. 58.
48. Llewellyn, D.; Vaddey, S. *Upper Rio Grande Impact Assessment*; Bureau of Reclamation: Columbia, MI, USA, 2013; p. 169.
49. Cook, B.I.; Ault, T.R.; Smerdon, J.E. Unprecedented 21st Century Drought Risk in the American Southwest and Central Plains. *Sci. Adv.* **2015**, *1*, e1400082. [CrossRef]

50. Cayan, D.R.; Das, T.; Pierce, D.W.; Barnett, T.P.; Tyree, M.; Gershunov, A. Future Dryness in the Southwest US and the Hydrology of the Early 21st Century Drought. *Proc. Natl. Acad. Sci. USA* **2010**, *107*, 21271–21276. [CrossRef]

51. Rumsey, C.A.; Miller, M.P.; Sexstone, G.A. Relating Hydroclimatic Change to Streamflow, Baseflow, and Hydrologic Partitioning in the Upper Rio Grande Basin, 1980 to 2015. *J. Hydrol.* **2020**, *584*, 124715. [CrossRef]

52. Steele, C.; Elias, E.; Reyes, J. Recent Streamflow Declines and Snow Drought in the Upper Rio Grande Tributary Basins. *Univ. Counc. Water Resour.* **2019**, 1.

53. Paredes-Tavares, J.; Gómez-Albores, M.A.; Mastachi-Loza, C.A.; Díaz-Delgado, C.; Becerril-Piña, R.; Martínez-Valdés, H.; Bâ, K.M. Impacts of Climate Change on the Irrigation Districts of the Rio Bravo Basin. *Water* **2018**, *10*, 258. [CrossRef]

54. Duran-Encalada, J.A.; Paucar-Caceres, A.; Bandala, E.R.; Wright, G.H. The Impact of Global Climate Change on Water Quantity and Quality: A System Dynamics Approach to the US–Mexican Transborder Region. *Eur. J. Oper. Res.* **2017**, *256*, 567–581. [CrossRef]

55. Corona, D.S.; Solis, S.S.; Peraza, E.H.; Hidalgo, H.S.; Herrera, C.Á. Aproximación e impacto directo de ciclones tropicales a la cuenca del río Conchos, Chihuahua, México. *Investig. Y Cienc. De La Univ. Autónoma De Aguascalientes* **2017**, *25*, 53–61. [CrossRef]

56. Scurlock, D. *From the Rio to the Sierra: An Environmental History of the Middle Rio Grande Basin*; US Department of Agriculture, Forest Service, Rocky Mountain Research Station: Fort Collins, CO, USA, 1998.

57. Wozniak, F.E. *Irrigation in the Rio Grande Valley, New Mexico: A Study and Annotated Bibliography of the Development of Irrigation Systems*; U.S. Department of Agriculture, Forest Service, Rocky Mountain Research Station: Fort Collins, CO, USA, 1998.

58. Ward, F.A.; Booker, J.F.; Michelsen, A.M. Integrated Economic, Hydrologic, and Institutional Analysis of Policy Responses to Mitigate Drought Impacts in Rio Grande Basin. *J. Water Resour. Plan. Manag.* **2006**, *132*, 488–502. [CrossRef]

59. Warman, A. La Reforma Agraria Mexicana: Una Visión de Largo Plazo. In *Land Reform, Land Settlement and Cooperatives—Réforme Agraire, Colonisation et Coopératives Agricoles—Reforma Agraria, Colonización y Cooperativas*; FAO Rural Development Division: Mexico City, Mexico, 2006.

60. U.S. Congress Pecos River Compact. *Congress of the United States*; U.S. Congress Pecos River Compact: Santa Fe, NM, USA, 1949.

61. Heard, T.; Perkin, J.; Bonner, T. Intra-Annual Variation in Fish Communities and Habitat Associations in a Chihuahua Desert Reach of the Rio Grande/Rio Bravo Del Norte. *West. North Am. Nat.* **2012**, *72*, 1–15. [CrossRef]

62. Contreras-Balderas, S.; Edwards, R.J.; de Lourdes Lozano-Vilano, M.; García-Ramírez, M.E. Fish Biodiversity Changes in the Lower Rio Grande/Rio Bravo, 1953–1996. *Rev. Fish Biol. Fish.* **2002**, *12*, 219–240. [CrossRef]

63. Contreras-Balderas, S.; Ruiz-Campos, G.; Schmitter-Soto, J.J.; Díaz-Pardo, E.; Contreras-McBeath, T.; Medina-Soto, M.; Zambrano-González, L.; Varela-Romero, A.; Mendoza-Alfaro, R.; Ramírez-Martínez, C.; et al. Freshwater Fishes and Water Status in México: A Country-Wide Appraisal. *Aquat. Ecosyst. Health Manag.* **2008**, *11*, 246–256. [CrossRef]

64. Wong, C.; Williams, C.; Pittock, J.; Collier, U.; Schelle, P. World's Top 10 Rivers at Risk. Working Papers. 2007. Available online: esocialsciences.com (accessed on 10 August 2022).

65. Garreaud, R.D.; Boisier, J.P.; Rondanelli, R.; Montecinos, A.; Sepúlveda, H.H.; Veloso-Aguila, D. The Central Chile Mega Drought (2010–2018): A Climate Dynamics Perspective. *Int. J. Climatol.* **2020**, *40*, 421–439. [CrossRef]

66. Walker, B.; Carpenter, S.; Anderies, J.; Abel, N.; Cumming, G.; Janssen, M.; Lebel, L.; Norberg, J.; Peterson, G.D.; Pritchard, R. Resilience Management in Social-Ecological Systems: A Working Hypothesis for a Participatory Approach. *Conserv. Ecol.* **2002**, *6*, 14. [CrossRef]

67. Gunderson, L.H.; Holling, C.S. *Panarchy: Understanding Transformations in Human and Natural Systems*; Island Press: Washington, DC, USA, 2002; ISBN 978-1-55963-857-9.

68. Becken, S. Developing A Framework for Assessing Resilience of Tourism Sub-Systems to Climatic Factors. *Ann. Tour. Res.* **2013**, *43*, 506–528. [CrossRef]

Article

IrrigTool—A New Tool for Determining the Irrigation Rate Based on Evapotranspiration Estimated by the Thornthwaite Equation

Cristian Ștefan Dumitriu [1], Alina Bărbulescu [2,*] and Carmen Elena Maftei [2]

1   S.C. Utilnavorep S.A., 55 Aurel Vlaicu Av., 900055 Constanța, Romania; cristian.dumitriu@unitbv.ro
2   Faculty of Civil Engineering, Transilvania University of Brașov, 5, Turnului Str., 900152 Brașov, Romania; cemaftei@gmail.com
*   Correspondence: alina.barbulescu@unitbv.ro

Abstract: In the context of climate change, irrigation has become a must for ensuring crop production because in some regions, the drought episodes became more frequent. The decision to efficiently allocate water resources should be made quickly, based on tools that provide correct information with a low computational effort. Therefore, we propose a new user-friendly tool—IrrigTool—for assessing the irrigation rate considering the precipitation, temperature, evapotranspiration, soil type, and crop. IrrigTool implements the Thornthwaite equations and can be used to identify weakness due to drought stress and as an educational tool. Apart from the computation, it provides a graphical representation of the results and possible comparisons of the output for two locations. The application is built in Microsoft Excel for graphics and Visual Basic VBA. The user does not have programming knowledge to use it. Data on monthly precipitation and temperature data must be introduced in the specified fields, and after pressing the run button, the results are automatically displayed. The article exemplifies the functioning on data series from Romania's Dobrogea region.

Keywords: evapotranspiration; irrigation; soil water reserve; water balance

check for updates

Citation: Dumitriu, C.Ș.; Bărbulescu, A.; Maftei, C.E. IrrigTool—A New Tool for Determining the Irrigation Rate Based on Evapotranspiration Estimated by the Thornthwaite Equation. *Water* **2022**, *14*, 2399. https://doi.org/10.3390/w14152399

Academic Editor: William Frederick Ritter

Received: 17 June 2022
Accepted: 1 August 2022
Published: 2 August 2022

Publisher's Note: MDPI stays neutral with regard to jurisdictional claims in published maps and institutional affiliations.

Copyright: © 2022 by the authors. Licensee MDPI, Basel, Switzerland. This article is an open access article distributed under the terms and conditions of the Creative Commons Attribution (CC BY) license (https://creativecommons.org/licenses/by/4.0/).

1. Introduction

Water scarcity is one of the world's most critical issues since water resources are not uniformly distributed and are affected by pollution. Moreover, people from some regions have no access to fresh water. Industrial activities, intensive chemical fertilizers utilization, vehicle exhaust, acidic rain, and climate change are among the most critical threats to groundwater quality [1,2]. The water quality for agricultural use is questionable in many situations, while in others, the water composition is not appropriate for such use [3]. Thus, water availability has become a restrictive factor for irrigation in different world regions [4].

In climate change conditions, this issue becomes more acute [5]. Since a population of about 7 billion people must be fed, an increasing concern for ensuring food necessities has manifested. Scientists and professionals should cooperate to find new approaches to ensure food production [6]. Different solutions have been proposed, each based on the particular climate of the region of interest. Rational irrigation, influenced by climate, crop, and soil type, aiming at ensuring the humidity necessary for plants' development and minimizing water consumption, is one of them [7].

While the climate influences irrigation necessity, it seems that, in its turn, irrigation impacts the climate of the regions where it is applied [8–10].

Efficient irrigation should consider the new technology and water management practices that may increase the performance of the available irrigation systems [11,12].

Different authors presented the procedure of "irrigation interval" scheduling [13] available only for particular regions. For example, the Michiana Irrigation Scheduler program schedules irrigation for some crops using the Stress Day Index and provides the estimated final yield throughout the season as the soil becomes too dry. The KanSched benefit by a network of weather stations and evapotranspiration (*ETRM*) is found on the Internet, but data must be input by hand. The Computerized Irrigation Scheduling by the Checkbook method (North Dakota State University) uses an earlier bulletin on checkbook irrigation scheduling. The Cropflex 2000 developed by the Colorado State University aims at managing irrigation and fertility. The Woodruff Chart maker developed by the University of Missouri uses the historical hydro-meteorological data for drawing an accumulative water-use curve for crops, emergence date, and weather. The curve is a graphical tool for scheduling the irrigations. Unfortunately, these software are of particular use for some USA regions only.

Water-requirement tables have been built by other scientists [14,15] based on the FAO documents [16,17]. Different software have been proposed to compute the evapotranspiration, an essential component of the water balance equation, mostly based on the Penman–Monteith equation [18,19], which is the official method adopted by FAO. CROPWAT 8.0 [20], WTRBLN [21], EVAP [22], and Excel worksheets have been proposed over time for water-balance calculation based on the Thornthwaite–Mather methodology [23] or estimating the *ETRM* by the Hamon equation [24].

WTRBLN [21] that computes the water balance has its initial version in Basic 3.0, and the new one in MATLAB needs a license to be used as well as computational skills. The Excel 2000 worksheet announced by Armiraglio et al. [23] is available on request, and that of Mammoliti et al. [25] computes only the water balance. EVAP [22] is a FORTRAN program that implements the Thornthwaite equation for the ETRM computation similar to [20,21,23] but not the irrigation rate.

Related studies using remote sensing (RS) are those of Droogers et al. [26], Olivera-Guerra et al. [27], Brocca et al. [28], Bretreger et al. [29], and Foster et al. [30].

Droogers et al. [26] proposed an irrigation scheme based on a two-step modeling approach. The remote-sensing *ETRM* was used for optimizing two parameters of the SWAP model, and then, a forward–backward algorithm was applied to assess the accuracy of remotely sensed actual *ETRM*. Olivera-Guerra et al. [27] proposed a new methodology to estimate the timing and the irrigation rate based on the Landsat-7/8 data in the following stages: deriving the crop water stress coefficient from the Landsat land surface temperature, estimating the daily root zone soil moisture, retrieving irrigation at the Landsat pixel scale, and aggregating pixel-scale irrigation estimates at the crop field scale.

Through the inversion of the soil-water balance equation and by using satellite soil moisture products as input, the amount of water entering into the soil and hence irrigation is determined by Brocca et al. [28]. The study of Bretreger et al. [29] utilized the Landsat satellites (5–8) for monitoring the crop situation based on the vegetation index for assessing crop development through a crop coefficient. Soil parameters were provided by the digital soil maps and in situ observations. The results were in concordance with the recorded irrigation time series. Foster et al. [30] presented an analysis of the relative accuracy of different satellite-based irrigation water-use monitoring approaches, with evidence of large uncertainties when water-use estimates are validated against in situ irrigation data at both the field and regional scale.

In this context, this study introduces a new tool—IrrigTool—aiming to optimize the water supply necessary for crop growth based on the average monthly temperatures and precipitation, the soil characteristics, and the type of agricultural crop. It uses the Thornthwaite equation for the *ETRM* computation. This tool can be utilized for learning and practical purposes. It is can also be used without deep knowledge of computer science.

2. Methodology

This section contains the basic formulas employed for the implementation purposes necessary to understand the implementation.

The irrigation rate (*M*) is the water quantity used to irrigate the surface of 1 ha cultivated with a specific type of plant. Hence, it represents the total water quantity that must be applied to a crop during and outside the vegetation period to fill the soil moisture deficit up to the value of the potential evapotranspiration [31].

The water application rate (m) is the quantity of water necessary per 1 ha of the crop during a unique watering operation. Therefore,

$$M = \sum m \tag{1}$$

and the watering number is

$$n = M/m. \tag{2}$$

The watering number is rounded up to the closest natural number, and the irrigation rate will be corrected accordingly.

The first element that must be considered for computing the irrigation rate is the water loss by evapotranspiration. The monthly water consumption by the plants' evapotranspiration and evaporation from the soil surface, denoted in the following by ETRM (m^3/ha), is computed here by the Thornthwaite equations, with a correction that depends on the crop and geographical zone.

For a month j, $ETRM$ is calculated by [32]:

$$ETRM_j = 160 \times \left(10t_j/I\right)^a \times K_1 \times K_p \tag{3}$$

where:

t_j—the medium temperature of the j-th month, $j = 1, \ldots , 12$.
I—the annual thermic index:

$$I = \sum_{j=1}^{12} \left(\frac{t_j}{5}\right)^{1.514} \tag{4}$$

$$a = 6.75 \times 10^{-7} \times I^3 - 7.71 \times 10^{-5} \times I^2 + 1.792 \times 10^{-2} \times I + 0.49239, \tag{5}$$

where K_1—a coefficient specific to each latitude,
K_p—a coefficient that depends on the crop type.
The second element to be estimated is the water soil storage (m^3/ha).

Outside the vegetation period, the depth taken into account for computing the water available to the plant is the maximum thickness of the storage layer from which the plant can utilize the water (H). $H = 1.5$ m for deep soils, H = the real soil depth for soils with short profiles, and H = the specific depth for soils with heavy and compacted argillic layers [32].

Outside the vegetation period, the soil water reserve varies in the interval [$R_{min, H}$, $R_{max, H}$], where [32]:

$$R_{min,H} = 100 \times H \times DA_H \times C_{0,H}, \tag{6}$$

$$R_{max, H} = 100 \times H \times DA_H \times C_{c,H}, \tag{7}$$

- $R_{min,H}$ (m^3/ha)—the minimum water reserve in the soil, corresponding to the wilting point, at the depth H,
- $R_{max,H}$ (m^3/ha)—the maximum water reserve in the soil, corresponding to the water field capacity, at the depth H,
- DA_H (t/m^3)—the soil bulk density corresponding to the depth H,
- $C_{c,H}$ (%)—the water field capacity corresponding to the depth H,
- $C_{0,H}$ (%)—the wilting coefficient corresponding to the depth H.

During the vegetation period, the water soil storage computation considers the depth of the active layer (where the principal mass of roots is developed at the maturity stage) h and the soil type. Therefore, for this period, the maximum and minimum water reserve in soil $R_{max,h}$ ($R_{min,h}$) are given by [32,33]:

$$R_{max,h} = 100 \times h \times DA_h \times C_{c,h} \left[m^3/ha\right], \tag{8}$$

$$R_{min,h} = 100 \times h \times DA_h \times P_{min,h} \left[m^3/ha \right],\tag{9}$$

with

- DA_h (t/m^3)—the soil bulk density corresponding to the depth h,
- $C_{c,h}$ (%)—the water field capacity corresponding to the depth h.
- $P_{min,h}$(%)—the minimum moisture level, defined as the limit under which the soil humidity should not decrease for ensuring normal conditions for the plant growth. It is computed at the depth h by [34]:

$$P_{min.h} = \begin{cases} C_{0,h} + 1/3(C_{c,h} - C_{0,h}), & \text{for light textural type} \\ C_{0,h} + 1/2(C_{c,h} - C_{0,h}), & \text{for medium textural type} \\ C_{0,h} + 2/3(C_{c,h} - C_{0,h}), & \text{for heavy textural type} \end{cases}\tag{10}$$

where $C_{0,h}$ (%)—the wilting coefficient, corresponding to the depth h.

$P_{min,h}$ depends on the soil texture and takes values between the field capacity and the wilting coefficient.

The water balance equation for a month is [34]:

$$R_f = R_i - ETRM + P_u - A_f\tag{11}$$

where:

ETRM—the monthly water consumption by the plants' evapotranspiration and evaporation from the soil surface,

P_u (m^3/ha)—the water intake from precipitation in that month,

R_i (m^3/ha)—the initial water reserve in soil at the considered depth, at the beginning of the month,

R_f (m^3/ha)—the final water reserve in soil for the considered depth at the month's end,

A_f—the groundwater supply. A_f is positive only for deep-rooted plants (alfalfa, corn, beets, soybeans, trees, etc.). Otherwise, $A_f = 0$.

In areas where irrigation is applied before sowing, the initial water reserve is taken equal to the field capacity $C_{c,H}$.

In rainy years during the vegetation period, when $P_u > ETRM + R_f - R_i$, irrigation is not necessary because the amount of rainfall completely covers the water need.

The water balance computation starts from 1st of October (October is the first month of the agricultural year). The water reserve at the end of a month j (R_{f_j}) becomes the initial water reserve for the next month ($R_{i_{j+1}}$).

From the above, it results that for the months October–March (months I–VI of the agricultural year), the water balance is calculated at H, whereas for the vegetation period (April–September), it is calculated at the depth h [34,35].

To compute the irrigation rate, the following rules should be followed:

a. The water reserve in soil must be lower than $R_{max,H}$ ($R_{max,h}$) in the cold (vegetation) season. When it is higher, the quantity that exceeds the above limits is considered lost by infiltration and cannot be used by plants;
b. In the cold period, the water reserve can decrease without limitation except for the crops subject to water provision.
c. In the vegetation period, the water reserve cannot decrease under $R_{min,h}$. If the reserve decreases under this value, watering must be applied.

If the groundwater level varies during the year, then the groundwater supply will be considered resulting from hydrological forecasts. If the groundwater level does not vary during the year, then one may use the data from [18].

The water application rate is computed by [36]:

$$m = 100 \times h \times DA_h \times (C_{c,h} - P_{min}).\tag{12}$$

The flowchart of the study is presented in Figure 1.

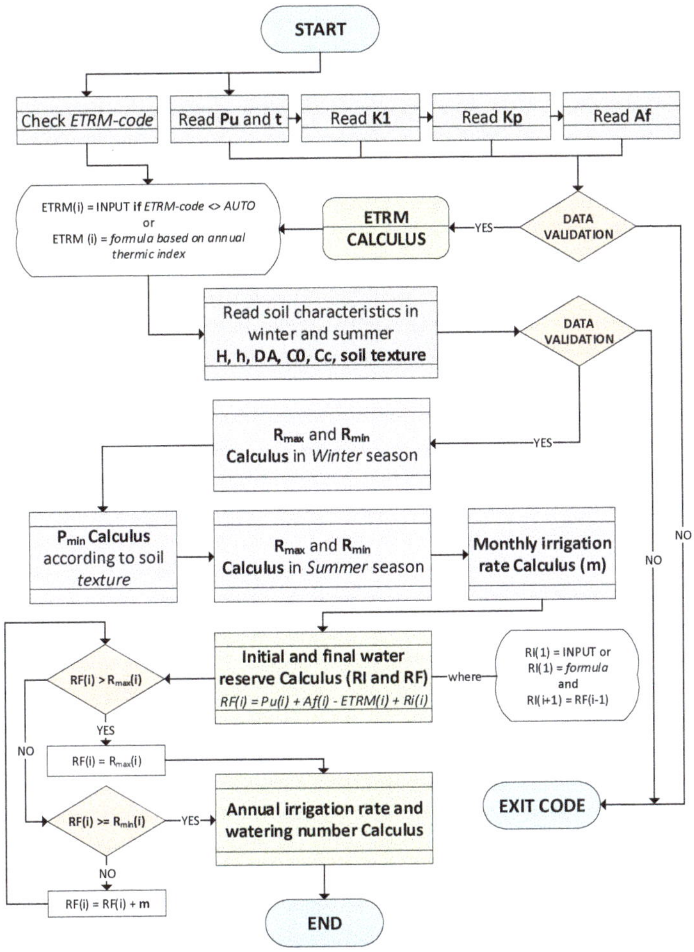

Figure 1. Flowchart (the variables are denoted in concordance with the coding).

The analysis steps (according to Figure 1) are:

(a) Input data from the Excel files where they have been previously introduced. They are: monthly precipitation (P_u in the flowchart) and temperature series ($t-$ in the flowchart), the series of the monthly coefficients K_1 and K_p, and the A_f series. The input data are validated by checking if the cells are filled in with numerical values. If not, the algorithm stops. Otherwise, it passes to the next step, as follows:

(b) Read the soil characteristics, namely H, h, DA, C_0, and C_c, corresponding to h and H from the worksheets where they were previously introduced. If all the values are numerical, and none is absent, the algorithm passes to step (c). Otherwise, it stops.

(c) Compute R_{min} and R_{max} for winter;

(d) Compute P_{min};

(e) Compute R_{min} and R_{max} for summer;

(f) Compute the monthly irrigation rate;

(g) Compute the initial and final water reserve for each month;

(h) Compute the water application rate and the annual irrigation rate;

(i)   Display the results and the graphical representation.

3. Data Series

The climate in Romania is continuously changing, so the drought regions are becoming more arid. The Dobrogea region (Figure 2), situated in the south-eastern part of the country, is experiencing a more extended period without precipitation followed by short periods with high rainfalls. It was shown that the mean annual temperatures were augmented by 0.8 °C after 1997, and many rivers are drying up in summer. Increasing trends are noticed for the region's maximum and minimum annual, summer, winter, and spring temperature series [37–41]. Therefore, the necessity of crop irrigation becomes more relevent than ever.

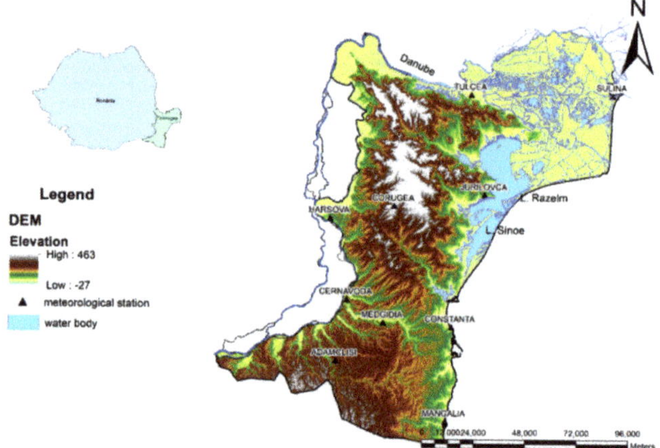

Figure 2. The map of Dobrogea [41].

The monthly average temperatures and precipitation series recorded at Constanța (CT) and Tulcea (TL) during the agricultural years 1998–2019, which were employed to exemplify the proposed tool, are presented in Figure 3. The basic statistics of these series are given in Table 1.

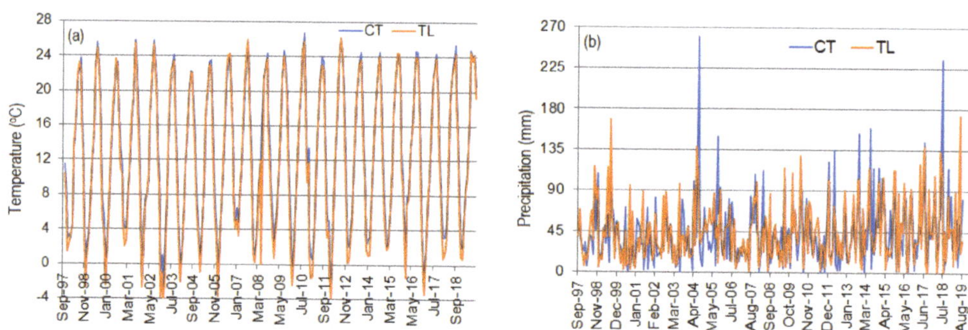

Figure 3. Monthly average (**a**) temperature and (**b**) precipitation series recorded at Constanța (CT) and Tulcea (TL).

Table 1. Basic statistics of the CT and TL monthly series used for exemplification.

Series	Max	Min	Mean	Median	Std. Dev.	Coef. of Variation	Skewness	Kurtosis
CT precipitation	259.20	0.30	43.27	36.00	36.53	0.84	1.96	6.95
TL precipitation	173.00	0.00	44.83	37.35	33.91	0.76	1.06	0.85
CT temperature	26.70	−0.20	12.83	12.60	8.17	0.64	−0.03	−1.29
TL temperature	26.20	−4.50	12.08	11.65	8.60	0.71	−0.06	−1.28

4. Implementation

The application was built in Microsoft Excel for graphics and VBA (Supplementary Materials S1 and S2) because these environments have also been employed for hydrological analysis, and the environment is easy to use [42–46].

The application's structure allows the irrigation rate calculation for two locations (S1 and S2) and the comparison of results.

The main worksheets—"Irrigation S1" (Figure 4) and "Irrigation S2"—contain the work data for each site, the numerical results obtained, and the afferent charts obtained after running the algorithm.

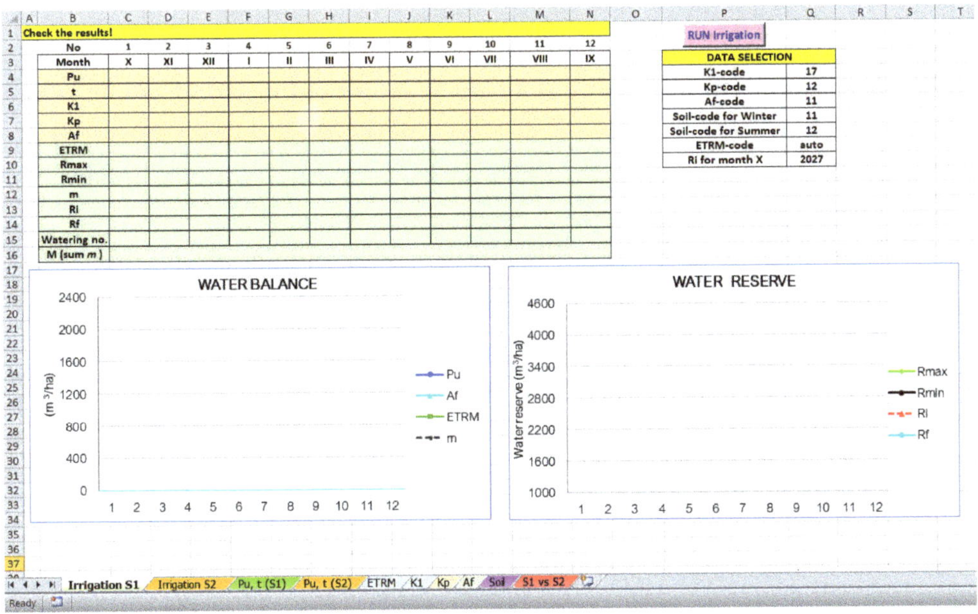

Figure 4. The "Irrigation S1" worksheet.

Before running, the input data is entered in seven worksheets ("Pu, t (S1)", "Pu, t (S2)", "ETRM", "K1", "Kp", "Af", and "Soil") that are simple and easy to populate by the user with data according to its need. They will be presented in the following section.

The software selects the values of interest through numerical codes, which correspond to the data sets entered by the user. In the "Irrigation S1" and "Irrigation S2" worksheets, the user indicates the codes (settable by him) in the table entitled "DATA SELECTION".

The user will introduce into the cells Q4–Q11 the following codes:

— in Q4—a code selected from the first column of the worksheet "K1";
— in Q5—a code selected from the first column of the worksheet "Kp";
— in Q6—a code selected from the first column of the worksheet "Af";

— in Q7—a code selected from the first column of the worksheet "Soil" corresponding to the winter season;

— in Q8—a code selected from the first column of the worksheet "Soil" corresponding to the summer season;

— in Q9—a code selected from the first column of the worksheet "ETRM" or "auto";

— in Q10—"0" or a value selected by the user.

Details will be provided in the following section.

The software takes the data corresponding to the codes from the worksheets "Pu, t (S1)" (Figure 5), "ETRM" (Figure 6), "K1" (Figure 7), "Kp" (Figure 8), "Af" (Figure 9), and "Soil" (Figure 10).

	A	B	C	D	E		H	I	J	K	L	M	N	O	P	Q	R	S	T
1	From year...	To year...	Pu (mm)	t (°C)			22 years	Check the results!											
2	1	22	38.8	10.4			Month	X	XI	XII	I	II	III	IV	V	VI	VII	VIII	IX
3		Find Averages	68	6.2			Pu$_{Avg}$												
4			40.7	1.4			t$_{Avg}$												
5			20	2.2															
6			5.8	3.4															
7			20.2	4.4															
33			3.5	18.2															
263			173	23.6															
264			22	24.3															
265			35	19.3															
266																			

Sheet1 | Irrigation S1 | Irrigation S2 | Pu, t (S1) | Pu, t (S2) | ETRM | K1 | Kp | Af | Soil | S1 vs S2

Figure 5. The "Pu, t (S1)" sheet.

	A	B	C	D	E	F	G	H	I	J	K	L	M	N	O
1															
2															
3	ETRM-Code	References	Month	X	XI	XII	I	II	III	IV	V	VI	VII	VIII	IX
4	11		ETRM	340	172	56	20	48	130	392	830	1186	1892	1657	782
5															
6															
7															
8															

Irrigation S1 | Irrigation S2 | Pu, t (S1) | Pu, t (S2) | ETRM | K1 | Kp | Af | Soil | S1 vs S2

Figure 6. The "ETRM" worksheet.

	A	B	C	D	E	F	G	H	I	J	K	L	M	N	O	P	Q
1																	
2																	
3	K1-Code	Latitude	Month	X	XI	XII	I	II	III	IV	V	VI	VII	VIII	IX		
4	11	0	K1	1.04	1.01	1.04	1.04	0.94	1.04	1.01	1.04	1.01	1.04	1.04	1.01		
5	12	10	K1	1.02	0.98	0.99	1	0.91	1.03	1.03	1.08	1.06	1.08	1.07	1.02		
6	13	20	K1	1	0.93	0.94	0.95	0.9	1.03	1.05	1.13	1.11	1.14	1.11	1.02		
7	14	30	K1	0.98	0.89	0.88	0.9	0.87	1.03	1.08	1.18	1.17	1.2	1.14	1.03		
8	15	35	K1	0.97	0.86	0.85	0.87	0.85	1.03	1.09	1.21	1.21	1.23	1.16	1.03		
9	16	40	K1	0.97	0.86	0.81	0.84	0.83	1.03	1.11	1.24	1.25	1.27	1.18	1.04		
10	17	45	K1	0.94	0.79	0.75	0.8	0.81	1.02	1.13	1.28	1.29	1.31	1.21	1.04	*used in the example	
11	18	46	K1	0.94	0.79	0.74	0.79	0.81	1.02	1.13	1.29	1.31	1.32	1.22	1.04		
12	19	47	K1	0.94	0.78	0.73	0.77	0.8	1.02	1.13	1.3	1.32	1.33	1.22	1.04		
13	20	48	K1	0.93	0.77	0.72	0.76	0.8	1.02	1.14	1.31	1.33	1.34	1.23	1.05		
14	21	49	K1	0.93	0.76	0.71	0.75	0.79	1.02	1.14	1.32	1.34	1.35	1.24	1.05		
15	22	50	K1	0.92	0.76	0.7	0.74	0.78	1.02	1.15	1.33	1.36	1.37	1.25	1.06		
16																	
17																	

Irrigation S1 | Irrigation S2 | Pu, t (S1) | Pu, t (S2) | ETRM | K1 | Kp | Af | Soil | S1 vs S2

Figure 7. The "K1" worksheet. Data are taken from [34,47].

Initially, the cells from the rectangle C4–N4–N12–C12 in the "Irrigation S1" and "Irrigation S2" are empty. They are automatically filled in, concomitantly with the charts "WATER BALANCE" and "WATER RESERVE", after running the code.

The monthly precipitation and temperature series are introduced in the worksheets "Pu, t (S1)" and "Pu, t (S2)" (Figure 5). The data are entered consecutively for an unlimited number of years for precipitation—in column C—and temperature—in D column D—separately. The worksheets (Figure 5) implement a 1–12 monthly average calculation module for both precipitation and temperature series. For performing the computation, the user can select all the years entered or a specific interval (for example, from year 5 to year 10) and press the button "Find Averages".

Kp-Code	Description	Month	X	XI	XII	I	II	III	IV	V	VI	VII	VIII	IX
11	corn (Constanta)	Kp	0.7	0.85	0.9	0.9	0.85	0.6	0.81	0.75	0.85	1.33	1.26	0.77
12	alphaalpha (Constanta)	Kp	1	1	1	1	1	1	1.6	1.11	1.01	1.07	1.15	0.97
13	corn (Tulcea)	Kp	0.7	0.85	0.9	0.9	0.85	0.6	0.94	0.61	0.98	1.29	1.28	0.81
14	alphaalpha (Tulcea)	Kp	1	1	1	1	1	1	1.53	1.32	1.33	1.27	1.3	1.19

Figure 8. The "Kp" worksheet. Data are reproduced from [35], p. 41.

Af-Code	Description	Month	X	XI	XII	I	II	III	IV	V	VI	VII	VIII	IX
11	pesimistic	Af (m^3/ha)	0	0	0	0	0	0	0	0	0	0	0	0

Figure 9. The "Af" worksheet.

Soil-code	Location (zone)	Season	Culture (crop)	Soil texture data	Soil characteristics			
					H or h (m)	DA(t/m3)	C0(%)	Cc(%)
11	Dobrogea	Winter	Corn	medium	1.5	1.29	6.9	23.2
12	Dobrogea	Summer	Corn	medium	0.75	1.3	7.5	24.15
13	Dobrogea	Winter	Sunflower	medium	1.5	1.29	6.9	23.2
14	Dobrogea	Summer	Sunflower	medium	1	1.28	7.4	23.9

Figure 10. The "Soil" worksheet.

The average values obtained are automatically displayed in the table next to the series: the cells I3–T3 for the average monthly precipitation and I4–T4 for the average monthly temperature. Cells are also automatically filled in the table from the main worksheets—"Irrigation S1" (for the series introduced in "Pu, t (S1)") and "Irrigation S2" (for the series introduced in "Pu, t (S2)")—namely the cells C4–N4 (for the average monthly precipitations) and C5–N5 (for the average monthly temperatures). These values are then employed to calculate the irrigation rates.

When running the program, the user may introduce the corresponding code (11 in the cell Q6 in the worksheets "Irrigation S1" and "Irrigation S2" from Figure 6) for ETRM. If the evapotranspiration is known (such as in Figure 6), and data were previously filled in the worksheet "ETRM", the balance Equation (11) is computed using these values. If ETRM is unknown, the user must introduce the code "auto" in the cell Q6 in the worksheets "Irrigation S1" and "Irrigation S2", and the ETRM is estimated using Equation (3).

To not restrict the use of the tool for some specific soil types, the series of coefficients K_1, K_p, and A_f are entered by the user in separate worksheets with the same names.

The "K1" worksheet (Figure 7) already contains some K_1 values corresponding to different northern latitudes, among which is the 45° north latitude utilized in the exemplification. The table can be filled with values for other latitudes and the corresponding K1 code (assigned by the user). Two different latitudes must have different K1 codes.

The "Kp" worksheet (Figure 8) contains some examples of K_p values but can be filled with other values according to the crop and the region where it is cultivated. The user can assign the Kp code in column A and introduce any necessary description in column B.

In Figure 8 there are two types of crops—corn and alfalfa—and the corresponding coefficients for each location—Constanța and Tulcea. The values are filled in according to [35] for the two locations used for exemplification. For other locations and other crops, some values are provided in [33].

The K1 code and Kp code must be filled in the cells Q4 and Q5, respectively, from the main worksheets. When pressing the button "RUN Irrigation" from these worksheets, the corresponding K_1 and K_p values are filled in automatically in the cells C6–N6 and C7–N7, respectively, in the main worksheets.

The "Af" worksheet (Figure 9) contains the groundwater's supply values. The user can introduce them, and the code is filled in the cell Q6 from the main worksheets. When running the code, the values are automatically filled in the table from the "Irrigation S1" and "Irrigation S2" worksheets (the cells C8–Q8).

Figure 9 refers to the case when there is no water supply from the groundwater.

The "Soil" worksheet (Figure 10) contains the values of $H, h, DA_H, DA_h, C_{0,H},$ and $C_{0,h}$. It must be filled with the corresponding values for each soil type and each season. Since the soil characteristics are specific to the season, crop, and soil type, the user should specify (in column B) the location, the crop (column D), and season (column E).

The irrigation rate and watering numbers (abbreviated by "Water no." in the charts) are computed by pressing the command button in the "Irrigation S1" and "Irrigation S2" worksheets using Equations (1) and (2). The results and graphs are updated (the previous results are overwritten) both in the two main worksheets and in the comparison one ("S1 vs. S2") (Figure 11).

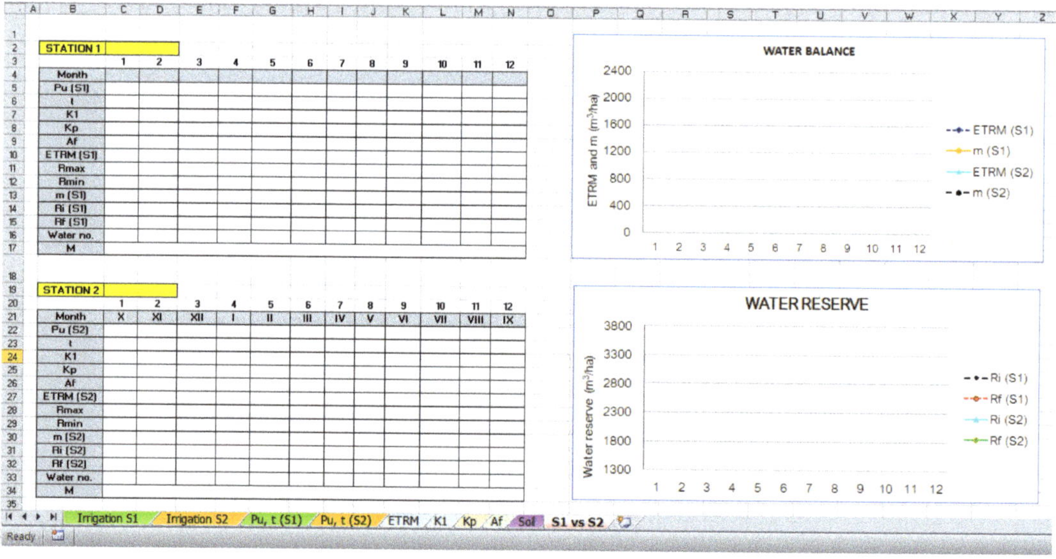

Figure 11. The "S1 vs. S2" worksheet.

5. Results

In the following, we exemplify how IrrigTool works, giving as input the monthly temperature and precipitation at Constanța and Tulcea (Romania) for 22 agricultural years (1998–2019). Given that the agricultural year starts in October (1 or month X in all the charts), the data series are introduced in the "Pu, t (S1)" and "Pu, t (S2)" worksheets. The chosen crop was corn.

Considering the geographical position of Romania, the K1 code represents the values of K1 corresponding to 45° northern latitude (Figure 7—C10).

The Kp code was chosen as 11 and 13 for CT and TL, respectively (Figure 8).

The water supply from groundwater was $A_f = 0$ because the region is arid, so the Af code was set to 11 (Figure 9).

The Soil code (Figure 10) for winter (summer) was the same, 11 (12), given the same soil characteristics. The ETRM code was set to "auto", so the ETRM was computed using Equation (3).

The initial water reserve in soil for October (Ri for month X) must be provided to start the computation. The user may introduce the value in the cell corresponding to "Ri for month X" from "Irrigation S1" and "Irrigation S2" or allow the software to compute it based on the formula (6) if "0" is inserted in that cell. The results obtained when the Ri for October is computed according to the methodology presented above are shown in Figure 12 (for CT) and Figure 13 (for TL), whereas their comparison is provided in Figure 14.

Row 15 in Figures 12 and 13 contains the watering numbers. One watering must be applied in June, August, and September and two in July for both locations. The water application rate is $m = 812$ m^3/ha (cells K12, M12, and N12, in Figures 12 and 13). The value 1623 from cell L12 in Figures 12 and 13 is equal to $2m$. The irrigation rate is 4058 in both cases (row 16 in the left-hand side table in Figures 12 and 13).

The "Water balance" charts (Figures 12 and 13) illustrate the mean monthly precipitation, ETRM, A_f, and m, whereas the second ones contain the water reserve (initial and final, maximum and minimum). The charts from Figure 14 summarize the findings for both locations.

The ETRMs are comparable at CT and TL. The initial and final water reserve is slightly higher at TL. The irrigation rate is the same at TL and CT. The watering should be applied the same month and in the same amount.

From Figures 12–14, row 30, it results that during the first eight months of the agricultural year (October–May), it is not necessary to apply irrigation. In June, August, and September, one watering should be applied (812 m^3/ha). In July, watering should be applied twice (a total of 1624 m^3/ha).

This aspect is emphasized in Figures 12 and 13, the bottom left chart, by the black dotted curves and in Figure 14, the top chart, by the yellow continuous curve (for CT) and the black dotted curve (for TL).

Given that the watering distribution depends on the initial soil reserve in October, the user may run the algorithm many times, at each run considering the initial water reserve in soil in October to be equal to the final water reserve in September in the previous run (and introducing manually this value in the cell "Ri for month X").

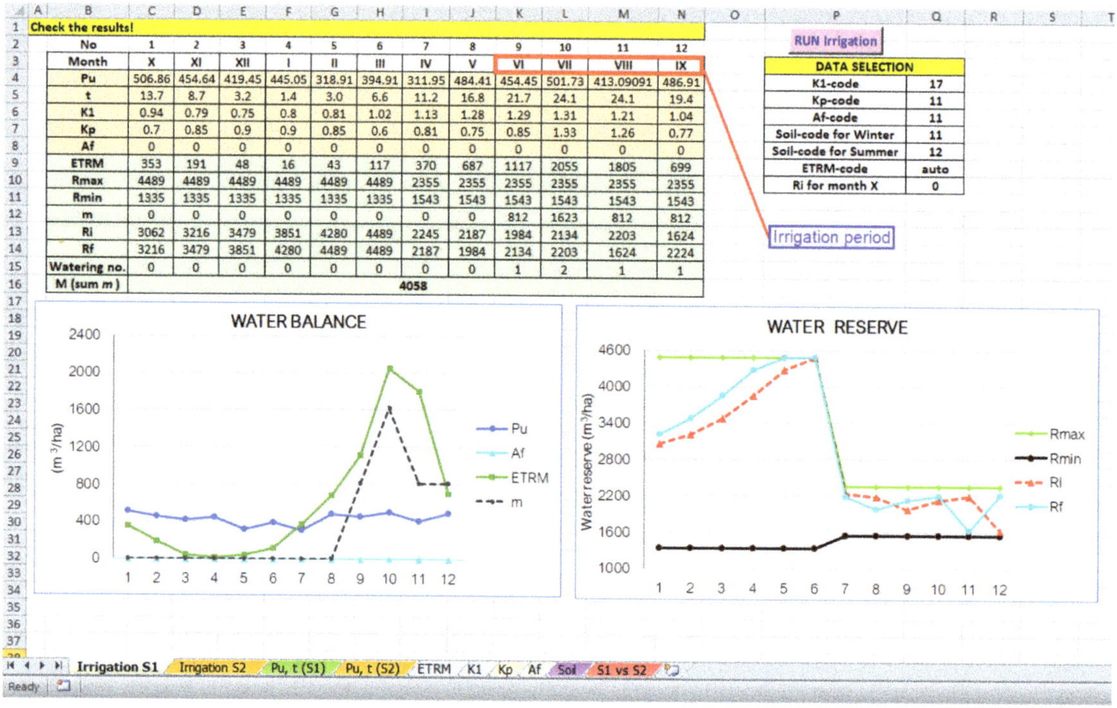
Check the results!
No 1 2 3 4 5 6 7 8 9 10 11 12
Month X XI XII I II III IV V VI VII VIII IX
Pu 506.86 454.64 419.45 445.05 318.91 394.91 311.95 484.41 454.45 501.73 413.09091 486.91
t 13.7 8.7 3.2 1.4 3.0 6.6 11.2 16.8 21.7 24.1 24.1 19.4
K1 0.94 0.79 0.75 0.8 0.81 1.02 1.13 1.28 1.29 1.31 1.21 1.04
Kp 0.7 0.85 0.9 0.9 0.85 0.6 0.81 0.75 0.85 1.33 1.26 0.77
Af 0 0 0 0 0 0 0 0 0 0 0 0
ETRM 353 191 48 16 43 117 370 687 1117 2055 1805 699
Rmax 4489 4489 4489 4489 4489 4489 2355 2355 2355 2355 2355 2355
Rmin 1335 1335 1335 1335 1335 1335 1543 1543 1543 1543 1543 1543
m 0 0 0 0 0 0 0 0 812 1623 812 812
Ri 3062 3216 3479 3851 4280 4489 2245 2187 1984 2134 2203 1624
Rf 3216 3479 3851 4280 4489 4489 2187 1984 2134 2203 1624 2224
Watering no. 0 0 0 0 0 0 0 0 1 2 1 1
M (sum m) 4058
RUN Irrigation
DATA SELECTION
K1-code 17
Kp-code 11
Af-code 11
Soil-code for Winter 11
Soil-code for Summer 12
ETRM-code auto
Ri for month X 0
Irrigation period
WATER BALANCE
(m 3/ha)
Pu
Af
ETRM
m
WATER RESERVE
Water reserve (m 3/ha)
Rmax
Rmin
Ri
Rf
Irrigation S1
Irrigation S2
Pu, t (S1)
Pu, t (S2)
ETRM
K1
Kp
Af
Soil
S1 vs S2
Ready

Figure 12. Computation results for CT.

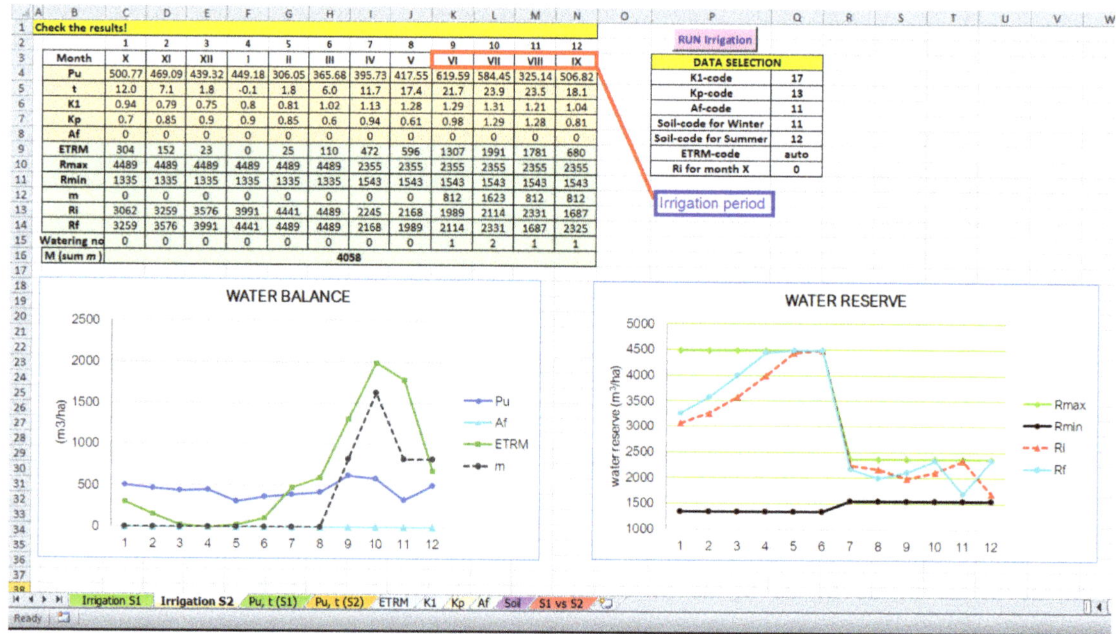

Check the results!
1 2 3 4 5 6 7 8 9 10 11 12
Month X XI XII I II III IV V VI VII VIII IX
Pu 500.77 469.09 439.32 449.18 306.05 365.68 395.73 417.55 619.59 584.45 325.14 506.82
t 12.0 7.1 1.8 -0.1 1.8 6.0 11.7 17.4 21.7 23.9 23.5 18.1
K1 0.94 0.79 0.75 0.8 0.81 1.02 1.13 1.28 1.29 1.31 1.21 1.04
Kp 0.7 0.85 0.9 0.9 0.85 0.6 0.94 0.61 0.98 1.29 1.28 0.81
Af 0 0 0 0 0 0 0 0 0 0 0 0
ETRM 304 152 23 0 25 110 472 596 1307 1991 1781 680
Rmax 4489 4489 4489 4489 4489 4489 2355 2355 2355 2355 2355 2355
Rmin 1335 1335 1335 1335 1335 1335 1543 1543 1543 1543 1543 1543
m 0 0 0 0 0 0 0 0 812 1623 812 812
Ri 3062 3259 3576 3991 4441 4489 2245 2168 1989 2114 2331 1687
Rf 3259 3576 3991 4441 4489 4489 2168 1989 2114 2331 1687 2325
Watering no 0 0 0 0 0 0 0 0 1 2 1 1
M (sum m) 4058
RUN Irrigation
DATA SELECTION
K1-code 17
Kp-code 13
Af-code 11
Soil-code for Winter 11
Soil-code for Summer 12
ETRM-code auto
Ri for month X 0
Irrigation period
WATER BALANCE
(m3/ha)
Pu
Af
ETRM
m
WATER RESERVE
water reserve (m^3/ha)
Rmax
Rmin
Ri
Rf
Irrigation S1
Irrigation S2
Pu, t (S1)
Pu, t (S2)
ETRM
K1
Kp
Af
Soil
S1 vs S2
Ready

Figure 13. Computation results for TL.

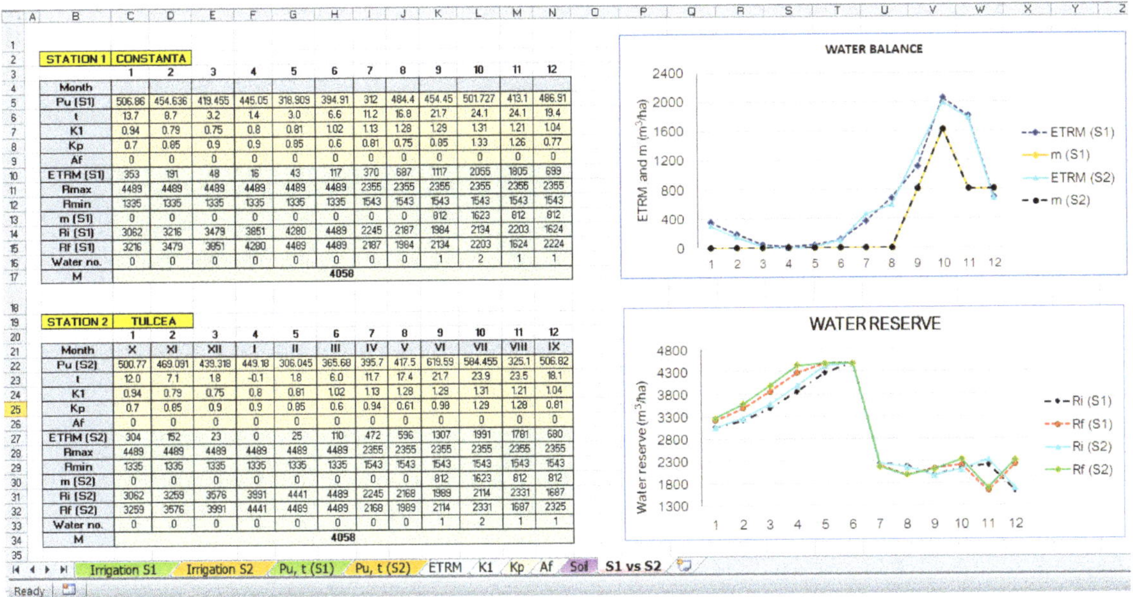

STATION 1	CONSTANTA											
	1	2	3	4	5	6	7	8	9	10	11	12
Month												
Pu (S1)	506.86	454.636	419.455	445.05	318.909	394.91	312	484.4	454.45	501.727	413.1	486.91
t	13.7	8.7	3.2	1.4	3.0	6.6	11.2	16.8	21.7	24.1	24.1	19.4
K1	0.94	0.79	0.75	0.8	0.81	1.02	1.13	1.28	1.29	1.31	1.21	1.04
Kp	0.7	0.85	0.9	0.9	0.85	0.6	0.81	0.75	0.85	1.33	1.26	0.77
Af	0	0	0	0	0	0	0	0	0	0	0	0
ETRM (S1)	353	191	48	16	43	117	370	687	1117	2055	1805	699
Rmax	4489	4489	4489	4489	4489	4489	2355	2355	2355	2355	2355	2355
Rmin	1335	1335	1335	1335	1335	1335	1543	1543	1543	1543	1543	1543
m (S1)	0	0	0	0	0	0	0	0	812	1623	812	812
Ri (S1)	3062	3216	3479	3851	4280	4489	2245	2187	1984	2134	2203	1624
Rf (S1)	3216	3479	3851	4280	4489	4489	2187	1984	2134	2203	1624	2224
Water no.	0	0	0	0	0	0	0	0	1	2	1	1
M	4058											

STATION 2	TULCEA											
	1	2	3	4	5	6	7	8	9	10	11	12
Month	X	XI	XII	I	II	III	IV	V	VI	VII	VIII	IX
Pu (S2)	500.77	469.091	439.318	449.18	306.045	365.68	395.7	417.5	619.59	584.455	325.1	506.82
t	12.0	7.1	1.8	-0.1	1.8	6.0	11.7	17.4	21.7	23.9	23.5	18.1
K1	0.94	0.79	0.75	0.8	0.81	1.02	1.13	1.28	1.29	1.31	1.21	1.04
Kp	0.7	0.85	0.9	0.9	0.85	0.6	0.94	0.61	0.98	1.29	1.28	0.81
Af	0	0	0	0	0	0	0	0	0	0	0	0
ETRM (S2)	304	152	23	0	25	110	472	596	1307	1991	1781	680
Rmax	4489	4489	4489	4489	4489	4489	2355	2355	2355	2355	2355	2355
Rmin	1335	1335	1335	1335	1335	1335	1543	1543	1543	1543	1543	1543
m (S2)	0	0	0	0	0	0	0	0	812	1623	812	812
Ri (S2)	3062	3259	3576	3991	4441	4489	2245	2168	1989	2114	2331	1687
Rf (S2)	3259	3576	3991	4441	4489	4489	2168	1989	2114	2331	1687	2325
Water no.	0	0	0	0	0	0	0	0	1	2	1	1
M	4058											

Figure 14. Comparison of the results for CT and TL (from the worksheet "S1 vs. S2").

6. Discussion

First, we compared the irrigation rate obtained using the Thornthwaite equation with the results obtained using the measured ETRM in the neighborhood of Constanţa (at Mamaia). We mention that the procedure followed in Romania is that the ETRM is measured only for nine months, from March to November, so the rest of the values are considered to be zero. Therefore, IrrigTool was run with the ETRM code 12 in the worksheet "ETRM" (Figure 15). The output is shown in Figure 16.

	A	B	C	D	E	F	G	H	I	J	K	L	M	N	O
3	ETRM-Code	References	Month	X	XI	XII	I	II	III	IV	V	VI	VII	VIII	IX
5	12	Mamaia	ETRM	325	204	0	0	0	104	322	692	1130	2055	1855	785

Figure 15. Average ETRM values in the neighborhood of CT, at Mamaia (5 km from CT).

There are no significant differences between the computed and measured ETRM values—cells C9–E9 in Figures 12 and 16 and the watering numbers are the same (one in June, August, and September and two in July).

In the previous section, we indicated that the irrigation rate depends on the crop. To illustrate this assertion, we considered the crop alfalfa with the Kp code 12 (alfalfa). In this case, the soil code must be modified as well. In this case, it will be 15 for summer and 16 for summer (Figure 17).

The watering number has a different distribution: one in June and July and two in August. The irrigation rate is 4224 m^3/ha because, in this case, $m = 1056$ m^3/ha.

It was also mentioned that the irrigation rate depends on the initial water reserve in the soil. To exemplify, let us consider the same settings as in Figure 17, with Ri in the month 1 = Rmin in the month IX (2614 m^3/ha), and run the algorithm. In this case, one watering should be applied in from May to July. The irrigation rate will remain the same: 4224 m^3/ha (Figure 18).

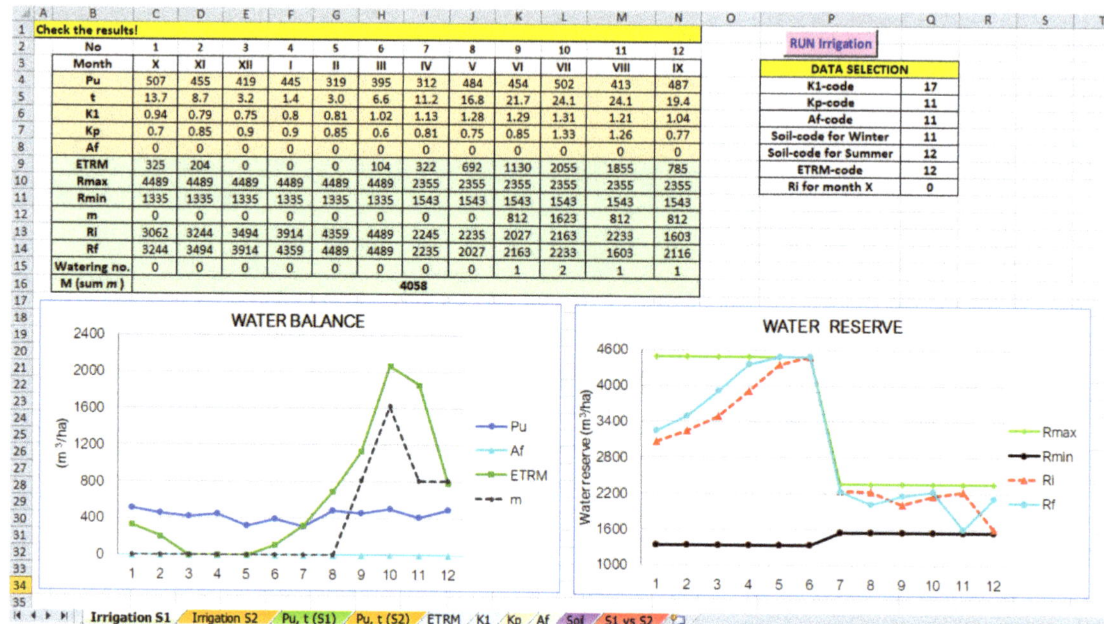

Figure 16. Output of running the algorithm using ETRM recorded at Mamaia (5 km from CT).

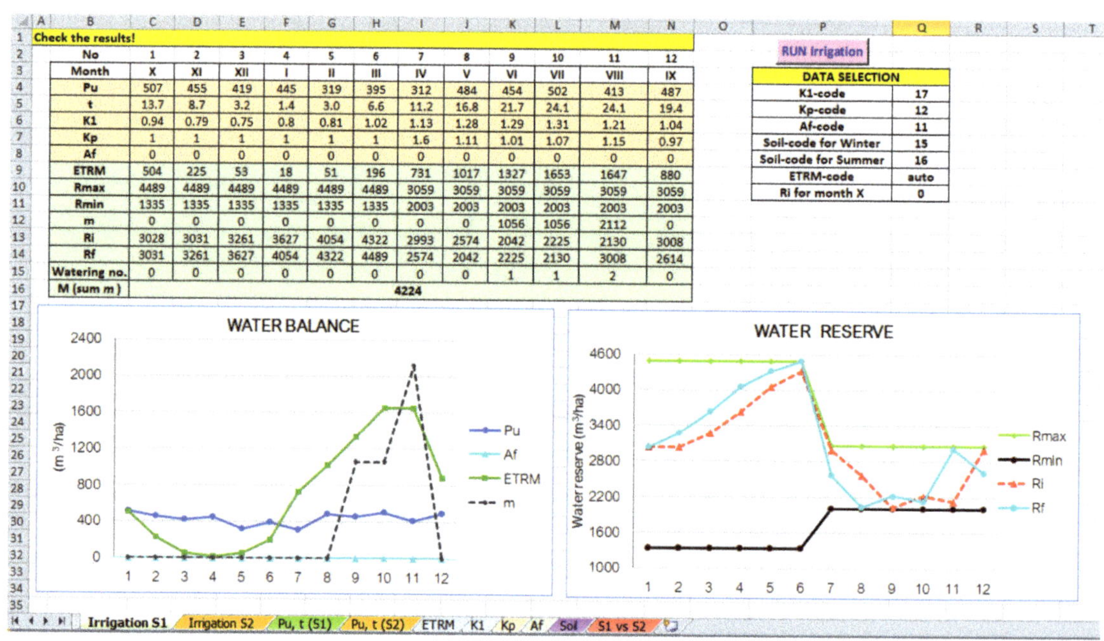

Figure 17. Output for alfalfa at CT, computed using the Thornthwaite equation.

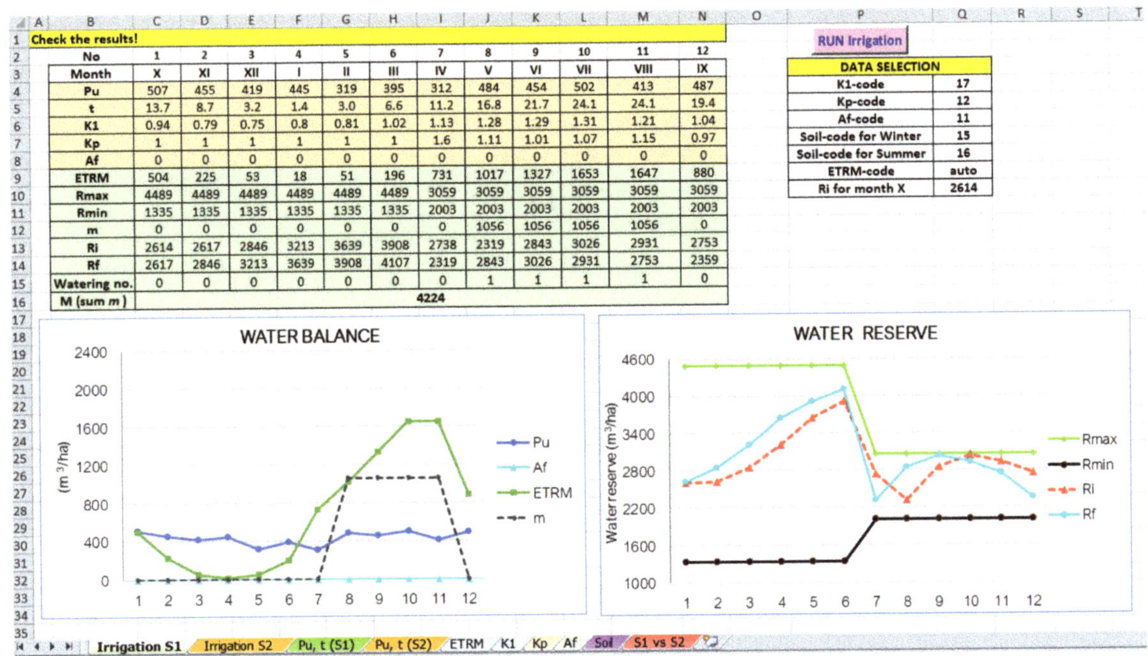

No	1	2	3	4	5	6	7	8	9	10	11	12
Month	X	XI	XII	I	II	III	IV	V	VI	VII	VIII	IX
Pu	507	455	419	445	319	395	312	484	454	502	413	487
t	13.7	8.7	3.2	1.4	3.0	6.6	11.2	16.8	21.7	24.1	24.1	19.4
K1	0.94	0.79	0.75	0.8	0.81	1.02	1.13	1.28	1.29	1.31	1.21	1.04
Kp	1	1	1	1	1	1	1.6	1.11	1.01	1.07	1.15	0.97
Af	0	0	0	0	0	0	0	0	0	0	0	0
ETRM	504	225	53	18	51	196	731	1017	1327	1653	1647	880
Rmax	4489	4489	4489	4489	4489	4489	3059	3059	3059	3059	3059	3059
Rmin	1335	1335	1335	1335	1335	1335	2003	2003	2003	2003	2003	2003
m	0	0	0	0	0	0	0	1056	1056	1056	1056	0
Ri	2614	2617	2846	3213	3639	3908	2738	2319	2843	3026	2931	2753
Rf	2617	2846	3213	3639	3908	4107	2319	2843	3026	2931	2753	2359
Watering no.	0	0	0	0	0	0	0	1	1	1	1	0
M (sum *m*)						4224						

DATA SELECTION	
K1-code	17
Kp-code	12
Af-code	11
Soil-code for Winter	15
Soil-code for Summer	16
ETRM-code	auto
Ri for month X	2614

Figure 18. Output for alfalfa at CT, computed by the Thornthwaite equation, with $R_i = 2614$ m^3/ha.

If the water reserve in the soil is known, it is indicated to use its value to better estimate the irrigation rate.

7. Conclusions

This article presented a new tool, IrrigTool, designed for computing the irrigation rate based on the Thornthwaite equation for ETRM. It is implemented in VBA and Excel and is easy to use without knowledge of the formula implementation. The user can introduce the data series in different worksheets whereby, during the computation, the programs import the data into the main worksheets. The output for two different data sets can be compared as series of numbers and graphically.

This tool's main advantage is that it permits the user to use either the Thornthwaite equation for ETRM computation or to manually fill in the recorded values of ETRM. Manually selecting or automatically computing the initial water reserve for the first month of the agricultural year is also permitted based on the available data. Moreover, there is no limitation related to the climatic conditions. The tool is user-friendly because it does not require programming skills but only basic knowledge of Excel.

A disadvantage of this tool is that the irrigation rate is computed only using the Thornthwaite equation (when ETRM is not known) and does not provide a comparison with another method. Therefore, in future work, we have the following goals: (a) to add a module that computes the ETRM by the Penman–Monteith equation and (b) to add a module to compare the irrigation rates computed based on the Thornthwaite and the Penman–Monteith equations.

Supplementary Materials: The following supporting information can be downloaded at: https://www.mdpi.com/article/10.3390/w14152399/s1, Excel–VBA application: IrrigTool. S1. The VBA code, S2. The VBA–Excel Application.

Author Contributions: Conceptualization, A.B. and C.E.M.; methodology, A.B. and C.E.M.; software, C.Ş.D.; validation, A.B. and C.Ş.D.; formal analysis, A.B. and C.Ş.D.; investigation, A.B.; data curation, A.B.; writing—original draft preparation, A.B.; writing—review and editing, A.B. and C.Ş.D.; visualization, C.Ş.D.; supervision, A.B.; project administration, A.B.; funding acquisition, A.B. All authors have read and agreed to the published version of the manuscript.

Funding: This research received no external funding.

Data Availability Statement: Not applicable.

Conflicts of Interest: The authors declare no conflict of interest.

References

1. Bărbulescu, A.; Dumitriu, C.Ş. Assessing the water quality by statistical methods. *Water* **2021**, *13*, 1026. [CrossRef]
2. Bărbulescu, A.; Nazzal, Y.; Howari, F. Assessing the groundwater quality in the Liwa area, the United Arab Emirates. *Water* **2020**, *12*, 2816. [CrossRef]
3. Gómez, V.M.R.; Gutiérrez, M.; Haro, B.N.; López, D.N.; Herrera, M.T.A. Groundwater quality impacted by land use/land cover change in a semiarid region of Mexico. *Groundw. Sustain. Dev.* **2017**, *5*, 160–167. [CrossRef]
4. Towards a Water and Food Secure Future. Critical Perspectives for Policy-Makers, Food and Agriculture Organization of The United Nations Rome. Revised Reprint World Water Council Marseille. 2015. Available online: http://www.fao.org/3/i4560e/i4560e.pdf (accessed on 9 October 2021).
5. United Nations World Water Assessment Programme. *The United Nations World Water Development Report 2015: Water for a Sustainable World*; UNESCO: Paris, French, 2015. Available online: http://www.unesco.org/new/en/natural-sciences/environment/water/wwap/wwdr/2015-water-for-a-sustainable-world/ (accessed on 9 October 2021).
6. Hargreaves, G.H.; Riley, J.P.; Sikka, A. Influence of climate on Irrigation. *Can. Water Resour. J.* **1993**, *18*, 53–59. [CrossRef]
7. Li, Y.; Li, J.; Gao, L.; Tian, Y. Irrigation has more influence than fertilization on leaching water quality and the potential environmental risk in excessively fertilized vegetable soils. *PLoS ONE* **2018**, *13*, e0204570. [CrossRef] [PubMed]
8. Haddeland, I.; Lettenmaier, D.P.; Skaugen, T. Effects of irrigation on the water and energy balances of the Colorado and Mekong river basins. *J. Hydrol.* **2006**, *324*, 210–223. [CrossRef]
9. de Rosnay, P.; Polcher, J.; Laval, K.; Sabre, M. Integrated parameterization of irrigation in the land surface model ORCHIDEE: Validation over Indian Peninsula. *Geophys. Res. Lett.* **2003**, *30*, 1986. [CrossRef]
10. Lobell, D.; Bala, G.; Mirin, A.; Phillips, T.; Maxwell, R.; Rotman, R. Regional Differences in the Influence of Irrigation on Climate. *J. Clim.* **2009**, *22*, 2248–2255. [CrossRef]
11. Schaible, G.D.; Aillery, M.P. Water Conservation in Irrigated Agriculture: Trends and Challenges in the Face of Emerging Demands. *Econ. Inform. Bull.* **2012**, *99*, 67. Available online: https://www.ers.usda.gov/publications/pub-details/?pubid=44699 (accessed on 20 September 2021). [CrossRef]
12. Huang, Q.; Xu, Y.; Kovacs, K.; West, G. Analysis of factors that influence the use of irrigation technologies and water management practices in Arkansas. *J. Agric. Appl. Econ.* **2017**, *49*, 159–185. [CrossRef]
13. Henggeler, J. Software Programs Currently Available for Irrigation Scheduling. Available online: https://www.irrigation.org/IA/FileUploads/IA/Resources/TechnicalPapers/2002/SoftwareProgramsCurrentlyAvailableForIrrigationScheduling.pdf (accessed on 9 October 2021).
14. Bos, M.G.; Kselik, R.A.L.; Allen, R.G.; Molden, D.J. *Water Requirements for Irrigation and the Environment*; Springer: Dordrecht, The Netherlands, 2009.
15. CRIWAR 3.0. Available online: https://www.bos-water.nl/links-downloads (accessed on 10 July 2022).
16. Allen, R.G.; Pereira, L.S.; Raes, D.; Smith, M. Crop Evapotranspiration-Guidelines for Computing Crop Water Requirements-FAO Irrigation and Drainage Paper 56, FAO—Food and Agriculture Organization of the United Nations Rome. 1998. Available online: http://www.fao.org/3/x0490e/x0490e00.htm (accessed on 9 October 2021).
17. Allen, R.G.; Pereira, L.S.; Raes, D.; Smith, M. FAO Irrigation and Drainage Paper no. 56, Evapotranspiration (Guidelines for Computing Crop Water Requirements). Available online: http://www.climasouth.eu/sites/default/files/FAO%2056.pdf (accessed on 10 May 2021).
18. Raes, D. The ETo Calculator. Evapotranspiration from a Reference Surface. Available online: https://www.ipcinfo.org/fileadmin/user_upload/faowater/docs/ReferenceManualV32.pdf (accessed on 8 May 2021).
19. Penman-Montieth, E.T. Worksheet with Solar Estimate and Equation Instructions. Available online: https://extension.umaine.edu/blueberries/factsheets/irrigation/penman-montieth-et-worksheet-with-solar-estimate/ (accessed on 8 May 2021).
20. CROPWAT 8.0 for Windows. Available online: https://www.fao.org/land-water/databases-and-software/cropwat/en/ (accessed on 5 May 2021).
21. Donker, N.H.W. WTRBLN: A computer program to calculate water balance. *Comput. Geosci.* **1987**, *13*, 95–122. [CrossRef]
22. Sellinger, C.E. Computer Program for Estimating Evapotranspiration Using the Thornthwaite Method. In *NOAA Technical Memorandum ERL GLERL-101*; United States Department of Commerce: Washington, DC, USA, 1996.

23. Armiraglio, S.; Cerabolini, B.; Gandellini, F.; Gandini, P.; Andreis, C. Calcolo informatizzato del bilancio idrico del suolo. *Nat. Bresciana. Ann. Mus. Civ. Sci. Nat. Brescia* **2003**, *33*, 209–216.
24. McCabe, G.J.; Markstrom, S.L. *A Monthly Water-Balance Model Driven by A Graphical User Interface*; Open-File Report 2007-1088; U.S. Geological Survey: Reston, VA, USA, 2007; 6p. [CrossRef]
25. Mammoliti, E.; Fronzi, D.; Mancini, A.; Valigi, D.; Tazioli, A. WaterbalANce, a WebApp for Thornthwaite–Mather water balance computation: Comparison of applications in two European watersheds. *Hydrology* **2021**, *8*, 34. [CrossRef]
26. Droogers, P.; Immerzeel, W.W.; Lorite, I.J. Estimating actual irrigation application by remotely sensed evapotranspiration observations. *Agric. Water Manag.* **2010**, *97*, 1351–1359. [CrossRef]
27. Olivera-Guerra, L.; Merlin, O.; Er-Raki, S. Irrigation retrieval from Landsat optical/thermal data integrated into a crop water balance model: A case study over winter wheat fields in a semi-arid region. *Remote Sens. Environ.* **2020**, *239*, 111627. [CrossRef]
28. Brocca, L.; Tarpanelli, A.; Filippucci, P.; Dorigo, W.; Zaussinger, F.; Gruber, A.; Fernández-Prieto, D. How much water is used for irrigation? A new approach exploiting coarse resolution satellite soil moisture products. *Int. J. Appl. Earth Obs. Geoinf.* **2018**, *73*, 752–766. [CrossRef]
29. Bretreger, D.; Yeo, I.Y.; Hancock, G. Quantifying irrigation water use with remote sensing: Soil water deficit modelling with uncertain soil parameters. *Agric. Water Manag.* **2022**, *260*, 107299. [CrossRef]
30. Foster, T.; Mieno, T.; Brozović, N. Satellite-Based monitoring of irrigation water use: Assessing measurement errors and their implications for agricultural water management policy. *Water Resour. Resear.* **2020**, *56*, e2020WR028378. [CrossRef]
31. CAWATERinfo. Irrigation and Water Application Rate. Available online: http://www.cawater-info.net/bk/4-2-1-1-3-3_e.htm (accessed on 7 June 2022).
32. Roșu, L. *Land Ameliorations*; Ovidius University Press: Constanta, Romania, 1999. (In Romanian)
33. Pleșa, I.; Florescu, G.; Mușat, I. *Land Amelioration*; Editura Didactică și Pedagogică: Bucharest, Romania, 1967. (In Romanian)
34. Ștefan, V.; Bechet, S.; Tomiță, O.; Titz, L. *Land Amelioration*; Editura Didactică și Pedagogică: Bucharest, Romania, 1981. (In Romanian)
35. Cismaru, C.; Gabor, V. *Irrigation. Arrangements, Rehabilitations and Modernizations*; Editura POLITEHNIUM: Iasi, Romania, 2004. (In Romanian)
36. Cazacu, E.; Dorobantu, M.; Georgescu, I.; Sarbu, E. *Irrigations Arrangements*; Ceres: Bucharest, Romania, 1982. (In Romanian)
37. Bărbulescu, A. Models for temperature evolution in Constanta area (Romania). *Rom. J. Phys.* **2016**, *61*, 676–686.
38. Bărbulescu, A.; Deguenon, J. About the variations of precipitation and temperature evolution in the Romanian Black Sea Littoral. *Rom. Rep. Phys.* **2015**, *67*, 625–637.
39. Bărbulescu, A.; Dumitriu, C.S.; Maftei, C. On the Probable Maximum Precipitation Method. *Rom. J. Phys.* **2022**, *67*, 801.
40. Bărbulescu, A.; Maftei, C. Statistical approach of the behavior of Hamcearca River (Romania). *Rom. Rep. Phys.* **2021**, *73*, 703.
41. Maftei, C.; Bărbulescu, A.; Rugină, S.; Nastac, C.D.; Dumitru, I.M. Analysis of the arbovirosis potential occurence in Dobrogea, Romania. *Water* **2021**, *13*, 374. [CrossRef]
42. Bărbulescu, A.; Maftei, C.; Dumitriu, C. The modeling of the climatic process that participates at the sizing of an irrigation system. *Appl. Comput. Math.* **2002**, *2048*, 11–20.
43. Halford Tools. Available online: https://halfordhydrology.com/tools/ (accessed on 29 June 2022).
44. GR2M Model. Available online: https://webgr.inrae.fr/en/models/monthly-model-gr2m/ (accessed on 29 June 2022).
45. GR4J_FR. Available online: https://webgr.inrae.fr/modeles/journalier-gr4j-2/ (accessed on 29 June 2022).
46. Niazkar, M. An Excel VBA-based educational module for bed roughness predictors. *Comput. Appl. Eng. Educ.* **2021**, *29*, 1051–1060. [CrossRef]
47. Miller, S. *Handbook for Agrohydrology*; Natural Resources Institute: London, UK, 1994. Available online: http://www.nzdl.org/cgi-bin/library.cgi?e=d-00000-00---off-0hdl--00-0----0-10-0---0---0direct-10---4-------0-1l--11-en-50---20-about---00-0-1-00-0-0-11-1-0utfZz-8-00&cl=CL3.37&d=HASH3b4d99e5f9716ab628b9b2>=2 (accessed on 7 June 2022).

MDPI
St. Alban-Anlage 66
4052 Basel
Switzerland
www.mdpi.com

Water Editorial Office
E-mail: water@mdpi.com
www.mdpi.com/journal/water

Disclaimer/Publisher's Note: The statements, opinions and data contained in all publications are solely those of the individual author(s) and contributor(s) and not of MDPI and/or the editor(s). MDPI and/or the editor(s) disclaim responsibility for any injury to people or property resulting from any ideas, methods, instructions or products referred to in the content.